CIVIL ENGINEERING AND ARCHITECTURE

COMPUTER AIDED BRIDGE ENGINEERING

(DETAIL DESIGN OF PRE-STRESSED CONCRETE I-GIRDER / BOX-GIRDER BRIDGES)

CIVIL ENGINEERING AND ARCHITECTURE

Additional books and e-books in this series can be found on Nova's website under the Series tab.

CIVIL ENGINEERING AND ARCHITECTURE

COMPUTER AIDED BRIDGE ENGINEERING

(DETAIL DESIGN OF PRE-STRESSED CONCRETE I-GIRDER / BOX-GIRDER BRIDGES)

SANDIPAN GOSWAMI

nova science publishers
New York

DOI: https://doi.org/10.52305/YVXV1963

Library of Congress Cataloging-in-Publication Data

ISBN: 978-1-68507-413-5

Published by Nova Science Publishers, Inc. † New York

To my wife 'Sangeeta Goswami',

and also to all my colleagues & friends in my professional career who helped me learn and gather knowledge on the subject.

Contents

PREFACE

The present book *Computer Aided Bridge Engineering* is for the design of pre-stressed concrete (PSC), I-girder (I-Beam), and PSC box-girder bridges. In this volume, the real project design calculations for a deck-girder superstructure is presented along with the design of an abutment and pier with pile foundation as the bridge substructure.

The book is proposed to be read in association with processing the design work by using the computer software ASTRA Pro as referred to in the book. The book describes two essential facets of the work, which are 'Analysis of the Grillage Model of the Deck-Girder Superstructure' and the subsequent 'Design of Deck Slab and PSC I-Girder'. Whereas the software provides three facets of the work, first is the 'Analysis of the Grillage Model of the Deck-Girder Superstructure', second is the 'Design of Deck Slab and PSC I-Girder, Abutment, Piers along with Pile Foundation' and the third is a 'Set of Sample Editable CAD Drawings for the work. The drawings may be modified as per the design work and be submitted as required for the construction. The drawings contain information on dimensions, structural detailing, bar-bending schedules, pre-stressing details and construction guides.

The computer applications for the design of bridge deck-girders, abutments, piers and pile foundations which are in AASHTO LRFD, are described in detail in the softcopy volume 'User's Guide for Hands on Practice with Computer Applications' for the book 'Computer Aided Bridge Engineering' and is available for download from web site 'www.roadbridgedesign.com'.

The structural analysis of the grillage model is to provide the basic forces to be used in the design. Other considerations are used in the exhaustive design worksheet programs, the various design inputs may be altered by the user directly in the design Excel Worksheet for instant results with the altered design input data.

The grillage model is a finite element model by using three-dimensional beam elements. The deck-girder superstructure is defined as a lumped mass model in which the relevant part of section properties and mass of the deck and girders are attributed to the surrounding beam elements. The boundary conditions with applicable restraints are defined relevant to supports on which the grillage model is resting.

Commonly, material properties are mentioned as single materials for the model. In a composite bridge the material properties are different for steel and concrete, the orthotropic analysis gives more correct results.

The grillage model is subjected to dead load (DL), super imposed dead load (SIDL) and live loads for moving vehicles. The applications of live loads are defined for various lane configurations and by following the applicable design standard.

The load-deflection stage analysis may also be carried out to get the non-linear effects of loads on the grillage structure.

The basis of design considers new concepts of actions & their combinations that gives ideas about various design situations like persistent, transient, accidental or seismic combinations that might occur during the life of the structure. This will help the bridge designer to design the bridge for the worst loading situations.

The ULTIMATE LIMIT STATES are introduced in design standards. They are 1. Limit state of equilibrium 2. Limit state of strength.

The SERVICEABILITY LIMIT STATES deals with 1. Limit state of internal stress 2. Limit state of crack control 3. Limit state of crack control 4. Limit state of vibration 5. Limit states of fatigue. However, limit state for vibration &limit state of fatigue are not covered i.e., specialized literature is suggested by the code.

The Design in AASHTO LRFD Standard

For the design in AASHTO-LRFD, the figures, tables, and equations are denoted by their home article number and an extension, for example, 1.2.3.4.5-1 wherever they are cited. In early editions, when they were referenced in their home article or its commentary, these objects were identified only by the extension. For example, in Article 1.2.3.4.5, Eq. 1.2.3.4.5-2 would simply have been called “Eq. 2.” The same convention applies to figures and tables. These objects are identified by their whole nomenclature throughout the text.

Please note that the AASHTO materials standards (starting with M or T) cited throughout the LRFD Specifications can be found in Standard Specifications for Transportation Materials and Methods of Sampling and Testing, adopted by the AASHTO Highway Subcommittee on Materials. The individual standards are also available as downloads on the AASHTO Bookstore, https://bookstore.transportation.org.

INTRODUCTION

The book *Computer Aided Bridge Engineering* is for detail design of PSC I-Girder and Box-Girder Bridge Superstructure, Integral Bridge Abutments, Piers along with Foundations.

By using the AASHTO-LRFD Standard, British Standard Eurocode II and IRC (Indian Roads Congress) Standard, the design of the deck slab and cantilever slab are done by calculating bending moments, shear forces, bending resistance in transverse direction, bending resistance in longitudinal direction, checking flexural cracking. The design of PSC I-girders is done for bending moments and shear forces by dead load, super imposed dead load (SIDL) and live loads (LL). The shrinkage strain, creep strain and effect of temperature rise and fall are also determined. The design is complete for pre-stressing cables, end anchorages, un-tensioned reinforcements, end cross girders, and shear connectors.

The book series 'Computer Aided Bridge Engineering' is intended to cover bridge design procedures for various bridge types in a series of volumes for various types of highway bridges. This is because the detail engineering design for every bridge type is significantly large.

The current volume of the book contains, *Detail Design of Pre-stressed Concrete I-Girder Bridge Deck-Girder Superstructure including the Detail Design of Integral Bridge Abutments on Pile/Open Foundations, Detail Design of Bridge Piers on Pile/Open Foundations, Detail Design of Pre-stressed Concrete Box Girder Bridge Superstructure.*

A question is seen in various technical forums like ResearchGate, Roads and Bridges etc. on the internet many times with engineers worldwide asking for complete, worked out, detailed engineering designs for various types of bridges, by referring to design standards of AASHTO-LRFD, BS Eurocode-II, IRC etc. which are the most widespread design standards around the world. Engineers commonly have their own personal worksheet programs that are developed by them over a long spell of time. But the less experienced engineers, university faculties, students are normally found less guided to choose the appropriate option from a wide range of options mentioned in the design standards. It therefore felt essential to provide the engineers worldwide with a very useful resource as 'Complete Step-wise Design Procedure on Real Project Data' by following the desired design standard.

The various market available textbooks only give steps and calculations of the design in a short-cut manner rather than following the detailed procedure. This resulted in the preparation of individual Excel worksheet programs for the design job. That may be possible although it is very tedious and time consuming, but the main difficulty was the availability of a finite element program to analyze the bridge model to get the forces for the design of individual components of the bridge. The books in the series 'Computer Aided Bridge Engineering' were therefore planned to provide the engineers with a textbook describing the entire procedure starting with a very powerful analysis followed by step wise design calculations in worksheets with the help of bridge and structure design software ASTRA Pro, which uses mSAP by University of California, Berkley, USA for finite element analysis of the grillage model of bridge deck-girder superstructure.

Software applications cater to bridge geometry, data section, analysis suit, design suit, graphical user interface, and a built-in high power 3D CAD system. The system opens a set of sample CAD construction drawings with various structural details specific to each bridge design. The drawings are in editable CAD format to make them applicable as required and can be saved with selected types acceptable to other CAD software. This significantly helps users understand the design requirement for thc construction of individual components (structural detailing of deck slab, long girders, cross girders, diaphragms, abutments, piers, bearings, truss fabrication, plate girder splicing, arrangements of stay cables, profiles of tendon cables etc.) for individual bridge types and structures.

The exhaustive detail analysis and design calculation Excel report referring to clauses and illustrations of relevant design standards shall help engineers to get well versed with the design procedure and deliver professional design work by following the design standards like AASHTO-LRFD or BS Eurocode 2 or IRC, which are most commonly used by the industry worldwide.

This is expected to meet the requirements by the teachers of the engineering institutions to train the students worldwide with the bridge design procedure. Once the students are thorough with the application they will come out from the institutions as 'Industry Ready' and shall be in the position to undertake the professional design jobs with the desired level of expertise. This will, in turn, help the young engineers to get jobs easily in the industry as bridge engineers, in various countries across the world and produce the deliverables as required for the project.

With the advent of computer technology, various software are available for the design of bridges. Still, SAP is considered the oldest and most authentic structural analysis program with a wide range of applications of the finite element method. SAP was developed by the University of California, Berkley, USA. The 'Normal Analysis', 'Stage Analysis' and 'Orthotropic Analysis' enable the users to decide and use the results of various analyses by applying the dead load, super imposed dead load and various lane combinations of moving vehicle live loads for the design of the bridge superstructure.

The design of the deck slab is commonly done with 'Transverse Analysis' by the 'Effective Width Method'. The structural analysis model for right-normal, skew and 3-span curved bridges are developed in ASTRA Pro with basic general arrangement of data. In every bridge design module, default data is provided which the user may change in the authorized versions of the program. The applications of load combinations are flexible by selecting from the standard loadings. In the worksheet design, some design data may be directly modified in the Excel worksheet for instant results.

The book series 'Computer Aided Bridge Engineering' is intended to provide the user a companion to understand and carry out the job of bridge design as a 'ONE-STOP' solution. The theory part is mentioned as minimum in the book as that is widely explained in the design standards and in numerous books available for the users to refer to. The book rather provides real life design procedures with step wise detail calculations, illustrated with figures and tables and referring to relevant clauses of the design standard.

The book series 'Computer Aided Bridge Engineering' provides the detail design for real project in AASHTO-LRFD standard with each topic in separate chapters.

The book series 'Computer Aided Bridge Engineering' is believed to serve as a unique companion for the professional engineers, consulting firms, government departments, university faculties, students all over the world. The current book 'Volume - I' is for the detail design of pre-stressed concrete I-girder &box-girder bridges with integral abutments and piers with foundations and bearings and is released to serve the users with a straightforward design application for composite bridges.

The design calculations are based on formulas and arc embedded in the Excel worksheet which users can see. It is therefore strongly suggested that the users download and install the ASTRA Pro software on the system and keep it open. During studying specific sections of the book, it should also be processed concurrently in ASTRA Pro. Users may download the software ASTRA Pro, the Design & Users Manuals and the tutorial videos from the web sites:

Visit www.roadbridgesoftware.com, Under menu 'Book Tutorials' for ASTRA Pro Uesrs' Guide is available for download. Any technical query may be sent on email to techsoftinfra@gmail.com and techsoft@consultant.com or website www.techsoft global.com. The tutorial videos may be watched first.

This book uses the AASHTO LRFD standard, however, if users want to design as per British Eurocode standard or IRC standard, they may write to sandipanmails@gmail.com for necessary advice.

ACKNOWLEDGEMENTS

We acknowledge the kind co-operation extended by Mr. Cullen Bob American Association of State Highway and Transportation Officials (AASHTO) by giving permission for using illustrations from,

AASHTO LRFD Bridge Design Specifications.

555 12th Street NW Suite 1000, Washington, DC 20004

Web: https://www.transportation.org

Mr. Lubin Gao, PhD, PE, Senior Bridge Engineer – Load Rating by giving permission for using illustrations from AASHTO LRFD Bridge Design Specifications., Federal Highway Administration (FHWA)

FHWA Office of Bridges and Structures

1200 New Jersey Ave., SE

Washington, DC 20590

Web: https://www.fhwa.dot.gov/bridge/loadrating/

We are highly thankful to Techsoft Engineering Services for allowing the use of analysis and design reports and drawings by Computer Aided Bridge Design in this book and also providing us the design tutorials of Bridge Design software ASTRA Pro by making them available for download from the web sites www.techsoftglobal.com and www.roadbridgedesign.com.

I also convey my heartfelt thanks to the wonderful people of Nova Science Publishers for their cooperation and publishing this work.

List of Abbreviations

AASHTO	American Association of State Highway and Transportation Officials
BR	Breaking Force
CAD	Computer Aided Design
CD	Chemical Deficiencies
CIP	Cast in Place
DL	Dead Load
DOF	Degrees of Freedom
FWS	Future wearing surface
IRC	Indian Roads Congress
LL	Live Loads
LRFD	Load Resistance Factor Design
mSAP	Microstructure Analysis Program
PD	Physical Deficiencies
POT-PTFE	Plate on Top-Poly Tetra Fluro Ethylene
PSC	Pre-stressed Concrete
PT	Post Tension
SAP	Structure Analysis Program
SIDL	Super Imposed Dead Loads
SLS	Serviceability Limit States
UDL	Uniformly Distributed Load
ULS	Ultimate Limit States
VB	Base wind velocity

Chapter 1

THE AIM OF THIS WORK

This book belongs to the book series of 'Computer Aided Bridge Engineering' for the design of pre-stressed concrete (PSC) I-girder (I-Beam) and PSC box-girder bridges. In this volume, the real project design calculations for deck-girder superstructure are presented along with the design of abutment and pier with pile foundation as the bridge substructure.

The book is proposed to be read in association with processing the design work by using the computer software ASTRA Pro as referred to in the book. The book describes two essential facets of the work, which are 'Analysis of the Grillage Model of the Deck-Girder Superstructure' and the subsequent 'Design of Deck Slab and PSC I-Girder'. The software provides three facets of the work, first is the 'Analysis of the Grillage Model of the Deck-Girder Superstructure', second is the 'Design of Deck Slab and PSC I-Girder, Abutment, Piers along with Pile Foundation' and the third is a 'Set of Sample Editable CAD Drawings for the work. The drawings may be modified as per the design work and be submitted as required for construction. The drawings contain information on dimensions, structural details, bar-bending schedules, pre-stressing details and construction guides.

The Computer Applications for the design of Bridge Deck-Girders, Abutments, Piers and Pile Foundations in AASHTO LRFD, are described in detail in the softcopy volume 'User's Guide for Hands on Practice with Computer Applications' for the book 'Computer Aided Bridge Engineering' which is available for download from the website 'www.roadbridgedesign.com'.

The structural analysis of the grillage model is to provide the basic forces to be used in the design. Other considerations are used in the exhaustive design worksheet programs, the various design inputs may be altered by the user directly in the design Excel Worksheet for instant results with the altered design input data.

The grillage model is a finite element model by using three-dimensional beam elements. The deck-girder superstructure is defined as a lumped mass model, in which the relevant part of section properties and mass of the deck and girders are attributed to the surrounding beam elements. The boundary conditions with applicable restraints are defined relevant to supports on which the grillage model is resting. Commonly, material properties are mentioned as single materials for the model. In a composite bridge, the material properties are different for steel and concrete, the orthotropic analysis gives more correct results.

The grillage model is subjected to dead load (DL), super imposed dead load (SIDL) and live loads for the moving vehicles. The applications of live loads are defined for various lane configurations and by following the applicable design standard.

The load-deflection stage analysis may also be carried out to get the on-linear effects of loads on the grillage structure.

The basis of design considers new concepts of actions & their combinations that gives idea about various design situation like persistent, transient, accidental or seismic combinations that might occur during the life of the structure. This will help bridge designer to design bridge for worst loading situations.

The ULTIMATE LIMIT STATES is introduced in design standards these are 1. Limit state of equilibrium 2. Limit state of strength

The SERVICEABILITY LIMIT STATES deals on 1. Limit state of internal stress 2. Limit state of crack control 3. Limit state of crack control 4. Limit state of vibration 5. Limit states of fatigue. However, Limit state for vibration & Limit state of fatigue are not covered i.e., specialized literature are suggested by the code.

1.1. The Design in AASHTO LRFD Standard

For the design in AASHTO-LRFD, the figures, tables, and equations are denoted by their home article number and an extension, for example 1.2.3.4.5-1 wherever they are cited. In early editions, when they were referenced in their home article or its commentary, these objects were identified only by the extension. For example, in Article 1.2.3.4.5, Eq. 1.2.3.4.5-2 would simply have been called "Eq. 2." The same convention applies to figures and tables. These objects are identified by their whole nomenclature throughout the text.

Please note that the AASHTO materials standards (starting with M or T) cited throughout the LRFD Specifications can be found in Standard Specifications for Transportation Materials and Methods of Sampling and Testing, adopted by the AASHTO Highway Subcommittee on Materials. The individual standards are also available as downloads on the AASHTO Bookstore, https://bookstore.transportation.org.

This book uses the AASHTO LRFD standard, however, if users want to design as per British Eurocode standard or IRC standard, they may write to sandipanmails@gmail.com for necessary advice.

1.2. The Design in British Standard Eurocodes

The British Eurocode standards provide common structural design rules for everyday use for the design of whole structures and component products of both a traditional and an innovative nature. Unusual forms of construction or design conditions are not specifically covered and additional expert consideration will be required by the designer in such cases.

The National Standards implementing Eurocodes comprise the full text of the Eurocode (including any annexes), as published by CEN, which may be preceded by a National title page and National foreword, and may be followed by a National annex.

The National annex may only contain information on those parameters which are left open in the Eurocode for national choice, known as Nationally Determined Parameters, to be used for the design of buildings and civil engineering works to be constructed in the country concerned, i.e.,

- values and/or classes where alternatives are given in the Eurocode,
- values to be used where a symbol only is given in the Eurocode,
- country specific data (geographical, climatic, etc.), e.g., snow map,
- the procedure to be used where alternative procedures are given in the Eurocode.

It may contain

- decisions on the application of informative annexes,
- references to non-contradictory complementary information to assist the user to apply the Eurocode.

1.3. The Design in IRC Standard

The IRC design is mostly similar to British Standard Eurocode 2 but different in respect to live load applications. Live load uses India Standard vehicle loading as per the width of the carriageway and lane configuration of the bridge deck. Users may refer to IRC 112-2011, Code of Practice for Concrete Road Bridges and IRC 6-2017, Standard

Specifications and Code of Practice for Road Bridges, Section: II, Loads and Stresses, Published by Indian Roads Congress.

References

ASTRA Pro User's Manual and Design Manual, Techsoft Engineering Services, Web site: www.techsoftglobal.com

Bridge Engineering Handbook, *Second Edition, Superstructure Design, Edited by Wai Fah Chen and Lian Duan*, Published by CRC Press, Taylor & Francis Group, 6000 Broken Sound Parkway, NW, Suite 300, Boca Raton, FL 33487, Web site: www.crcpress.com

British Standard, BS 5400-3: 2000, *Steel, concrete and composite bridges* Ð, Part 3: Code of practice for design of steel bridges

British Standard, BS 5400-4: 1990, *Steel, concrete and composite bridges* — Part 4: Code of practice for design of concrete bridges

British Standard, *Design Manual for Roads and Bridges*, Volume 1, Section 3, Part 14, BD 37/01, Loads For Highway Bridges, Published by: British Standards Institution.

British Standard, *Eurocode 2: Design of Concrete Structures*, BS EN 1992-1-1:2004, Published by: British Standards Institution.

Design Specifications, Customary U.S. Units, Sixth Edition 2012, American Association of State Highway and Transportation Officials. 444 North Capitol Street, NW, Suite 249, Washington, DC 20001, Phone 202-624-5800 / Fax 202-624-5806, Web site: www.transportation.org

IRC 112-2011, and IRC 6-2017, *Code of Practice for Road Bridges*, Published by Indian Roads Congress.

Chapter 2

GRILLAGE MODEL ANALYSIS OF BRIDGE DECK AND GIRDER BY MICROSAP

ABSTRACT

This chapter describes the grillage model with which the bridge decks are schematized. First, the theoretical principles on which this kind of modeling is based are recalled; the equivalent condition between three-dimensional Beam Elements and corresponding grillage models are imposed through the use of a kinematics and an energetic criterion. Secondly, the same technique is generalized to three-dimensional structures and specialized to the case of cellular decks. For this kind of deck, structural behaviors usually neglected by the current technical approaches, like shear lag, distortion and warping, are considered. The chapter presents the method introducing these effects in a grillage analysis; the method provides a series of criteria with which it's possible to define the rigidities of the equivalent model. These criteria are applied with finite element solutions. Finally, the application is executed in order to obtain the desired results for forces developed in the nodes/joints and beam elements/members.

2.1. GENERAL

SAP-IV was developed by the Earthquake Engineering Research Centre, College of Engineering, University of California, Berkeley, Ca., USA. It was the first complete structural analysis package which introduced the basic concepts like multiple element types, disk based large capacity equation solution, etc., used by all major structural analysis packages today. mSAP is the Structural Analysis Program for Microcomputers and is an adaptation of SAP-IV to the microcomputer environment. The package has been thoroughly edited and source code revised to conform to the Fortran-77 standard, and to utilize the facilities provided by MS-Windows operating environment. In this chapter the data file for the model is explained as it is created by software ASTRA Pro in user-friendly text file format, by taking various input from users, which is next formatted into SAP data and finally the SAP data is processed by running the mSAP program to obtain the analysis report.

The bridge deck analysis is done by the use of a grillage model, the assignment of rigidities to the grillage members is certainly the main phase of this pattern. The expression of the rigidities must be assigned to the beams, given by various manuals for more common types of deck. These estimations of the equivalent rigidities derive from theoretical considerations and experimental observations referred to only "principal" deformation modes, deflections and torsion of deck in longitudinal and transverse directions.

This chapter emphasizes introducing the typical process of designing deck-girder superstructure of pre-stressed concrete (PSC) 'I-Girder' concrete bridge, along with the consideration that a design engineer needs to go through each phase of the design process. A wide range of contents with respect to bridge design process are covered in this chapter. Upon the completion of structural analysis, the bridge will be designed in detail. The step-by-step procedure of determining the concrete section dimensions, pre-stressing tendon profile, reinforcing bar layout and material properties are specified in this chapter.

The analysis by 'Grillage Model' of deck-girder superstructure of the bridge consists of the sections as: Geometry with Cantilever Footpaths, Member Section Properties, Member Material Properties, Bearing Supports, Dead Loading (DL), Super Imposed Dead Loading (SIDL), Live Loading (LL) by moving vehicle loads, of various classes and applied lane wise on the bridge deck by following the relevant design standards.

2.2. Overview of Deck-Girder Grillage Model

2.2.1. Geometry

Grillages are most widely used to analyze beam/girder and slab type bridge decks. Solid slab type decks are generally modeled using finite elements, however grillages may be used for this type of deck with reasonably accurate results.

Longitudinal grillage members are arranged to represent the main beams with transverse members representing the deck slab and diaphragm beams.

The spacing of transverse grillage members is chosen to be about 1.5 times the spacing of the main longitudinal members, but may vary up to a limit of 2:1. Transverse members are required at the diaphragm positions and, in order to achieve a member at mid span, there needs to be an odd number of members.

In the case of skew-deck, the transverse members should be arranged orthogonally to the main members to achieve the correct magnitude for moments and deflections.

However, this arrangement can be impractical for small skew angles (below 35°) and a skew mesh is usually adopted. The skew mesh will tend to slightly overestimate the magnitude of moments and deflections and is therefore considered to be a safe solution. The section properties of transverse members in skew meshes should be calculated using the orthogonal spacing.

2.2.2. Member Section Properties

The bending inertia and torsion inertia are required for all the members in the grillage model.

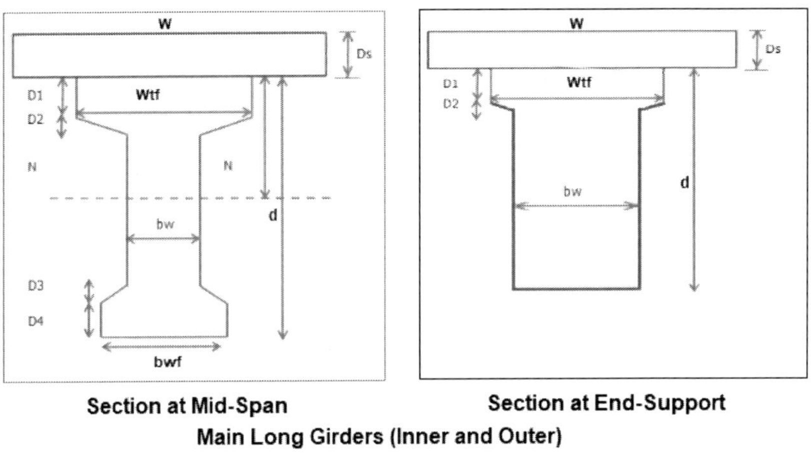

(a) Main Long Girders (Inner and Outer)

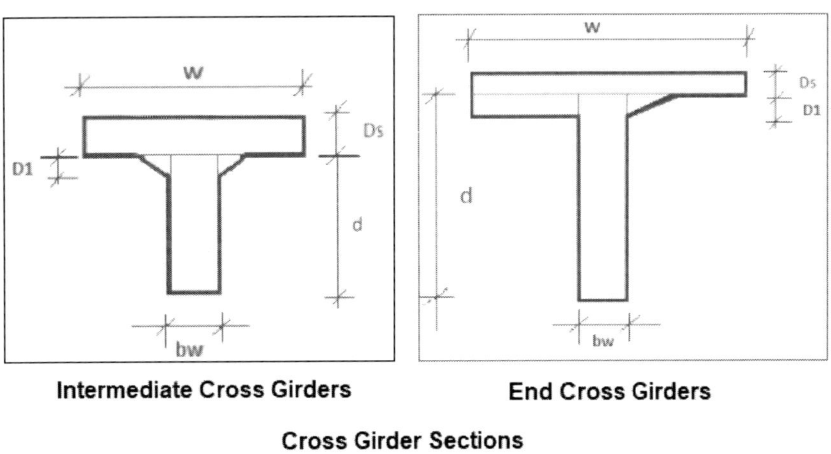

(b) Cross Girder Sections

Figure 2.1. (a) & (b) - Member Section Properties.

The bending inertia for the composite sections can be calculated using a suitable performance. Member properties for the grillage are calculated for the local member axis as shown and care will be needed to ensure you are using the correct notation. If the spread sheet is used, then 'Ixx' from the spread sheet will be input as 'Iy' for the local member axis shown. Similarly, 'Iyy' from the spread sheet will be input as 'Iz' and the torsion inertia 'J' will be input as 'Ix'.

The torsion inertia for a rectangle is given by $J=k_1b^3b_{max}$

where:

b is the length of the short side

b_{max} is the length of the long side

$k_1=\{1-0.63(b/b_{max})(1-b^4/12b^4_{max})\}/3$

The torsion inertia of the section is the summation of the inertias of the individual rectangles. As the deck slab is used in both the longitudinal members and transverse members then the inertia of this rectangle is halved. This only relates to the Torsion Inertia, the full section is used to calculate the Bending Inertia.

2.2.3. Cantilever Footpaths

When the deck is cantilevered from the edge of the main deck beams it is convenient to extend the grillage to the parapet beam. This will simplify the application of footway loading and accidental wheel loading.

However, the torsion effects in the edge beam can be significantly overestimated in a beam and slab type deck if the members are not modeled correctly.

Most of the cantilever effects should induce bending in the deck slab with secondary torsion effects being induced in the supporting edge beams.

If a simple two-dimensional (2D) grillage is used then the bending effects from the cantilever will be resisted by torsion in the edge beam. This will overestimate the torsion in the edge beam and underestimate the bending in the deck.

Some grillage programs allow the centroids of the members to be offset from the 2D plane (as shown in Figure 2.3). Alternatively, a three-dimensional model may be used by including rigid, vertical, dummy members into the model, although this solution does make the geometry more complicated.

2.2.4. Bearing Support

Most grillage analyses allow the supports to be modeled either as free, or rigid or sprung. Spring supports are used to model the elastic deformation either of the bearing or of the support structure. Rubber bearings will distort under load and have a significant effect on the distribution of loads throughout the deck. Even the elastic deformation of concrete columns can have an effect on the distribution of loads in a continuous deck.

A simple line beam analysis will give an approximate magnitude for the reactions. This will enable a suitable bearing to be chosen for the grillage model. Alternatively, the grillage analysis can be carried out with rigid vertical supports and modified later.

Bearings may be chosen as either of:

- 'Free and sliding-guided bearings' will be fixed or sprung in the vertical direction and releases applied for rotational directions.
- The 'Fixed' bearing will be fixed or sprung in all directions.

When using spring supports it is usual to have to fix one bearing in the vertical direction to achieve a stable solution. A grillage model is not analyzed for in-plane loading as any longitudinal or transverse restraint cannot be modeled.

2.2.5. Loading

All loading is proportioned to the grillage members and grillage joints (nodes) before the moments, shears and torsions are calculated. Many programs have the facility for applying patch loads and point loads which do not necessarily coincide with joints or members. The program will distribute these loads to the members before calculating the moments, shears and torsion effects.

There are a number of ways of proportioning the loads to the joints if the program does not have this facility. The 48kN point load shown acting within a 600 square grid can be proportioned to a pair of opposite members and then again to the joints. This distribution will achieve sufficiently accurate results.

It is convenient to apply all loads to the structure as nominal loads. Load factors can be applied to the combination cases to avoid inputting numerous load cases. Consequently, load cases should not be too complex. For example, carriageway construction 150mm thick is considered in BD21 as 100mm surfacing with 50mm fill and should be applied as two load cases as different load factors are applied to fill than to surfacing.

2.2.6. Dead Loading

Dead load or DL is applied to the main longitudinal members. Some programs automatically generate dead load by applying a density to the cross-sectional area of the member. Care is needed to avoid double accounting for the weight of the deck slab.

Superimposed dead load or SIDL (carriageway surfacing, footpath fill and surfacing and parapets) are input as uniformly distributed loads along the length of the longitudinal grillage members. Some programs have the facility of applying patch loads which can be used for the surfacing providing it is of constant thickness.

2.2.7. Live Loading

Live load or LL can consist of HA (udl + kel) load, HB load, pedestrian load, accidental wheel load and wind load. Collision load on parapets is only included if high containment parapets are required. Horizontal loads such as traction or braking and skidding are generally not included as the deck is very stiff for resisting horizontal loads compared with vertical loads. Unless the deck has a very high superelevation or steep longitudinal gradient then the load component in the vertical direction for skidding and centrifugal loads will be negligible.

2.2.8. Results

It is always good practice to carry out approximate checks of the output as the job proceeds. One simple check is to obtain the total reactions for each load case to see if they match with an estimate of the total load applied in each load case.

Standard procedure for most bridges deck. The grillage analogy method for analyzing bridge superstructures has been in use for quite some time. Various studies were done on grillage idealization by mesh layout of the structure, together with the relevant background information. The bridge deck is analyzed by both grillage analogy as well as by finite element method. In grillage analysis the three-dimensional beam elements are used, which are also a member of finite element library. By calling the finite element method it is meant the usage of plate elements in modeling the bridge deck. By lumped mass technique, the deck-girder structure of the bridge is more correctly defined in grillage model by using beam elements. In finite element method the model is difficult idealize the actual deck-girder structure, by using the plate elements. For slab bridges without any girder, the modeling by using plate elements may be more appropriate. Bridge deck analysis for the grillage model by using beam elements is many times compared with finite element model with plate elements. Though the finite element method gives lesser values for bending moment in deck as compared to grillage analysis, the later method seems to be easy to use and comprehend.

2.3. Grillage Modeling of Slab-on-Girders Bridge

The grillage model of a deck-girder is created by idealization of the beam and slab bridge where an assembly of interconnected linear beam elements forms a grid in the horizontal plane. This is more acceptable to structural engineering than a grillage model for a slab bridge, where a model by plate elements is more realistic. The T- beams and I-beams are the most commonly adopted type of bridge decks consisting of longitudinal girders at definite spacing, with or without connected by cross girders to support the deck slab. Commonly, the diaphragms are used to connect the longitudinal girders, the diaphragms are provided at the supports.

The longitudinal grid lines in the grillage model for T-beam or I-beam decks are made to coincide with the centre lines of physical girders and these longitudinal members are given the material and section properties of the girders plus associated portions of the slab which they represent. Additional dummy grid lines between physical girders and the edges of the bridge deck or at a suitable distance from the edge are also set in the grillage model for extra nodes to obtain force information and also to improve the accuracy of the result. To consider the bridge footpaths, one extra longitudinal grid line along the centre-line of each footpath slab is also provided in the grillage model. The above procedure for setting longitudinal grid lines is applicable to both normal and skew decks.

When intermediate cross-girders are considered in the actual deck, the transverse grid lines represent the properties of cross girders and associated deck slabs. The grid lines are set-in along the centre-lines of cross-girders. If after considering the effective flange widths of these girders, some portions of the slab are left out then, additional dummy grid lines are also placed in between these transverse physical cross-girders. If the spacing of transverse grid line is still greater than two times the spacing of longitudinal grid lines, the left-over slabs are to be replaced by two or more grid lines so that the above requirement for spacing is fulfilled.

In the actual deck, if a diaphragm is considered on thc supports, the grid lines should also be placed by coinciding with these diaphragms. A typical grillage lay-out of Deck-Girder system is shown in Figure 2.2.

When no intermediate diaphragms are provided, the transverse medium i.e., deck slab is assumed as a number of transverse strips, replaced by transverse grid lines. The spacing of transverse grid lines should be about 1/8 of effective span for making the grid convenient. As a guideline, the ratio of spacing of transverse and longitudinal grid lines may be maintained between 1 and 2 and the total number of lines should be odd. This spacing ratio should reflect the span-width ratio of the deck. Therefore, for square and wider decks, the ratio tentatively becomes 1 and for long and narrow decks it is tentatively 2.

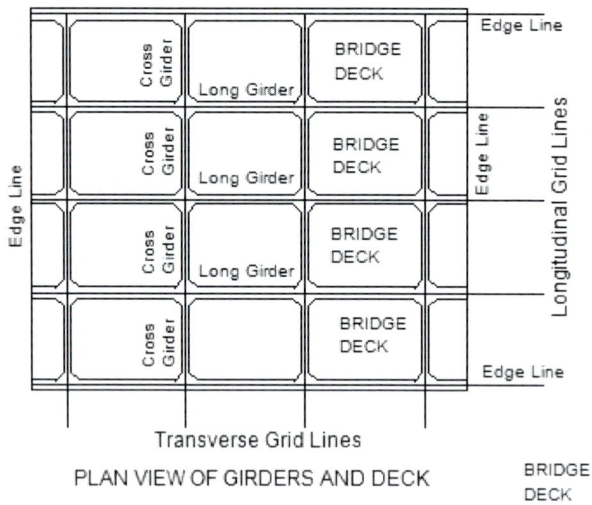

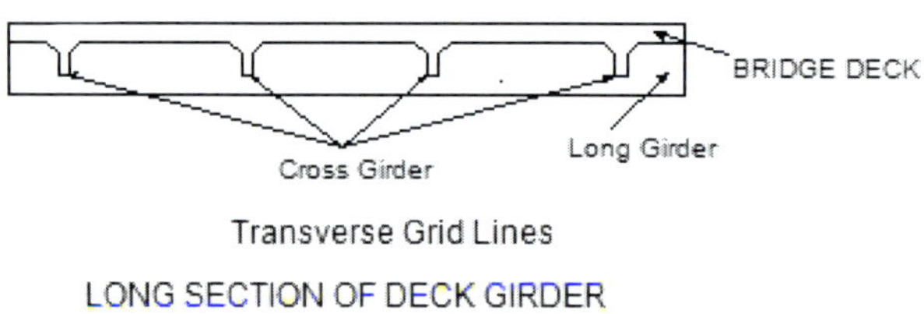

Figure 2.2. Grillage Layout of Deck-Girder System.

At abutments, the transverse grid lines are placed by joining the centers of bearings and minimum seven transverse grid lines are recommended, including transverse edge lines. In cases where cross-girders are not provided, the transverse grid lines should be normal to the longitudinal lines. The transverse grid lines are to be extended up to the extreme longitudinal grid lines, at the edges.

In skew bridges with small skew angle up to 15° and with no intermediate diaphragms, the transverse grid lines are made parallel to the support lines as shown in Figure 2.3 (a). Additional transverse grid lines are provided in between these support lines with their spacing not exceeding twice the spacing of longitudinal lines, which is the same as normal bridges.

In skew bridges with high skew angles in between 15° to 45°, the transverse grid lines are set along abutments, at either end. The interior transverse grid lines are not parallel to the end transverse lines, but are set by following the skew angle of the bridge, as shown in Figure 2.3 (b).

The Deck-Girder system may be either precast I-sections with in-situ reinforced concrete slab or cast-in-place T-beam where the girder is cast as monolithic with the deck slab. For the first case, the ratio of modulus of elasticity of slab and precast beam materials are to be mentioned properly in the analysis. This is referred to as a two-stage construction.

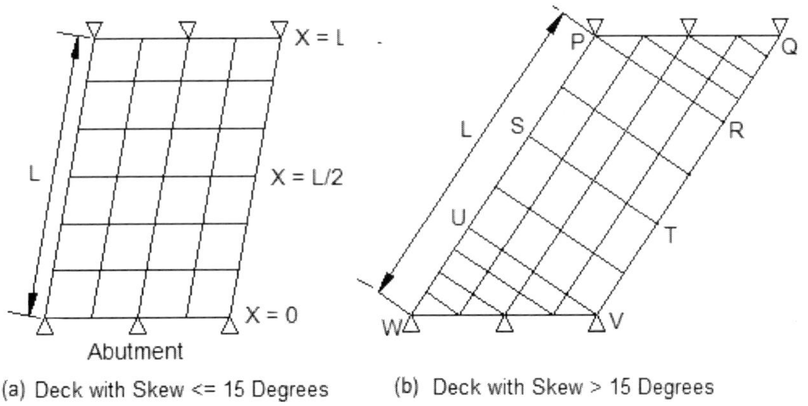

Figure 2.3. Arrangements of Grillage Members in Skew T-Beam Bridge without Cross Girders.

2.4. Grillage Modeling of Box-Girder Bridge

Idealization of a box-girder bridge as grillage model is similar to that of slab-on-girders type of deck systems but there is some behavior difference between them. The box-girder bridge usually has problems of shear deformations or shear lag due to usually wide flanges of the deck and distortions of the cells, if intermediate transverse diaphragms are not provided in sufficient numbers. The grillage analogy in this structural system has the added advantage in respect of time taken for the analysis and simple to comprehend.

The method is suitably adopted for multi-cell rectangular box-girder decks (Figure 2.4), where the effects of shear deformations and cell-distortions are negligible and could be ignored. However, it can also be used for decks with one or two cells only. The outer webs may be vertical or inclined, as shown in Figure 2.5.

In rectangular box girders, the longitudinal grid lines are placed by coinciding with webs of the actual structure (Figure 2.4). For box girders with sloping end webs (Figure 2.5), the grillage simulation is to be done by engineering judgment for placing the longitudinal members. However, a grid line is placed at the junction of the inclined web with slab as shown in Figure 2.5. Longitudinal grid lines are placed along the edges of the side cantilevers with nominal stiffness for making the grillage model convenient for analysis. For bridges with footpaths, additional longitudinal grid lines are placed at the centres of the footpaths in the similar way of slab and T-beam bridges, discussed earlier.

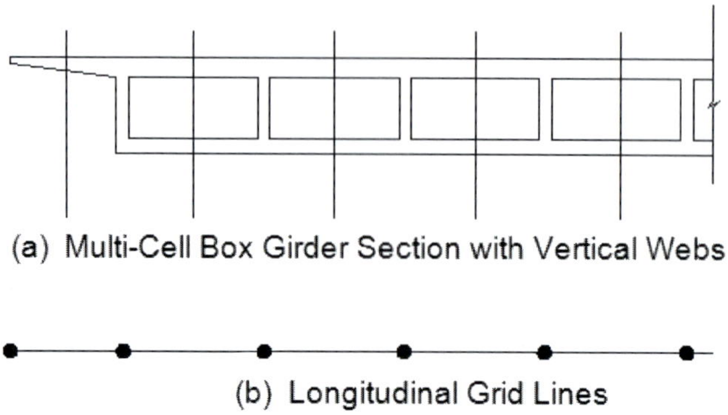

Figure 2.4. Longitudinal Grid Lines for Multi-Cell Box Girder Deck with Vertical Webs.

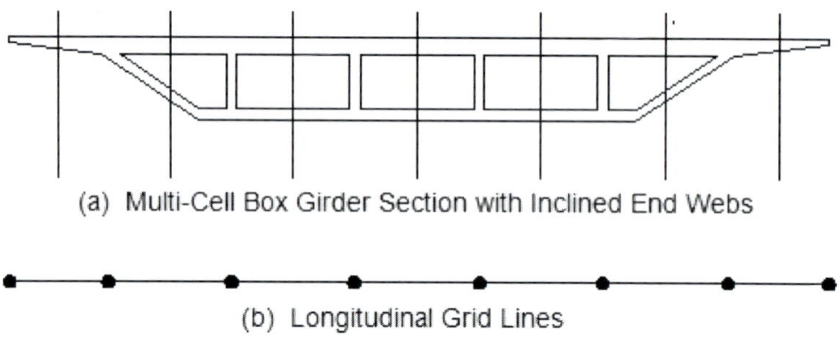

Figure 2.5. Longitudinal Grid Lines for Multi-Cell Box Girder Deck with Inclined Webs.

The top and bottom slabs with no diaphragms are represented by equally spaced transverse grid lines along the span. The number of grid lines and their spacing are similar to an as-adopted slab-on-girder bridge. If the deck is provided with diaphragms, the transverse grid lines are placed along each intermediate diaphragm and at supports. Additional grid lines, if needed, to meet the minimum requirements of transverse grid lines, may also be placed to account for top and bottom slabs and are placed in between the diaphragms. A closer spacing of transverse grid lines in the grillage model gives analysis results of more continuous structural behavior and will provide more details of forces and displacements.

For skew box-girder bridges, the procedure for placing of longitudinal and transverse grid lines is the same as that in the case of slab-on-girders bridges, discussed earlier.

2.5. Evaluation of Equivalent Elastic Properties

Once the bridge structure is simulated into the equivalent grillage model, by a set of longitudinal and transverse grid lines meeting at discrete nodes, the next step for grillage analogy method is to assign appropriate elastic properties. The evaluation of flexural and torsional stiffness of each member of the idealized grillage model needs the computation of equivalent flexural moment of inertia 'I' and torsional inertia 'J' for the members.

This is accomplished by considering isolated sections of the deck by considering them as individual beams. The inertias are calculated for each section and attributed to the related grillage members representing that section.

2.5.1. Flexural Moment of Inertia, 'I'

For different individual components like slab, 'T' or 'I' beams, box-girders etc. the computation of flexural moment of inertia 'I' is straight forward and needs no elaboration. However, in beams having the slab cast monolithically with the web of the beam, effective flange-width of the associated slab is considered.

2.5.2. Torsional Inertia, 'J'

The torsional inertia 'J', is generally not a simple geometric property of the cross-section like the flexural moment of inertia 'I' and needs proper consideration. There is no accurate analytical method for obtaining 'J'. Therefore, for determining 'J' for different cross-sections, the approximate method which is based on the elastic theory of torsion of prismatic beams is discussed here.

Saint-Venant derived an approximate expression for computing the torsional inertia 'J', of open sections and this is applicable to all cross-sectional shapes without having re-entrant corners as below;

$$J = \frac{A^4}{40\, I_p}$$

where 'A' is the area of cross-section and 'I_P' is the polar moment of inertia.

For a rectangle with sides b and d, the expression for 'J' becomes;

$$J = \frac{3b^3d^3}{10(b^2 + d^2)}$$

For a thin rectangle where b > 5d, the value of 'J' is more accurate by the expression;

$$J = \frac{bd^3}{3}$$

If the cross-section has re-entrant corners, then 'J' is quite less than that given by the first of the above three equations, therefore, the value of 'J' is obtained by conceptually sub-dividing the section into rectangular shapes avoiding re-entrant corners and summing the values of 'J' of these sub-divisions. The value of 'J' of a sub-divided part with conceptual cuts on two opposite faces may be computed as if the sub-division is a part of the thin and wide strip for which 'J' = $bd^3/3$. A typical T-section with re-entrant corners and its sub-division is described in Figure 1.6. For the portions 1, 2, 3 and 4 the 'J' values are designated as J_1, J_2, J_3 and J_4 respectively, then;

$$J_1 = \frac{1}{2}\left(\frac{b_1 d_1^{\,3}}{3}\right) \quad J_2 = \frac{1}{3}\left(\frac{b_2 d_2^{\,3}}{3}\right)$$

$$J_3 = \frac{3 b_3^{\,3} d_3^{\,3}}{10\left(b_3^{\,2} + d_3^{\,2}\right)} \quad J_4 = \frac{3 b_4^{\,3} d_4^{\,3}}{10\left(b_4^{\,2} + d_4^{\,2}\right)}$$

and the value of 'J' for the beam section as a whole, $J = J_1 + J_2 + J_3 + J_4$

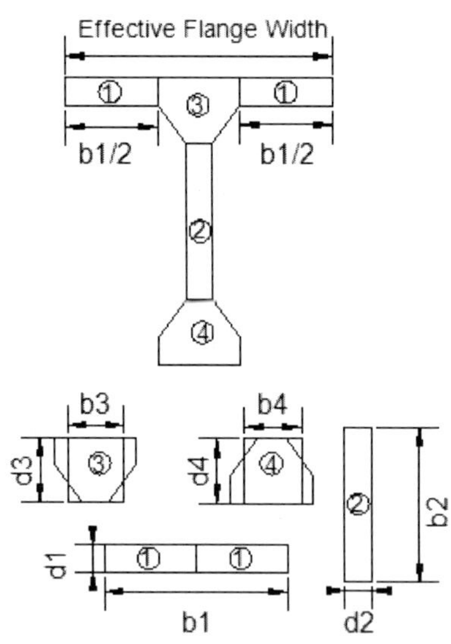

Figure 2.6. Sub-divided t-Section with Re-entrant Corners.

It may be noted that the value of 'J' of the part of deck slab forming the flange is to be divided into two halves to maintain the continuity in the other direction. Widths b_3 and b_4 of segments 3 and 4 are so adjusted to make the areas $b_3 \times d_3$ and $b_4 \times d_4$ same as original areas of the corresponding segments.

It is to be noted that the load distribution is likely to be more sensitive to the value of flexural moment of inertia 'I' rather than to that of torsional inertia 'J', and therefore, a small error in computing 'J' is not likely to affect much the final results. But an incorrect sub-division of a section will invariably lead to an under-estimation of 'J' and this may result a conservative design to a small extent.

2.6. Application of Loads, Analysis, Force Responses and Interpretations

The loads consisting of dead load, superimposed dead load, live and impact loads acting on the bridge deck-girder superstructure are to be appropriately distributed to the nodes of the grillage model. In the following sections, the application of various types of loads and their placement on the deck are discussed.

2.6.1. Dead Load

The dead loads (DL), including the superimposed dead loads (SIDL) of the deck-girder superstructure, is comprised of its own weight, wearing course, parapet, kerb etc. which are permanent and stationery in nature. The dead loads are acting as distributed loads on the deck surface and distributed on longitudinal grid members by idealizing into equivalent nodal loads. This is more relevant when the distributed load is non-uniform.

If the load is uniform all along the length of the linear grid members, then it is not necessary to calculate the equivalent nodal load, rather it can be treated as a uniformly distributed load (UDL). If the dead load is 'UDL', but its centroid does not coincide with the grid member then it is to be applied as vertical 'UDL' combined with torsion 'UDL'.

Figure 2.7 shows a solid slab with non-uniform spacing of grid lines of the grillage model. The equivalent vertical load along each grid line is calculated in the form of vertical UDL by considering its tributary area. For grid line 'A' the applied load is un-symmetric, therefore the equivalent load will consist of a combination of vertical and torsional UDL. Whereas for grid line 'B', the applied load is un-symmetric, therefore the equivalent load will be the vertical load only.

The self-weight of cross beam members and diaphragms need additional considerations. These beams located at specific intervals are actually small discrete loads on the grid members. However, for simplicity in the computation of the total weight of all the cross beams, each span is calculated and equally divided in the form of distributed loads to various grid members of the grillage model. The weights of railings, curbs, footpaths etc. are applied as a lumped mass on the edge member of the grid.

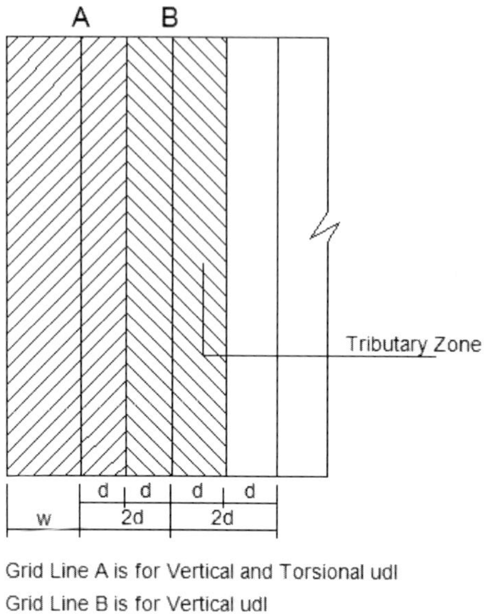

Figure 2.7. Longitudinal Grid Lines with ‘udl’ where ‘d’ is not equal to ‘w’.

2.6.2. Live Load

The live load on highway bridges is determined by the moving vehicle loads. The design standards of AASHTO-LRFD, BS Eurocode 2-BD 37/01, IRC 6 recommend different types of hypothetical loadings based on lane configurations of the carriageway on the bridge deck, for which a bridge is to be designed. The above design standards may be referred to for detail elaborations.

The vehicle live loads consist of a set of wheel loads. These are distributed over small areas of contacts of wheels forming patch loads. These patch loads are treated as concentrated loads acting at the centres of contact areas. This is a conservative assumption and facilitates the analysis. This assumption has very negligible effects on the results and does not make any significant difference in the design.

For example, a standard live load by a vehicle class of 10 axles and 20 wheel loads, each of 3.5 tonnes may be idealized as 20 point loads of 3.5 tonnes each, 10 loads on each track. The total load of the vehicle in this case is 70 tonnes.

For any standard live load by a vehicle class of only one vehicle loading per two lanes of the carriageway is considered. For a three lane bridge, the live loading may be either by three lane loading by one same class of vehicle or two lanes of loading by two different classes of vehicles. It is specified by loading standards that vehicles are not allowed to go close to the kerb for some specified distances. The distances are different for different types of loading.

The wheel loads of vehicles are placed either in the panels formed by the longitudinal and transverse members of the grid or directly on the nodes of the grid. The wheel loads falling in the panels are transferred to the surrounding nodes of the panels to facilitate the analysis. The procedure of distributing the point wheel loads to the four surrounded nodes of the panels are explained in the next section.

To obtain the worst force on the nodes by the wheel loads different positions of each type of loading system are to be tried on the grillage model, which is on the bridge deck. This is achieved by placing the wheel loads of a vehicle on the bridge and by moving it longitudinally and transversely in small increments by occupying a large number of different positions on the bridge deck.

2.6.3. Impact Load

The impact load on the bridge deck-girder superstructure is a major loading caused by the vibrations by the vehicle moving on the bridge. It is commonly considered as a percentage of the moving live load. Some design standards consider that the impact load varies with type of live load, span length and bridge type whether concrete or steel. The impact load is provided as a factor that is applied on the live load, to include the effect of impact.

2.6.4. Panels in the Grillage Model

The longitudinal and transverse members cross each other and panels are formed either of rectangular or parallelogram or triangular shapes, in normal skew bridges. All the wheels of a vehicular loading may not come directly on the nodes but mostly they come on the inside the panels of surrounding nodes. These wheel loads acting on the panels are transferred to the contiguous nodes forming the panel. Next, the grid is analyzed as a grillage model. It is therefore important to identify the panels in the idealized grillage model in which a particular wheel load is coming.

The identification of the panel includes its type as rectangular or parallelogram or triangular. In the grillage model the longitudinal members are parallel to the length of the bridge and transverse members are at right angles to the long members. The panels in a normal bridge are rectangular as shown in figure 2.8 (a), whereas, the panels in a skew bridge has triangular panels near the ends and rectangular panels in the central zone as shown in figure 2.8 (b).

In case of skew bridges where the skew angle is small, there the transverse members are considered not at right angles with the long members, rather these are parallel to the supports which are at an angle with the long members. This results the forming of parallelogram panels and are shown in Figure 2.8 (c).

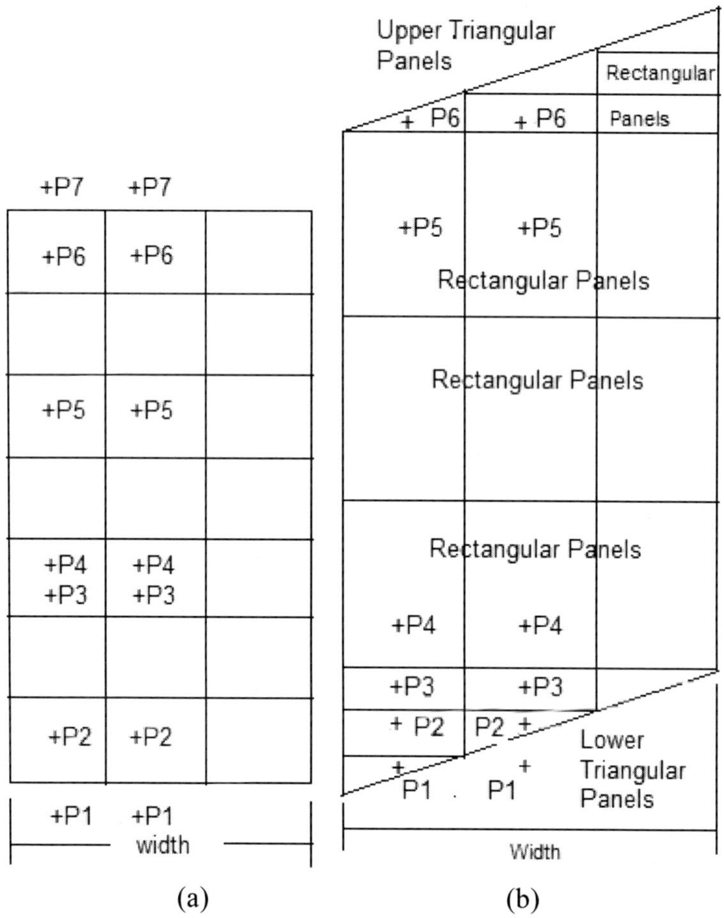

Figure 2.8. (a) & (b) - Rectangular Panels in Normal Bridge Deck.

In skew decks, it may so happen that, one wheel of an axle may fall either on the triangular panel or on the parallelogram panel and other wheels are outside the bridge deck, similar to the loads P1 and P6 in figure 2.8 (b) and 2.8 (c). In case of normal right bridges, all the wheels of an axle are either on a panel in the grid or outside it or outside the span, as shown in figure 2.8 (a). This happens when the axle loads are either entering a span or leaving a span. The grillage model is analyzed by taking into account only those wheel loads which are present in between either end supports of the grillage model.

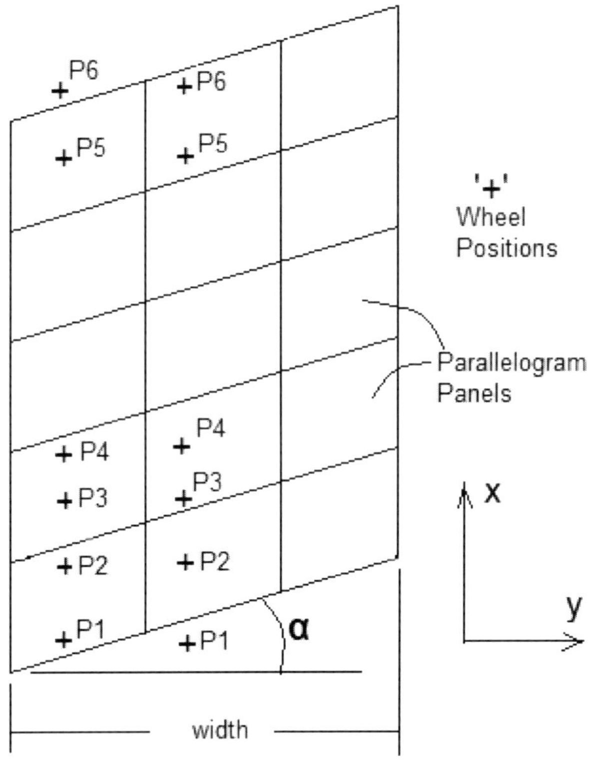

(c)

Figure 2.8. (c) - Triangular and Parallelogram Panels in Skew Bridge Deck.

At the time the wheels of vehicular loading system enters the bridge deck, the positions of the wheels are determined along with identifying the encountered panels one by one. The co-ordinates of the nodes and the grillage geometry are considered for such identification. Next, the load in the panel is transferred as equivalent loads to the contiguous nodes of the panels.

A panel may be subjected to one or more wheel loads. Each load is transferred to the nodes individually and the effects of all equivalent loads are algebraically summed up at each node for all the wheel loads located inside the panel. The nodes will also receive the loads from adjacent panels around them and the loads directly coming on the nodes.

2.6.5. Transfer of Loads to the Nodes

The grillage analysis is done by transferring the wheel loads in the panels to the nodes as equivalent loads. The equivalent nodal loads are computed by using any of the following approaches:

(i) By static approach where the wheel load is apportioned as equivalent vertical shear by considering that the panel in the grillage model is simply supported along its boundary.

(ii) By the approach where the equivalent load consists of vertical shear and bending moments by considering that the panel in the grillage model has fixed supports with applied degrees of freedom along its boundary.

The first approach is simpler, but it neglects the fixed end moments causing some error, which is not significant in longitudinal direction but such neglect in transverse direction may result incorrect transverse moments. The effect is more significant in the following situations:

(iii) When the wheel loads are applied on the cantilever part of the deck slab grillage model,

(iv) When the spacing of the girders are large,

(v) When very few wheel loads are acting on the deck.

If static division of the wheel load is intended for simplicity then it is required to keep smaller transverse spacing of the grid members to distribute the loads effectively in longitudinal and transverse directions. In such a case, additional (dummy) longitudinal grid lines are introduced in the grillage model. These dummy grid members are assigned zero section inertia values.

In the second approach, the wheel loads are distributed in the form of vertical shear and bending moments. Although the procedure is tedious but is superior in respect of results. Good computer software takes care of such tedious computations and is reliable for providing better results. Both approaches are used in the industry and for small grillage models, both give almost similar results. For the large size grillage model for a larger span of the bridge deck in between the supports at either end, the second approach is recommended.

2.6.5.1. Transfer of Dead Loads

The dead loads act as UDL on longitudinal grid members. These are distributed to the nodes of the grillage model in the following two ways:

(i) By considering that the longitudinal grid members are simply supported at nodes, the vertical load at each node of the longitudinal grid member due to UDL is obtained statically. The total equivalent vertical load on each node is obtained by summing up the loads coming from the adjacent members connected to the node.

(ii) By considering that the longitudinal grid members are fixed at nodes, the equivalent nodal loads will consist of a vertical load and a moment. Thus, if the UDL due to dead load is 'W' per unit length and is acting on any longitudinal grid line as shown in Figure2.9, the vertical force 'V' and moment 'M' at node 2 is given by;

$V = W \times (L_1 + L_2) / 2$ and $M = (WL_2^2 - WL_1^2) / 12$

In the similar way, the torsion 'T' per unit length due to the transverse eccentricity of loads is also distributed on the nodes treating the concerned longitudinal grid lines as fixed. For sign convention, downward vertical force is considered positive and for moments, the right hand screw rule is followed.

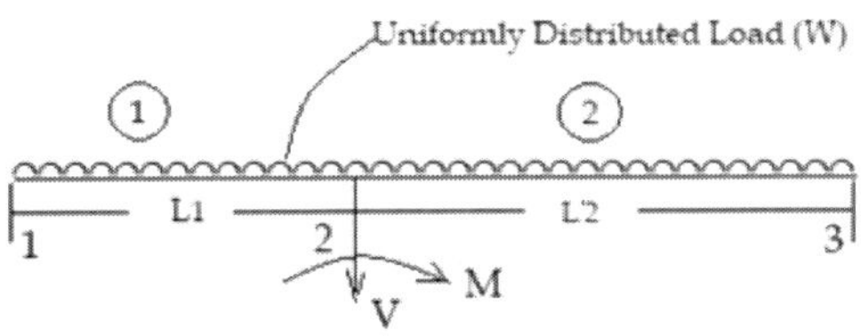

Figure 2.9. Transfer of Dead Load to the Nodes.

2.6.5.2. Transfer of Live Loads

The live loads are considered on footpaths, if provided, and on the bridge deck, by vehicular loading systems moving, in the form of UDL. These live loads are suitably transferred to the nodes of the grillage model for the analysis.

Footpath Live Load

If footpaths are provided on the bridge deck, the live loads carried by the footpaths are to be considered for the analysis. The footpath live load consists of a uniformly distributed load on its area over part or full length. Its magnitude etc. is considered as specified in the applicable design standard. The footpath live load may be distributed to the longitudinal grid members situated in their vicinity according to the tributary area of each grid member, and further transfer of this live load to the nodes of the grillage will also be done in the similar manner as dead load. The bridge structure is analyzed for footpath live load either on one side or both sides of the deck applied over part length and the developed forces are added to the forces due to other live loadings. Such combination may result in increase of forces.

Vehicular Live Load

The vertical wheel loads acting inside the panels of the grillage model are distributed in equivalent loads to the nodes of the concerned panel. This is done by following any one of the approaches as discussed above in (i) or (ii) in paragraph 2.5.2. Both approaches are applicable for rectangular, triangular and parallelogram panels.

The concentrated wheel load acting in the panel of the grillage model is transferred to the concerned nodes in two steps. First, the load is distributed, as equivalent forces, along the direction parallel to transverse grid members and then these forces are transferred to the adjacent nodes of the longitudinal grid members.

Case I: Equivalent Vertical Nodal Load

It is assumed that the panel is simply supported along its boundary.

In a rectangular panel (Figure 2.10), the equivalent vertical nodal loads are obtained by simple static division and are given by the following equations:

P1 = a x d x P / (Lx x Ly)
P2 = a x c x P / (Lx x Ly)
P3 = b x c x P / (Lx x Ly)
P4 = b x d x P / (Lx x Ly)

For a triangular panel at the lower side of the model, with load 'P' as shown in Figure 2.11, the equivalent nodal vertical loads are obtained by (i) distributing the load on longitudinal 1-2 and (ii) end transversal 1-3 at edges E and F respectively, in a direction parallel to the transversal 2-3. The equivalent vertical loads at E and F and on nodes 1, 2 and 3 are given by,

PE = (s – d) x P / s and PF = d x P / s
P1 = (c x PE / Lx) + (c x PF / Lx) = (c x P / Lx)
P2 = (a x PE / Lx) = (a x Ly - d x Lx) x P / (Lx x Ly)
P3 = a x PF / Lx = (d x P / Ly)

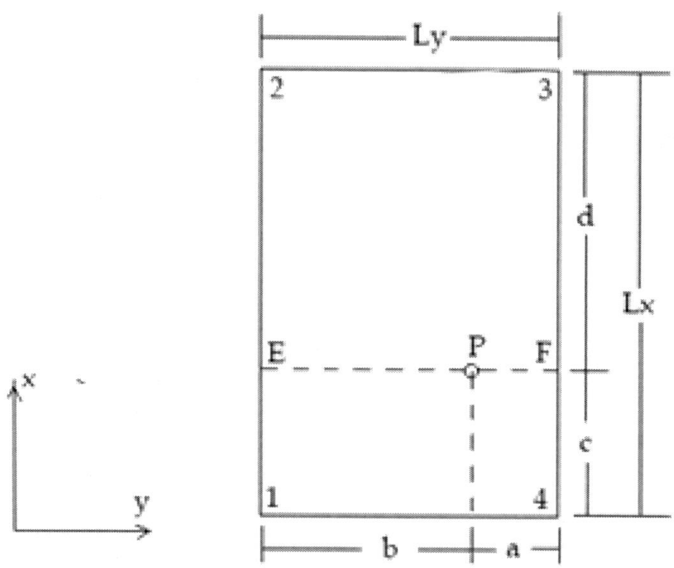

Figure 2.10. Static Distribution of Loads in Rectangular Panels.

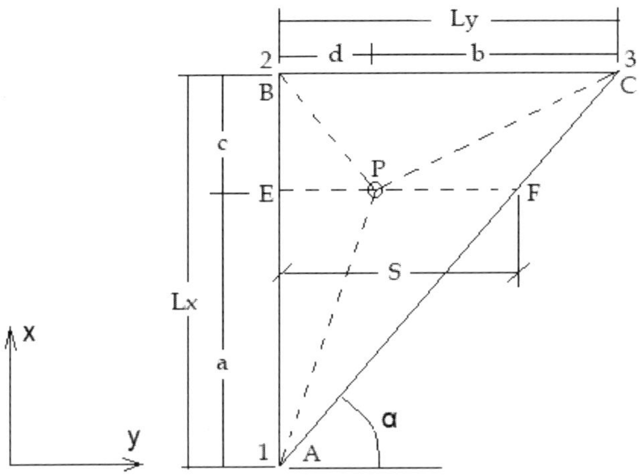

Figure 2.11. Static Distribution of Loads in Triangular Panels.

The nodal forces can also be obtained alternatively, by dividing the panel load *P* in ratios of the equivalent corresponding areas of triangles formed and the total area of the triangular element. The expressions for these may be as given below,

P_1 = (Area of Triangle BPC) x P / (Area of Triangle ABC)
P_2 = (Area of Triangle APC) x P / (Area of Triangle ABC)
P_3 = (Area of Triangle APB) x P / (Area of Triangle ABC)

The nodal loads P_1,P_2 and P_3 obtained from the first set of equations above will be the same as given by the second set of equations above, by substituting the dimensions in the figures 2.10 and 2.11.

The equivalent nodal loads in triangular elements at the upper side of the grillage model (Figure 2.8b) will be same in magnitude and direction as in lower triangular element given above. In panels having parallelogram shapes and containing the wheel load *P* (Figure 2.12), the equivalent nodal vertical loads at *E* and *F* and on nodes 1, 2, 3 and 4 are similarly evaluated and given in equations below.

$P_E = a \times P / L_y$ and$P_F = b \times P / L_y$
$P_1 = a \times d \times P / (L_x x L_y)$
$P_2 = a \times c \times P / (L_x x L_y)$
$P_3 = b \times c \times P / (L_x x L_y)$
$P_4 = b \times d \times P / (L_x x L_y)$

Case II: Equivalent Vertical Nodal Load and Moments

The concentrated wheel load located in a panel of the grillage model is distributed to the corresponding nodes of the panel in the form of vertical shear and moments. The panel is treated as bound by the contiguous grillage members as fixed along the edges. Again, the transfer of wheel load is done in two steps, as in Case I, first the load is distributed on the longitudinal at its edge parallel to the transversals in the form of vertical loads and moments and next, these are again transferred to the nodes of the longitudinal as in a fixed beam subjected to concentrated load and moments on its span. The distributions of loads to the nodes of the rectangular, triangular and parallelogram panels are Illustrated and the resulting equations for vertical shear and two moments at the nodes of the panels in each case are derived.

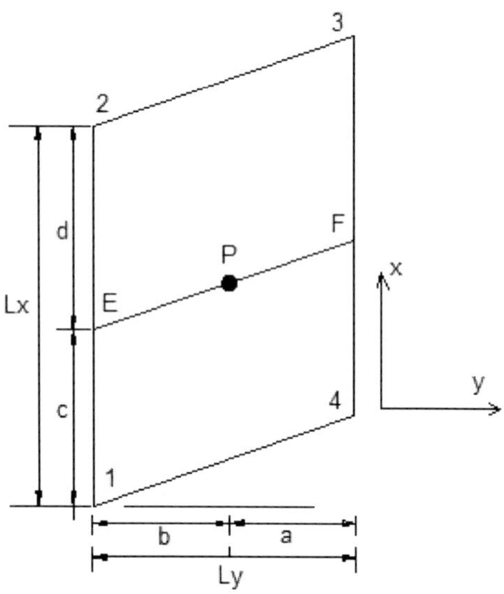

Figure 2.12. Static Distribution of Load in Parallelogram Panels.

Forces on the diagrams are shown in the positive direction of axes but proper signs are incorporated in the equations given. Negative signs preceding the equation indicate that the force is in the opposite direction to the vector drawn on the diagrams. Also, the equations pertaining to fixed end distribution of a point load, a bending moment and a torsional moment for a beam element are given below; these will be useful in subsequent discussions.

a) Fixed End Distribution of forces by the Point Load ‘P’ on beam AB are given by the expressions as below, and are illustrated in Figure 2.13 (a),
   $P_A = b^2 \text{ x } (3a + b) \text{ x } P/L^3$
   $P_B = a^2 \text{ x } (3b + a) \text{ x } P/L^3$
   $M_A = a \text{ x } b^2 \text{x } P/L^2$
   $M_B = - a^2 \text{ x } b \text{ x } P/L^2$

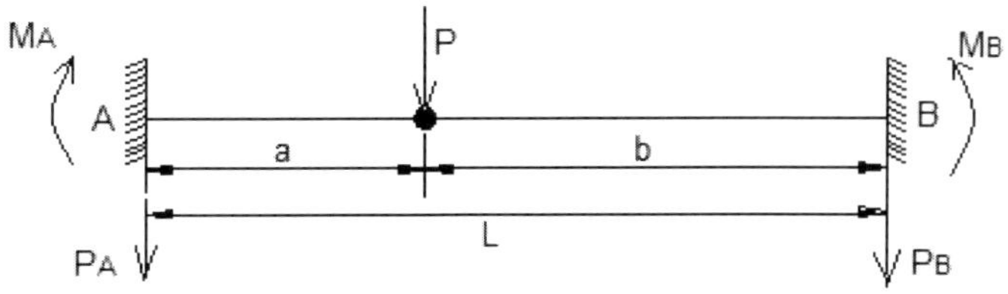

Figure 2.13. (a) - Distribution of Point Load.

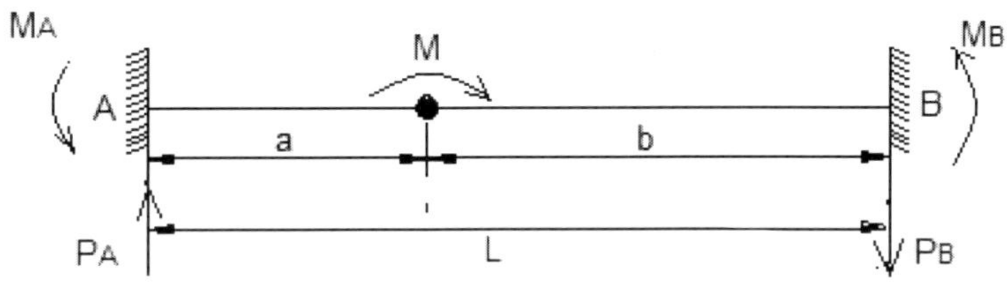

Figure 2.13. (b) - Distribution of Bending Moment.

b) Fixed End Distribution of forces by the Bending Moment 'M' on beam AB are given by the expressions as below, and are illustrated in Figure 2.13 (b),
$P_A = - P_B = - 6 \times a \times b \times M / L^3$
$M_A = - b \times (2a - b) \times M / L^2$ $M_B = - a \times (2b - a) \times M / L^2$

c) Fixed End Distribution of forces by the Torsional Moment 'T' on beam AB are given by the expressions as below, and are illustrated in Figure 2.13 (c),
$T_A = T \times b / L$
$T_B = T \times a / L$

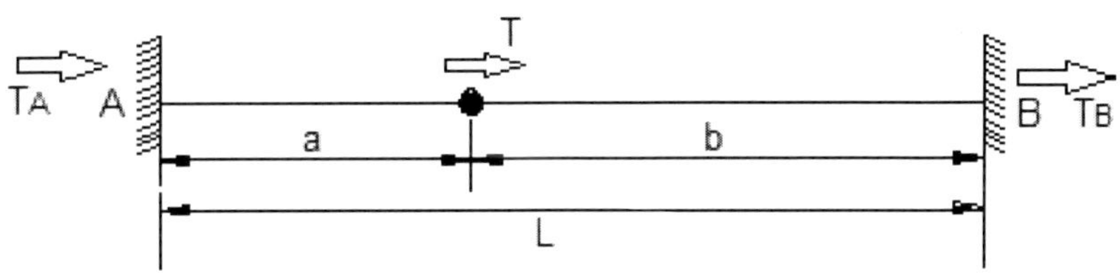

Figure 2.13. (c) - Distribution of Torsion.

Sign Convention: Clockwise Moment and Torsion and Downward Load are '+ve'

2.6.6. Forces Developed in a Rectangular Panel (Normal & Skew Bridges)

A wheel load 'P' is acting on a rectangular panel with nodes 1, 2, 3 and 4 as shown in Figure 2.14. The load 'P' is first transferred to points 'E' and 'F' lying on the longitudinal members 1-2 and 3-4. By considering the edges 'E' and 'F' as fixed, the vertical forces and moments about X-axis at 'E' and 'F' are given by the following expressions,

$P_E = a^2 x\,(3xb + a)\ x\ P / L_y^3$
$P_F = b^2 x\,(3xa + b)\ x\ P / L_y^3$
$M_{XE} = a^2\ x\ b\ x\ P / L_y^2$
$M_{XF} = -\ a\ x\ b^2\ x\ P / L_y^2$

By considering the member 1-2 as fixed at its ends, at node 1, 'P_E' will induce vertical force 'P_1' and moment 'M_{Y1}' and 'M_{XE}' will generate moment 'M_{X1}' in the directions shown in Figure 2.14 and the expressions are as follows:

$P_1 = d^2 x\,(3\ x\ c + d)\ x\ P_E / L_x^3$ $M_{x1} = d\ x\ M_{XE} / L_x$ $M_{y1} = -\ c\ x\ d^2\ x\ P_E / Lx^2$

In the similar way, the vertical force and the two moments, at other nodes of the panel can be calculated by the following expressions:

At node 2,
$P_2 = c^2 x\,(3\ x\ d + c)\ x\ P_E / L_x^3$
$M_{x2} = c\ x\ M_{XE} / L_x$
$M_{y2} = -\ c^2\ x\ d\ x\ P_E / Lx^2$

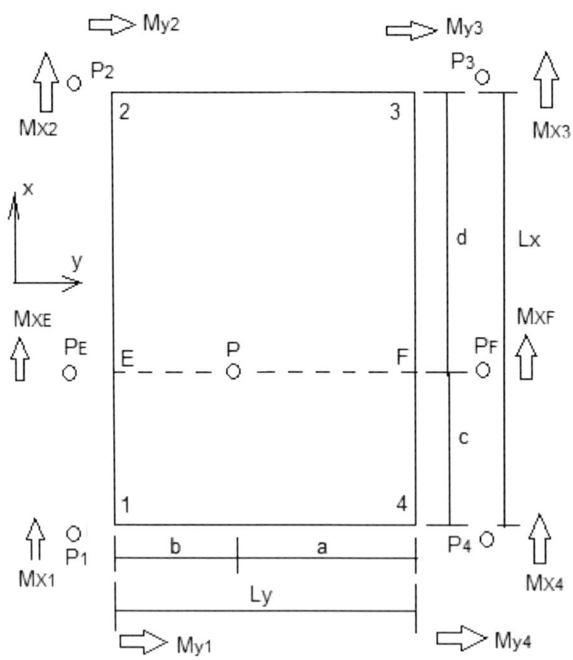

Figure 2.14. Distribution of Loads in Rectangular Panel.

At node 3,
$P_3 = c^2 x\,(3\ x\ d + c)\ x\ P_F / L_x^3$
$M_{x3} = c\ x\ M_{XF} / L_x$

$M_{y3} = - c^2 \times d \times P_F / Lx^2$

At node 4,

$P_4 = d^2 x (3 \times c + d) \times P_F / L_x^3$

$M_{x4} = d \times M_{XF} / L_x$

$M_{y4} = - c \times d^2 \times P_F / Lx^2$

2.6.7. Forces Developed in a Triangular Panel (Skew Bridges)

A lower triangular panel of the grillage model is described in Figure 2.15, acted upon by a load 'P'. The load 'P' is first distributed to members 1-2 and 1-3, parallel to the transversal 2-3 at 'E' and 'F' respectively. The vertical forces 'P_E' and 'P_F' and associated moments 'M_{XE}' and 'M_{XF}' at 'E' and 'F' respectively, are given by the following expressions,

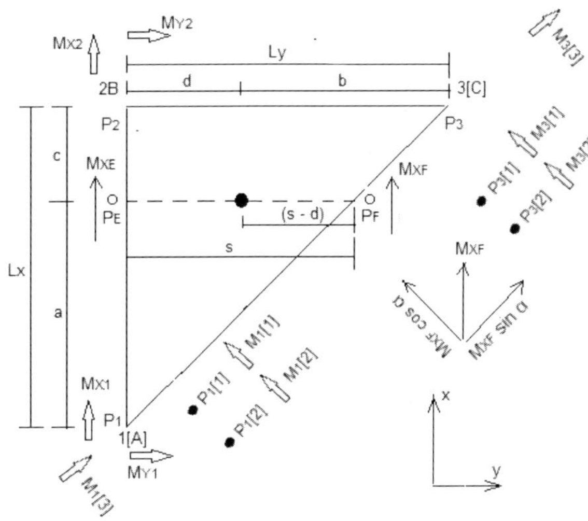

Figure 2.15. Load Distribution in Lower Triangular Panel.

Now, 'P_E' will result in vertical forces 'P_1' and 'P_2' and moments 'M_{Y1}' and 'M_{Y2}' and the moment 'M_{XE}' will result in moments 'M_{X1}' and 'M_{X2}' at nodes 1 and 2 respectively as shown. Similarly, 'P_F' on member 1-3 will result vertical forces '$P_1[1]$' and '$P_3[1]$' and moments '$M_1[1]$' and '$M_3[1]$' at nodes 1 and 3 respectively. The vertical forces and moments due to 'P_E', 'M_{XE}' and 'P_F' at nodes 1 and 3 are obtained by the expressions as described below,

At node 1 due to P_E,

$P_1 = c^2 \times (3 \times a + c) \times P_E / L_x^3$

$M_{x1} = c \times M_{XE} / L_x$
$M_{y1} = - c^2 \times a \times P_E / L_x^2$

At node 2 due to P_E,
P_2 or $PB = a^2 \times (3 \times c + a) \times P_E / L_x^3$
M_{x2} or $M_{XB} = a \times M_{XE} / L_x$
M_{y1} or $M_{YB} = - a^2 \times c \times P_E / L_x^2$

At node 1 due to P_F
$P_1[1] = c^2 \times (3 \times a + c) \times P_F / L_x^3$
$M_1[1] = c^2 \times a \times P_F / (L_x^2 \times \sin \alpha)$

At node 3 due to P_F
$P_3[1] = a^2 \times (3 \times c + a) \times P_F / L_x^3$
$M_3[1] = a^2 \times c \times P_F / (L_x^2 \times \sin \alpha)$

Next, 'M_{XF}' is acting in the direction of X-axis, and is resolved into components '$M_{XF} \sin \alpha$' and 'M_{XF}' $\cos \alpha$' in the directions parallel and perpendicular to member 1-3 respectively. 'M_{XF}' $\sin \alpha$ will result torsional moments '$M_1[3]$' and '$M_3[3]$' at nodes 1 and 3 and 'M_{XF}' $\cos \alpha$ will result bending moments '$M_1[2]$' and vertical force '$P_1[2]$' at node 1 and bending moment '$M_3[2]$' and vertical force '$P_3[2]$' at node 3. These resulting forces are obtained by the expressions as described below,

$M_1[3] = c \times M_{XF} \times \sin \alpha / L_X$
$M_3[3] = a \times M_{XF} \times \sin \alpha / L_X$
$P_1[2] = - 6 \times a \times c \times M_{XF} \times \sin \alpha \times \cos \alpha / L_X^3$
$M_1[2] = - c \times (2xa - c) \times M_{XF} \times \cos \alpha / L_X^2$
$P_3[2] = 6 \times a \times c \times M_{XF} \times \sin \alpha \times \cos \alpha / L_X^3$
$M_3[2] = - a \times (2xc - a) \times M_{XF} \times \cos \alpha / L_X^2$

At node 1, the total vertical force 'P_A', moment about x-axis 'M_{XA}' and about y-axis 'M_{YA}' are given by the following equations,

$P_A = P_1 + P_1[1] + P_1[2] = [c^2 \times (3 \times a + c) \times P_E/L_X^3]+[c^2 \times (3 \times a + c) \times P_F/L_X^3]-[6 \times a \times c) \times M_{XF} \times \sin \alpha \times \cos \alpha/L_X^3]$

$M_{XA} = M_1[1] \times \cos \alpha + M_{X1} + M_1[2] \times \cos \alpha + M_1[3] \times \sin \alpha = [c^2 \times a \times P_F \times \cot \alpha/L_X^2]+[c \times M_{XE}/L_X]-[c \times (2xa - c) \times M_{XF} \times \cos^2 \alpha/L_X^2] +[cx\, M_{XF} \times \sin^2 \alpha/L_X]$

$M_{YA} = M_{Y1} - M_1[1] \times \sin \alpha - M_1[2] \times \sin \alpha + M_1[3] \times \cos \alpha = -[c^2 \times a \times P_E/L_x^2] -[c^2 \times a \times P_F/L_x^2] + [c \times (2xa - c) \times M_{XF} \times \sin \alpha \times \cos \alpha/L_x^2] +[cx\, M_{XF} \times \sin \alpha \times \cos \alpha/L_X]$

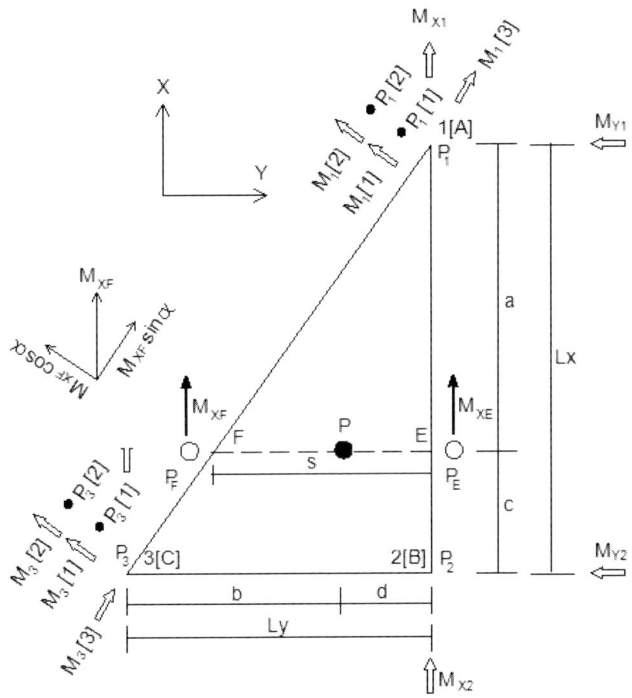

Figure 2.16. Load Distribution in Upper Triangular Panel.

Similarly at node 3, the total forces 'P_C', 'M_{XC}' and 'M_{YC}' are given by the following equations,

$P_C = P_3[1] + P_3[2] = [a^2 \times (3 \times c + a) \times P_F/L_X^3]+[6 \times a \times c) \times M_{XF} \times \sin\alpha \times \cos\alpha/L_X^3]$

$M_{XC} = M_3[1] \times \cos\alpha + M_3[2] \times \cos\alpha + M_3[3] \times \sin\alpha = -[a^2 \times c \times P_F \times \cot\alpha/L_X^2]-[a \times (2xc - a) \times M_{XF} \times \cos^2\alpha/L_X^2] +[a \times M_{XF} \times \sin^2\alpha/L_X]$

$M_{YC} = -M_3[1] \times \sin\alpha - M_3[2] \times \sin\alpha + M_3[3] \times \cos\alpha = -[a^2 \times c \times P_F/L_X^2] + [a \times (2xc - a) \times M_{XF} \times \sin\alpha \times \cos\alpha/L_X^2] +[a \times M_{XF} \times \sin\alpha \times \cos\alpha/L_X]$

Similar equations can be derived for an upper triangular panel also. The distributions of vertical forces, bending moments and torsion on nodes of the panel are explained in Figure 2.16.

The resultant vertical forces at joints A, B and C will remain same as described in the case of lower triangular panel. Also, the bending moments and torsion at the three nodes will be the same in magnitude but opposite in direction of that of the lower triangular panel.

2.6.8. Forces Developed in a Parallelogram Panel (Skew Bridges)

The load in a parallelogram panel can be distributed to its nodes in a similar manner as in the case of rectangular or triangular panels explained above. A parallelogram panel and the distribution of load to its nodes in the form of equivalent vertical loads and moments are shown in Figure 2.17.

The vertical loads and moments at edges E and F are given as

$P_E = [a^2 \text{ x } (3 \text{ x } b + a) \text{ x } P/L_Y^3]$
$P_F = [b^2 \text{ x } (3 \text{ x } a + b) \text{ x } P/L_Y^3]$
$M_E = [a^2 \text{ x } b \text{ x } P\text{x} \sec \alpha/L_Y^2]$
$M_F = - [a \text{ x } b^2 \text{x } P\text{x} \sec \alpha/L_Y^2]$
$M_{XE} = [a^2 \text{ x } b \text{ x } P/L_Y^2]$
$M_{XF} = - [a \text{ x } b^2 \text{x } P/L_Y^2]$
$M_{YE} = [a^2 \text{ x } b \text{ x } P\text{x} \tan \alpha/L_Y^2]$
$M_{YF} = - [a \text{ x } b^2 \text{x } P\text{x} \tan \alpha/L_Y^2]$

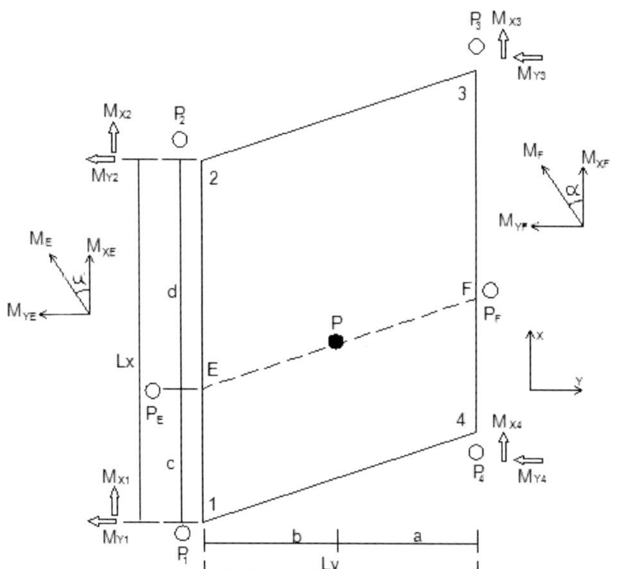

Figure 2.17. Load Distribution in Parallelogram Panel.

The forces at 'E' and 'F' are distributed to the nodes 1, 2, 3 and 4 and the resultant vertical loads, bending moments and torsions at each node are directly obtained by the following equations,

At node 1,
$P_1 = [d^2 \text{ x } (3 \text{ x } c + d) \text{ x } P_E/L_X^3] - [6 \text{ x } c \text{ x } d \text{ x } M_{YE}/L_X^3]$

$M_{X1} = [d \text{ x } M_{XE} / L_X]$
$M_{Y1} = - [c \text{ x } d^2 \text{ x } P_E / L_X^2] - [d \text{ x } (2 \text{ x } c - d) \text{ x } M_{YE} / L_X^2]$

At node 2,
$P_2 = [c^2 \text{ x } (3 \text{ x } d + c) \text{ x } P_E / L_X^3] + [6 \text{ x } c \text{ x } d \text{ x } M_{YE} / L_X^3]$
$M_{X2} = [c \text{ x } M_{XE} / L_X]$
$M_{Y2} = [c \text{ x } d^2 \text{ x } P_E / L_X^2] - [c \text{ x } (2 \text{ x } d - c) \text{ x } M_{YE} / L_X^2]$

At node 3,
$P_3 = [c^2 \text{ x } (3 \text{ x } d + c) \text{ x } P_F / L_X^3] + [6 \text{ x } c \text{ x } d \text{ x } M_{YF} / L_X^3]$
$M_{X3} = [c \text{ x } M_{XF} / L_X]$
$M_{Y3} = [c^2 \text{ x } d \text{ x } P_F / L_X^2] - [c \text{ x } (2 \text{ x } d - c) \text{ x } M_{YF} / L_X^2]$

At node 4,
$P_4 = [d^2 \text{ x } (3 \text{ x } c + d) \text{ x } P_F / L_X^3] - [6 \text{ x } c \text{ x } d \text{ x } M_{YF} / L_X^3]$
$M_{X4} = [d \text{ x } M_{XF} / L_X]$
$M_{Y4} = [c \text{ x } d^2 \text{ x } P_F / L_X^2] - [d \text{ x } (2 \text{ x } c - d) \text{ x } M_{YF} / L_X^2]$

The distribution of wheel loads acting in panels of different shapes to the associated nodes is done either as vertical forces only by static distribution or as a combination of vertical loads and moments. The decision is with the designer. A good bridge design software should use the transfer of loads, by replacing the vertical load lying in the panel by a combination of vertical loads and moments for better analysis results.

2.6.9. Grillage Analysis and Force Responses

The grillage model is given the geometry, material properties, section properties, support conditions and applied dead loads, super imposed dead loads and live loads. The computer program will transfer the live loads in the form of applied wheel loads to the nodes of the grillage model in the form of equivalent forces, and gives the results by determining the nodal deformations and member forces.

2.6.9.1. Analysis of Grillage Model

The grillage model is analyzed by using the direct stiffness method by computer programs. There are three possible displacements at each joint of the grillage model. The grillage model being in the X-Y plane is the concept followed in this book. The displacements in each node are joint rotations about X and Y axes and joint translation in Z-direction, normal to X-Y plane. The displacements in the same X-Y plane of grillage model and rotation about Z-axis are small and are ignored. However, in software ASTRA

Pro the plane of grillage model is considered in X-Z plane, where Y-axis is vertical which is perpendicular to the X-Z plane and is described in the relevant section of the user's manual of ASTRA Pro. The analysis of grillage model by the stiffness method has the steps as described below,

Step 1: Formulation of Stiffness Matrix

The first step in the analysis of grillage involves the formation of the stiffness matrix of the structure corresponding to the appropriate degrees of freedom. A 6 x 6 member stiffness matrix $[K_m]$ is created for each member of the grillage in terms of its geometric and elastic properties. The matrix $[K_m]$ is developed in terms of local degrees of freedom and is different for each member meeting at a joint. The matrix $[K_m]$ for a grillage member is therefore transformed into global degrees of freedom by using a 6 x 6 transformation matrix $[\lambda]$, by considering the orientation of the member. Hence, for each grillage member, the member global stiffness matrix $[K'_m]$ is obtained.

The assemblage of all the members meeting at that joint is considered, to satisfy the equilibrium conditions at the joint. The structure stiffness matrix [K] of the entire grillage model is obtained by assembling all member global stiffness matrices $[K'_m]$. This assembled matrix [K] is the addition of all internal forces which will be subsequently equated to the externally applied loads along the same degree of freedom.

The symmetry and banded nature of the matrix [K] is utilized in storing only the banded upper triangular portion of the matrix. This enables the computer program to analyze the grillage model of the bridge deck with a large number of nodes.

Step 2: Formulation of Load Vectors

The transferred external equivalent loads to the joints (nodes) of the grillage model constitute the load vector {B}, which are the moments about X and Y axes and vertical force along Z axis. The load vectors may either be formed separately for dead load, live load and impact load or their effects are added to form a single vector.

Step 3: Identification of Support Conditions

The stability of the structure is made by defining boundary conditions at the supports of the bridge deck grillage model. The support may be either on compressible (sinking) neoprene or elastomeric type bearings or rigid (non-yielding) steel or concrete (POT/PTFE) bearings. Depending on the type of bearings used, stiffness matrix developed above is modified accordingly. The supports are commonly defined in ASTRA Pro as 'PINNED' or 'FIXED BUT MX MZ', which releases or restrains the relevant degrees of freedoms of the node at the support and is based on considering the type of bearing proposed.

The elastomeric neoprene bearings have negligible rotational stiffness. The axial stiffness of the neoprene bearing is only considered in the stiffness matrix at the position corresponding to the vertical deflection of the supported node. Since bearing dimensions are not known in the beginning, a suitable size of the bearing is assumed initially and its axial stiffness is evaluated for analysis which may be modified, if reactions obtained are not as desired. Commonly rigid bearings are assumed initially and based on the reaction and rotations, as obtained from the analysis, the elastomeric neoprene bearing suitably dimensioned may be used as replacements and a revised analysis is carried out. The stiffness of the bearings has marked effects on the structural behavior in skew bridges, and it is very important to account for their effects properly.

The use of simple rigid type supports, need the removal of the rows and columns corresponding to the vertical deflections from the stiffness matrix. The load vector is also to be modified by removing the rows corresponding to the vertical deflection at the supported nodes. By solving the modified stiffness matrix using the modified load vector, the result gives the deformation vector and it does not contain the vertical deflections at the supported nodes. The support locations in the grillage model are defined at the same nodes where these actually exist.

For obtaining the reactions at the supports, the structure stiffness matrix [K] is partitioned into four sub-matrices pertaining to the free and restrained deformation vectors {DP} and {DR} and correspondingly load vectors is sub-divided into external load vector {P} and reaction vector {R}.

Step 4: Solution of Simultaneous Equations

A large number of simultaneous equations are created from the assembly of stiffness matrix considering equilibrium at each joint/node. The number of these equations depends upon the size of the grid. To solve these simultaneous equations the efficient techniques like Gauss-Elimination and Cholesky Factorization are used. The resulting deformation vector is used to compute member forces be using basic member material and section properties.

Step 5: Determination of Nodal and Member Deformations and Forces

The solution of simultaneous equations will produce nodal deformations of the structure. The member displacements in global as well as local coordinates can be evaluated by multiplying member stiffness matrix with the member displacement vector. The output consists of vertical deflection and rotations about X and Y axes at each node, shear force and bending moments at both the ends of each member and torsional moment in each member, along with reactions at the supports.

2.6.9.2. Force Responses

It is mentioned that the solution of equations results in nodal deformations i.e., deflection, slope and rotation at each end of the member. The shear force for a member, the bending moments at the two ends of the member, the torsional moment in a member and reactions at the supported nodes are usually produced as the output. Ordinarily the output is obtained for various longitudinal and transverse positions of different types of live loading. But, the output obtained is always very large. Scanning the analysis report for this output, for the worst member and joint forces obtained by analysis of a grillage model, even of moderate size, is a problem. Only, the critical values of the force responses are extracted from the analysis report for the design of various components of the bridge deck-girder structure.

For the design of any bridge structure we need the force envelope of various responses on it. The envelopes are the force response and are the largest values of responses picked up under dead load, SIDL and live load listed at each node of each grillage member along the beam/girder, from its start to middle to end.

This is achieved for the same dead load and SIDL but a particular live load by moving it over the deck in small increments both longitudinally and transversely. Each moving increment is considered as a separate load case for new wheel positions applying the loads, and the deck is analyzed. When the load moves from one position to the next position, for each load case, the force responses are again obtained for the new position of load and these values are computed and algebraically added with the previous force values, stored for each node. The largest values of each force responses like shear force bending moment and torsion for each grillage member shall be extracted from the output along corresponding to the load case. The process is repeated until the whole length and breadth of the bridge is moved by the live load.

The number of movements of loads in longitudinal and transverse directions by mentioning user given increment depends upon the factors like span, carriageway width, type of live loading, extent of accuracy desired, available computer time, etc. However, as a preliminary guidance, the minimum increment of movements of loads may be about 1/15th of span length or half the size of the mesh in longitudinal direction is chosen, the smaller the increment of movement will result in more load cases, hence a longer processing time. The movement of loads in transverse direction, for single or multi-lane loading of bridge deck by moving vehicles is very much limited due to the restrictions on wheels from coming closer to the curb by a specified distance. For multi-lane loading the lateral clearance between the vehicles in separate lanes also to be considered in mentioning the z-coordinates in ASTRA Pro, a-coordinate is always measured from the edge of the bridge deck in transverse direction. The value of x-coordinate is the length of the train of vehicles moving along the longitudinal direction of the bridge deck. At the start of the movement of vehicles, the x-coordinate is given as negative value equal to the length/span of the bridge deck, by considering the x-coordinates as zero at the start of the bridge deck. In the transverse analysis of bridge deck by effective width method, the loads may be moved transversely in five to seven equal intervals in a two lane bridge and in steps of about 750 mm in wider bridges.

2.6.10. Interpretation of Results of Grillage Analysis

The output or the result obtained from the analysis of grillage consists of vertical deflections and X and Y rotations of each node, shear force and torsional moment of each beam element as the grillage member, bending moments at the two ends of each beam element and reactions at each support.

The above results are to be judiciously used while designing a bridge deck. Since the deck has been initially idealized as a grillage model and the analysis has been performed on the idealized grid, the results depend on the boundary conditions of the supports. This may sometimes need modifications and proper interpretations before the result are finally used in design. Some of the important interpretations of the output and its modifications are required due to the local effects for slab bridges, slab-beam bridges and cellular/box-girder bridges are discussed below.

2.6.11. Grillage Model and Analysis for Slab Bridges

Modifications in the output results may be made, if necessary, due to local effects for slab bridges which are not considered earlier in the grillage analysis. The modified force responses may be used in the design for better accuracy. Some of the significant observations pertaining to force responses for slab bridges are described below.

The 'per unit force response' is the basis for the design of slabs. The computer output gives force response for the width which is particular for a grillage member. Hence, these force responses obtained from analysis output should be converted into 'per unit width', before these values are considered for design.

The analysis output gives various force responses with positive or negative values for each beam element. This needs careful consideration to the sign conventions adopted in the computer program. For example, a negative bending moment value at end 1 of a grillage member will denote a sagging moment while it indicates a hogging moment at end 2 of the same member.

When a grillage member is continued across a node, the values of moment at end 2 of one member and end 1 of the adjacent member in continuation are usually different.

This is because of the torsional moments in other adjacent members framing in other directions. To deal with such different values of moments at nodes along a line in the grillage model, the average value of the two moments may be taken, such pattern of "Saw Tooth" is described in Figure 2.18.

Only one value of the shear force for a grillage member is obtained from the output and may be used in design. Similarly, maximum reactions given in the analysis report, are taken as design values for reactions at supported nodes.

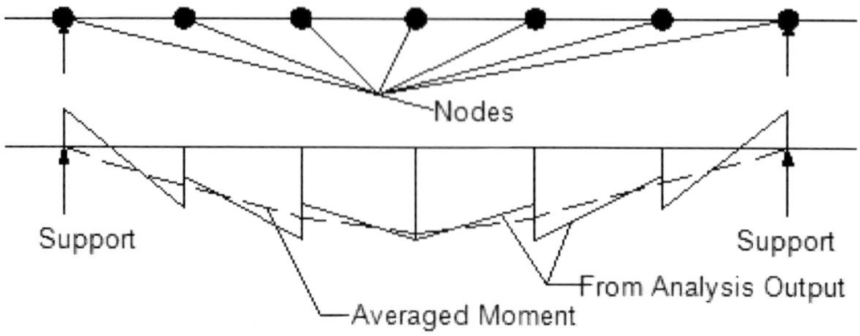

Figure 2.18. Saw-Tooth Pattern of Moment Diagram Using Grillage Analysis Results.

In an orthotropic slab with two different materials, the torque per unit width is same in orthogonal directions in the plane of the slab, however, it is often found different in the grillage analysis output. The torque per unit width at any point may be taken as the average of the two values.

The design of the section for the bending as shown in Figure 2.19 are based upon the principal moments m_l and m_2. The principal moments m_1, m_2 and their deviation at the point of interest is obtained from the values of bending moments per unit width m_X and m_Y in X and Y directions and the corresponding averaged torsional moment per unit width m_{xy} by using the equations as given below,

$$m_{1,2} = \frac{m_x + m_y}{2} \pm \left\{ \left(\frac{m_x - m_y}{2} \right)^2 + m_{xy}^2 \right\}^{1/2}$$

$$\tan 2\theta = \frac{2\, m_{xy}}{m_x - m_y}$$

The direction of reinforcement, in reinforced concrete bridges, may not always be along the direction of the principal moment. This is more common in slab bridges with skew. For such cases, the reinforcement component in the direction of each principal moment should be adequate.

At some a critical load combination, if at a bearing support has a net negative or downward reaction force as opposed to the usual positive or upward reaction force, then it indicates that under that combination of loading there is upward displacement at the bearing point. This may happen for bridges where the skew angle is large and there may be one or more such bearing support in the deck.

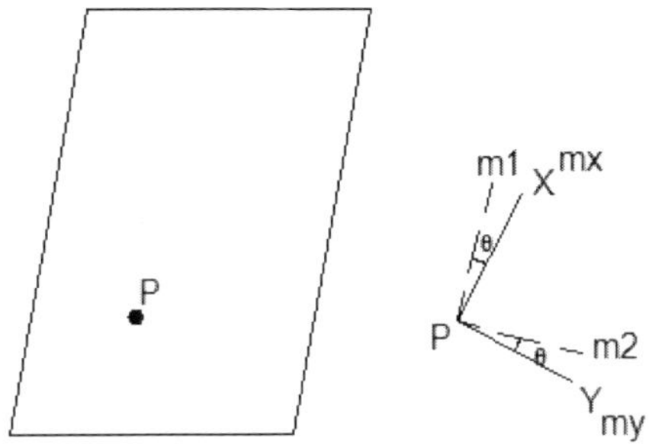

Figure 2.19. Principal Moments and their Deviation from X-Y Axis.

In the analysis, conditions of zero deflections are assumed at all the bearing points. Thus the behavior of the actual structure is not consistent with the assumed condition of the grillage model. In such case a repeat-analysis by force-deflection stage analysis is necessary. In the repeat-analysis, the displacements are imposed at the support nodes when analyzing under loading forming the above critical load combination. Displacements at the bearing supports are not desirable and should be prevented as far as possible. This could be done by providing tension POT/PTFE bearings at such bearing points as opposed to the usual compression bearings. In cases where a downward reaction force is small, lateral displacements in the position of the bearing supports can also change it to net upward reactions to achieve 'no lift'.

A slab bridge, can be idealized as an assembly of orthogonal beams only if Poisson's ratio of the slab material is zero. By ignoring the Poisson's ratio which is about 0.15 for concrete leads to under estimated moments. Such under estimation is usually negligible for longitudinal bending moments but considerable for transverse bending moments, as the curvature along the span is considerably large in the transverse direction. The following equations may be used to minimize errors resulting from ignoring the Poisson's ratio,

$$M_X = M_{xo} + YxM_{yo}$$
$$M_Y = M_{Yo} + YxM_{xo}$$

Where Mxo and MYo are the values obtained by grillage analysis for which Poisson's ratio = 0 and Mx and MY are the relevant corrected moments for an applicable value of Poisson's ratio.

For load dispersion, the actual area of application of load becomes larger than the contact area of the wheel with the slab. If such application area is larger than the panels of the grillage model, the load gets sufficiently dispersed for the grillage model to reproduce the distribution of moments throughout the slab. For such case no further modification of moments is necessary. On the other hand, if the application area of the load is small compared to the panels of the grillage model, no worthwhile information can be obtained about the local high values of force under the load, though the distributed moment field in the grillage model will simulate that in the deck.

2.6.12. Grillage Model and Analysis for Bridge Decks with Slab-on-Girders

The forces obtained from the analysis result may be considered for design in respect to certain assumption in analysis and actual structure. This may need some additional manual exercise to calculate the shear force, bending moments and torsion to be used in the structural design of the slab and girders of the bridge deck. In the following sections, some special considerations are discussed to make the analysis output useful for the structural design.

In decks with a beam and slab system, the stepping of moments in members on either side of a node also occurs. The difference in bending moments in two adjacent members meeting at a node will generally be large in outer girders, as exist in the grillage model. In the grillage model some grid members are actual beams as exist in the deck structure and some are dummy members. The dummy members are used to create additional nodes for getting forces at these nodes to facilitate the structural design. Where all the members meeting at the node are physical beams, the actual values of bending moments as obtained from the analysis result should be used. If at a node there are no physical beams in the other direction and the grid beam elements are dummy or represent a slab linear way, the average of the bending moments on either side of the node should be taken as there is no real beam of any significant torsional strength.

Design shear forces and torsions can be read directly from analysis output of the grillage model without any modifications.

In composite construction, where the stiffness of the member of the grillage model are calculated from two dissimilar material properties of slab and beam elements, the output force responses for shear, bending moment or torsion are attributed proportionally by considering the contribution to the particular stiffness by each element.

In the grillage model, when the longitudinal grid lines are not physically supported at the ends, the load carried by these lines are transferred toward the nearby supports through the end transversals. If this phenomenon is ignored, it will give a lower value of shear in supported longitudinal grid lines. To account for this under-estimation, the shear force of these un-supported beams may be added to the shear force of the adjacent physically supported beams. Similarly, in order to avoid under-estimation of the bending moments in supported longitudinal beams, the bending moments of unsupported grid lines also have to be considered for design of supported longitudinal beams.

Similarly, there may be an under-estimation of the bending moments in the supported longitudinal beams, as obtained by the analysis output. The bending moments of unsupported grid lines also have to be taken into account by adding the values to the bending moments in supported longitudinal beams. This will result in a safer design of supported longitudinal beams.

2.6.13. Grillage Model and Analysis for Box-Girder or Cellular Bridges

The grillage model of the box-girder is not by straight forward lumped mass technique. This is also a lumped mass technique but considers the contribution of the judiciously attributed section properties to the grid members of the grillage model. After obtaining the section properties for a cross sectional area, the Moment of Inertia about two principal axes for the whole section, the parts of these are assigned to the grid members of the grillage model, thus creating the idealized grillage model.

For this grillage model, the forces obtained from the analysis result may also be considered for design in respect to certain assumptions in analysis and actual structure. This may also need some additional manual exercise to calculate the shear force, bending moments and torsion to be used in the structural design of the slab and girders of the bridge deck. In the following sections, some special considerations are discussed to make the analysis output useful for the structural design.

The bending moment diagrams, if drawn by using the analysis output for box-girder or cellular deck, will also indicate the pattern of "saw tooth" as described in Figure 1.19 with large differences in bending moments at the nodes. This is the reason for the transfer of the torsional moment in the transverse members at each joint to bending moments and shear forces in the longitudinal member. The true design bending moment is obtained by taking the average value of the bending moments on the two sides of each joint. The stresses in the top and bottom slab are calculated from these average moments.

The transverse bending moment in the grillage member is equivalent to the opposed transverse compression of the top slab and tension of the bottom slab or the reverse. This is for the transverse flexure without distortion. In narrow decks, except in the diaphragms, the transverse bending moment is significantly small compared to the longitudinal bending moment. But in wide decks, the transverse bending moment may be large, especially near skew supports. From grillage analysis output, the transverse moment diagram also has a 'saw-tooth' pattern like the longitudinal moment diagram. The top and bottom slab stresses are calculated from the average moments.

The bending moments for slabs are derived from the shear force in the transverse grillage members. Some fractions of this shear force carried by the top and bottom slabs of the cell are assumed proportional to the flexural stiffness of the slabs. It is also assumed that the points of contraflexure lie midway between the webs and the moment at each end of a slab is simply the shear force it carries multiplied by half the distance between webs. The transverse moment in the cantilever slab can be taken directly from the output of the grillage analysis, because members defining the cantilever slab are not representing the cells of the box-girder. The total design moment in cellular box-girder deck is obtained by adding the slab moments of the cell, cantilever moments and the local moments.

The transfers of torsion-shear in the slabs of the box-girder must be calculated from the average torsion per unit width of transverse and longitudinal grillage model members. The shear force from grillage analysis output represents the total shear force in each web of the box-girder deck.

As the grillage model analysis ignores the effects of Poisson's ratio on the interaction of longitudinal and transverse moments, it causes neglect and a little error in narrow decks. In case of wide decks with little stiffness against cell-distortion, and while the moments are small and the Poisson's ratio is significant, the error is considerable in calculated transverse moments. However, since the concrete has relatively low Poisson's ratio of approximately 0.15, ignoring it may not result in any considerable error.

The grillage analogy method of analysis for bridge decks, is straight forward and is also acceptable to engineers without being well conversant in higher mathematics and numerical techniques. The setting out of the grid lines for physical as well as dummy members and the evaluation of elastic properties of the members can be handled by following the guidelines provided above. The determination of the dead loads and super imposed dead loads etc., is simple, as determined by the computer program by using the sectional area and density of materials relevant for the grid members. The transfer of live loads to nodes, formulation and inversion of matrix, solution of equations, evaluation of design responses etc. are done by an efficient computer software package. Preparation of input data is also simple as described in following section of this chapter, by referring to the versatile software ASTRA Pro and making it almost the same for many other market available software packages. The analysis input data can be done mechanically even without fully understanding the computer program. In ASTRA Pro, the input data for a set of analysis for various live loads is prepared by the software itself based on the general arrangement data for the bridge deck given by the user. The interpretations of results and its modifications due to local effects relevant to a particular type of deck may appear a bit difficult for the beginners, but after a few trials the users shall develop a better understanding of the analysis input, the analysis process and the structural behavior of the bridge deck obtained by the grillage analysis output report.

2.7. THE ANALYSIS DATA FOR THE DECK-GIRDER GRILLAGE MODEL

2.7.1. Title of the Analysis Data

The first line of the analysis input data has to start with the title, for example "Astra Floor PSC I Girder Bridge Deck Analysis with Moving Load." For understanding various data, a syntax user may refer to the relevant section of ASTRA Pro User's Manual.

2.7.2. Units for Linear Measurements and Loads of the Analysis Data

The analysis input data has to define the linear and load measurement units, for example 'UNIT FT KIP'. For understanding various data, a syntax user may refer to relevant section of ASTRA Pro User's Manual.

2.7.3. Joint Coordinates of Grillage Model

Referring to the AASHTO model in three dimensions, we have the joint co-ordinates data in the analysis file, the data mentions the section header as 'JOINT COORDINATES' followed by data as '10.0000.0000.000'.

There are four values in each line of data, these are joint/node numbers, X Co-ordinates, Y Co-ordinates and Z Co-ordinates. It may be observed that the 'Y co-ordinates' are of 'zero' values, for making the grillage model as in 'x-z' plane, where 'x' is measured along the length and 'z' is measured along the width of the bridge deck. In case the bridge is in skew, the co-ordinates are modified by applying the skew angle. For understanding the various data, a syntax user may refer to relevant section of ASTRA Pro User's Manual.

2.7.4. Elements/Members Connectivity of Grillage Model

The girders and edges of the deck slab along with cantilever slab are defined as linear beam members/elements. Every single beam element is defined between two nodes at the start and end of the element. The cross sections of the part of girders and part of deck supported by that part of the girder are considered to compute their masses. The masses are attributed to the beam elements forming the grillage model and the model is created as a 'Lumped Mass Model', for the finite element method of analysis. The data mentions the section header as 'MEMBER CONNECTIVITY' followed by data as '112'. There are three values in each line of data, the beam element/member number, start node/joint and end node/joint. For understanding the various data, a syntax user may refer to relevant section of ASTRA Pro User's Manual.

2.7.5. Section Properties of the Beam Elements/Members

For providing the section properties, the data mentions the section header as 'SECTION PROPERTIES', followed by the data values. The data values are of two alternative types, either the data type (i) as '1 TO 7 15 TO 84 92 TO 161 169 TO 223 320 TO 367 PRIS YD 8.0000 ZD 1.0', which mentions about the beam element/member numbers, width of beam element (YD) and depth of beam element/member (ZD) or the data type (ii) as '8 TO 14 162 TO 168 PRIS AX 783.2800 IX 105729.9933 IZ 259268.4480', which mentions about the beam element/member numbers, cross section area of beam element (AX), moment of inertia about x-axis (IX) and moment of inertia about x-axis (IZ). Any of the above two alternatives is used in the analysis data as per user's convenience. For understanding the various data, a syntax user may refer to relevant section of ASTRA Pro User's Manual.

2.7.6. Material Properties of the Beam Elements/Members

For providing the material properties, the data mentions the section header as 'MATERIAL CONSTANT', followed by the data values. The data values are the elastic modulus as 'E 4696 ALL', unit weight as 'DENSITY CONCRETE ALL' and the Poisson's ratio as 'POISSON CONCRETE ALL' applicable for all the beam element/member numbers. For understanding the various data, a syntax user may refer to relevant section of ASTRA Pro User's Manual.

2.7.7. Support at the Nodes/Joints of the Beam Elements/Members

For defining the supports, the data mentions the section header as 'SUPPORT', followed by the data either as '1 2 3 4 5 6 7 8 PINNED', which mentions the node/joint numbers for the support locations and the support type, or data type as '193 194 195 196 197 198 199 200 FIXED BUT FX MZ', which mentions the node/joint numbers for the support locations, the support type and removal of restraints along any desired degrees of freedom (DOF). There are three translational degrees of freedom (DOF) mentioned as FX-FY-FZ and three rotational degrees of freedom MX-MY-MZ, at each node/joint. For understanding the various data, a syntax user may refer to relevant section of ASTRA Pro User's Manual.

2.7.8. Applied Loads on the Beam Elements/Members

2.7.8.1. Dead Loads and Super Imposed Dead Loads

For defining the dead loads (DL) and super imposed dead loads (SIDL) on beam elements/members, the data mentions about the load name for the section header as 'LOAD 1 DEAD LOAD SELF WEIGHT1', followed by load type as 'MEMBER LOAD', followed by the applied load data as '248 249 270 TO 273 294 295 250 TO 269 274 TO 293 UNI GY -1.0229', which essentially contains the beam element/member numbers, the load type UNI for uniformly distributed loads, the sign -/+ for downward/upward direction of the load and the magnitude of the load in units as specified in paragraph 1.5.2 above. There are three translational degrees of freedom (DOF) mentioned as FXFY-FZ and three rotational degrees of freedom MX-MY-MZ, at each node/joint. For understanding the various data, a syntax user may refer to relevant section of ASTRA Pro User's Manual.

2.7.8.2. AASHTO-LRFD Class Live Loads

For defining the AASHTO-LRFD Live Loads (LL) on the grillage model, the data mentions about the section header as 'DEFINE MOVING LOAD FILE LL.TXT, followed by mentioning the applied load type as 'TYPE 1 LRFD_HL_93 1.10' where the live load type of AASHTO is mentioned along with the impact factor of '1.10', which is 10% additional to the live load. Various moving load details are mentioned and taken from the text file "LL.TXT" that contains the details of each live load class.

The next data is 'LOAD GENERATION 307', which mentions that the applied one-lane loading of 'TYPE 1' on the model is generated by moving the single train of loads on the grillage model for '307' times. The next data is 'TYPE 1 -43.620 0 6.500 XINC 0.5' which mentions start co-ordinates and increment for moving of the live load, by placing the start of the train of loads at 'x' co-ordinate '-43.620' feet and at a distance of '6.5' feet, which is the 'z' co-ordinate from the side edge of the bridge deck. The single train of load of length 43.620 feet will be moved along the longitudinal or 'x' direction of the model, at an increment 'XINC' of '0.5' feet for 307 times.

The data for start co-ordinates and increment for moving of the live load, may be mentioned as 'TYPE 1 -43.620 0 6.500 XINC 0.5' and next 'TYPE 1 -43.620 0 16.500 XINC 0.5', which defines that the applied two-lane loading of 'TYPE 1' on the model is generated by moving the two train of loads side-by-side on the grillage model for '307' times. The start of each of two trains of loads is placed at the 'x' co-ordinate '-43.620' feet, which is the longitudinal distance away from the start of the bridge deck at the 'x' co-ordinate '0.0'. The two trains of loads are at lateral distances of '6.5' and '16.5' feet respectively from the side edge of the bridge deck, which are the 'z' co-ordinates for the trains of loads. Each of the two trains of load of length 43.620 feet will be moved along the longitudinal or 'x' direction of the model, at an increment 'XINC' of '0.5' feet for 307 times. For understanding the various data, a syntax user may refer to relevant section of ASTRA Pro User's Manual.

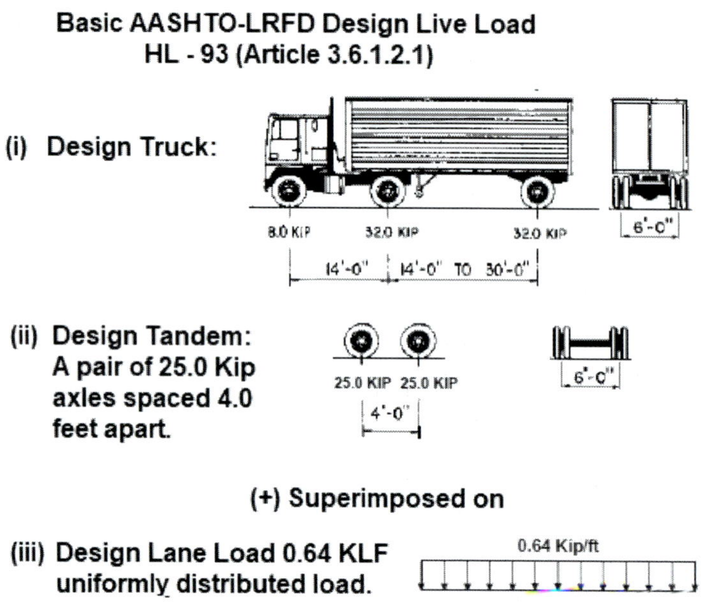

Figure 2.20. The AASHTO-LRFD Class Live Load applied as: [(i) + (iii)] or [(ii) + (iii)].

The various types of AASHTO-LRFD live loads are mentioned in FILE LL.TXT. There are four lines of data for each load type, the first line mentions for load type and vehicle load class, the second line mentions about axle loads in kip, the third line mentions about the spacing between axles and the fourth line mentions the width between wheels in the axle.

TYPE 1 LRFD_HTL57
10.5 10.5 10.5 10.5 10.5 4.5
1.6 4.572 4.572 1.6 4.572
1.800

TYPE 2 LRFD_HL93_HS20
4.0 16.0 16.0
4.2672 4.2672
1.800

TYPE 3 LRFD_HL93_H20
4.0 16.0
4.2672
1.800

TYPE 4 LRFD_H30S24
6.0 24.0 24.0
4.25 8.0
1.800

2.7.9. Analysis Specifications

The data for analysis specification is mentioned at the end of the analysis data as 'PRINT SUPPORT REACTIONS, PRINT MAX FORCE ENVELOPE LIST, PRINT MAX FORCE ENVELOPE LIST, PERFORM ANALYSIS and FINISH' each in separate lines. User is suggested not to modify the data, as the output may be affected in generating the essential details in the results.

2.8. Computer Applications for the Analysis of PSC I-Girder Bridge Grillage Model with AASHTO LRFD Live Load

The computer applications for the analysis of Bridge Deck-Girder Grillage model with Live Loads as per AASHTO LRFD. BS Eurocode2 BD 37/01 and IRC 112 / IRC 6 are described in detail in the softcopy volume 'User's Guide for Hands on Practice with Computer Applications' for the book 'Computer Aided Bridge Engineering' is available for download from web site 'www.roadbridgedesign.com'.

References

ASTRA Pro software, *User's Guide, Tutorial Videos, User's Manual and Design Manual*, Techsoft Engineering Services, Web site: www.techsoftglobal.com and www.roadbridgedesign.com

Bakht, B. And Jaeger, L. G., *Bridge Analysis Simplified*, McGraw Hill, New York, 1985.

Bridge Engineering Handbook, Second Edition, *Superstructure Design*, Edited by Wai Fah Chen and Lian Duan, Published by CRC Press, Taylor & Francis Group, 6000

Broken Sound Parkway, NW, Suite 300, Boca Raton, FL 33487, Web site: www.crcpress.com

British Standard, Design Manual for Roads and Bridges, Volume 1, Section 3, Part 14, BD 37/01, *Loads For Highway Bridges*, Published by: British Standards Institution.

British Standard, *Eurocode 2: Design of concrete structures*, BS EN 1992-1-1:2004, Published by: British Standards Institution.

Design Specifications, Customary U.S. Units, Sixth Edition 2012, American Association of State Highway and Transportation Officials. 444 North Capitol Street, NW, Suite 249, Washington, DC 20001, Phone 202-624-5800 / Fax 202-624-5806, Web site: www.transportation.org

Gere, M.J. and Weaver, W., *Matrix Analysis of Framed Structure*, D. Van Nostrand Co., USA, 1965.

Grillage Analogy in Bridge Deck Analysis, Prof. C. S. Surana and Prof. R. Agrawal, Narosa Publishing House, 6-Community Centre, Panchsheel Park, New Delhi 110017.

Hambly, E.C., *Bridge Deck Behaviour*, Chapman and Hall, London, 1976.

IRC 112-2011, *Code of Practice for Road Bridges*, Published by Indian Roads Congress.

IRC 6-2017, Standard Specifications and Code of Practice for Road Bridges, Section: II, *Loads and Load Combinations*, Seventh Revision, March 2017, Published by Indian Roads Congress, Kama Koti Marg, Sector-6, R. K. Puram, New Delhi 110022.

Pucher, A., *Influence surfaces of Elastic Plates*, Springer Verlag, Wien and New York, 1964.

Rusch, E.H.H. and Hergenroeder, A., *Influence Surfaces for Moments in Skew Slabs*, Werner-Verlag, Dusselforf, 1969.

Chapter 3

DESIGN OF PSC I-GIRDER BRIDGE DECK-GIRDER SUPERSTRUCTURE IN AASHTO-LRFD

ABSTRACT

By using AASHTO-LRFD standard, the design of the deck slab and cantilever slab are done by calculating bending moments, shear forces, bending resistance in transverse directions, bending resistance in longitudinal direction, checking flexural cracking. The design of PSC I-Girders is done for bending moments and shear forces by dead load, super imposed dead load (SIDL) and live loads (LL). The shrinkage strain, creep Strain and effect of temperature rise and fall are also determined. The design is complete for pre-stressing cables, end anchorages, un-tensioned reinforcements, end cross girders and shear connectors. The values in 'red color' are design input data by the user.

3.1. GENERAL

The span arrangement and configuration should be closely studied, first in respect to the site location of the bridge. A bridge crossing over a navigable waterway is very much dictated by the horizontal and vertical clearance required. It is also important to know the soil condition and landscape (e.g., over water, land, valley, or mountainous area). For segmental concrete bridges, uniformity of the span lengths is critical in order to maximize the benefit of pre-casting the segments. The more uniform the span distribution, the more economical the bridge. It is also preferable to have an uneven number of spans from the architectural point of view. A range of 20 to 50 metres for PSC I-Girder and 30 to 60 metres for incremental launched box girders bridges are recommended.

It is not recommended to place many piers with short span lengths. Therefore, the number of piers should be reduced and the span length increased. The shape and size of the piers are also important. For shorter span lengths, the lateral pier dimensions should be slender in order to reduce the wall view effect from an oblique view. For shallow valley, it is important to consider the L/H ratio of the opening between two piers, where L is the span length and H is the pier height. It is preferable to have an L/H ratio equal to or greater than 1.5. The end spans should be less than the typical span length (60% to 80% of typical span length) in order to achieve an efficient design.

The next step is to study the structural system of the bridge. The final bridge structural system is studied carefully, particularly the stability during construction and how the static system changes from stage to stage of the construction until the bridge is completed. This is the nature of segmental bridges, unlike other types of structures such as reinforced concrete or steel conventional types of bridges. The erection and stitching of the segment pieces require many stages and the structural system is constantly changing from time to time until completion.

Segmental bridge construction is not only popular for long span bridges over rivers and canyons but also in urban and city interchanges and for transit structures such as light rail and high-speed rail guide-way structures. Aesthetic value has become one of the most important parameters in selecting the bridge structure. The owners and stakeholders are placing more and more importance on aesthetically pleasing structures. Fortunately for segmental box girder bridges, the shape of the box girder is already inherently attractive. The architect should properly design the pier shape to ensure that the overall look of the bridge is compatible and suitable for a particular location.

The design forces obtained from the planer frame analysis and 3D live load influence lines may be combined in a spreadsheet using the LRFD Service Limit State and Strength Limit State combinations. The maximum tensile and compressive stresses at each predetermined section in the top slab are summarized and compared to the stress limits at service limit states specified in AASHTO-LRFD Bridge Design Specifications (AASHTO 2012).

The pre-stressing force is usually estimated in preliminary calculations, and then analyzed in a planer time-dependent run using the design software. All other loads are incorporated into the planer model, except live loads. The results are then compiled in a spreadsheet to check stresses. By varying the pre-stressing force, the combined stresses of service limit states are calculated. Using the selected tendon forces per unit length, the size and spacing of transverse tendons in the segment are determined.

The LRFD strength limit states may be also tabulated in a spreadsheet and an envelope of maximum and minimum values is determined for each chosen section. The values of the moment envelope can then be compared to the calculated bending capacities for each of the corresponding transverse components. Concrete service stresses were computed using several AASHTO standard specifications and the current LRFD specifications for comparison purposes.

3.1.1. Service Limit State Design (3.5.5)

When checking concrete stresses, only service limit state I is checked with a live load factor of 1.0 for tension as well as compression. Also, a linear temperature gradient of 10° F between interior and exterior surfaces of the box is used in service limit state I. The current LRFD specification does not specify this loading, leaving it up to the owner or designer to establish if it should be included on a project-by-project basis. This example is based on a load factor of 0.5 for transverse temperature gradient when accompanying live load. Also, in addition to service limit state I, LRFD requires a check for service load stresses due to dead load and full temperature gradient. This limit state can often govern at locations where live load influences are small.

In addition to service limit states under maximum loading, temporary stresses such as those prior to barrier placement and vehicular traffic should be checked to ensure that allowable stresses are not exceeded during the construction process. Service load combinations used in this example include the following:

Service I (Tension & Compression)

$$1.0(DC+DW+EL)+1.0(PT)+1.0(LL+IM)+1.0(CR+SH)+/-0.5(TG)$$

Segmental Load Combination (LRFD Equation 3.4.1-2)

$$1.0(DC+DW+EL)+1.0(PT)+1.0(CR+SH)+/-1.0(TG)$$

3.1.2. Strength Limit State Design (3.5.6.1 Flexural Strength Design Check)

For purposes of the transverse design, strength limit state IV is the same as strength limit state I without live load, with 25% more self-weight. This loading does not govern in this example.

For temperature gradient load factors, LRFD specifications suggest determining a load factor on a project-specific basis, with a recommendation of 0.0 in most instances. Since these loads are a result of restrained deformations, the loads should disappear if the reinforcement begins to yield at ultimate. In addition, the segmental guide specifications do not include this component in ultimate load combinations. For these reasons, the temperature gradient was not used in the strength limit state combinations below.

The LRFD specifications require minimum reinforcement equal to that required to resist 1.2 times the cracking moment. This requirement governs only the bottom slab (soffit) design. To satisfy the minimum steel requirement, the transverse bar spacing in the bottom soffit was decreased from 12 (205)to 8 in (203 mm), which represents an increase in reinforcement of 50%.

Also, under ultimate flexure, the amount of web steel reinforcing required for transverse bending should be calculated. This should be combined in an appropriate manner with the reinforcing required for longitudinal shear. Listed below is the ultimate load combination per LRFD: (Strength I)

$$\gamma_p DC + \gamma_p DW + 1.0EL + 1.75(LL + IM) + 0.5(CR + SH)$$

3.1.3. Shear Strength Design Check (3.5.6.2)

Traditionally, shear behavior has been ignored in the design of concrete decks for AASHTO bridges. Box girder decks are similar in this sense, but can often have large construction loads placed on them. In these special cases, both one-way and two-way action shear should be investigated.

3.1.4. Longitudinal Design (3.6.1 Design Methodology)

The following design example illustrates the longitudinal design process for a typical precast segmental bridge. The structure is assumed to be erected using the precast balanced cantilever construction method with a ground-based crane. Owing to changes in the static system during erection, as cantilevers are made continuous through cast-in-place closure joints, it is necessary to analyze the structure for time-dependent effects. Time-dependent analysis is a function of the segment casting date, times that the segments are incorporated into the structure, as well as dates associated with changes in the static system throughout the construction process.

It is customary to establish an assumed sequence of construction and to estimate a reasonable construction schedule. Casting and erection dates of the segments are established on the basis of the construction schedule and the segment production rate. Casting dates are a function of an assumed number of casting cells and the time required to cast each segment. For purposes of estimating these dates, the segment production rate is assumed as one typical segment per day per casting cell and one pier/expansion joint segment per week per casting cell. Segments are not to be erected earlier than one month after casting. During construction, when actual casting and erection dates become available, the stage-by-stage analysis should be rerun in order to obtain correct camber values.

Time-dependent properties of concrete are established on the basis of the environmental humidity and dimensions of the cross section, and can be adjusted for concrete composition (e.g., limestone aggregate), rate of hardening, and ambient temperature. Section properties are determined for each segment considering the effects of shear lag in the top and bottom slab.

The above information may be entered into a time-dependent analysis software. A stage-by-stage analysis may be performed using an assumed post-tensioning layout while carefully modeling appropriate boundary conditions for each step of the construction process. After the construction has been modeled, the structure may be stepped through time day 10,000 to allow all time-dependent effects to occur. It is also essential in statically indeterminate structures to sum up all locked-in forces that result from various stages of structural systems until day 10,000. Additional loads may be placed on the structure such as live load, temperature gradient, and support settlement, as appropriate, and analyzed for initial (at end of construction) and final conditions at day 10,000.

3.1.5. Tendon Layout and Envelope (3.6.2)

An approximate tendon layout can be based on preliminary calculations for construction loading of an atypical cantilever. Span continuity tendons can be estimated by preliminary design based on the final structure's approximate creep and shrinkage effects using load factor dead and live load combinations. The assumed layout can then easily be modified during the final design to satisfy all applicable LRFD limit state load combinations.

The preliminary design for this example indicates the need for twelve cantilever tendons and five bottom continuity tendons per web. On the basis of previous experience, two eight-strand continuity tendons were added in the top slab across the closure pour to control stresses resulting from temperature gradients. The final design resulted in an increase of one cantilever tendon and one bottom span continuity tendon at interior spans only.

The tendons used are based on a 12-strand system using 0.6 in (15.24 mm) diameter strands. Only 11 strands were used for bottom continuity tendons to provide space for 5% contingency post-tensioning as required for internal tendons. One out of 12 strands will provide approximately 8% of the contingency post-tensioning if needed. An empty duct was provided for the cantilever tendons combined with an anchorage on the last segment of the cantilever to allow for contingency post-tensioning. This empty duct should be grouted if no contingency tendons are required.

Note: The various formulas used in the design may be seen from the design Excel Worksheet of software ASTRA Pro by downloading from web site www.techsoftglobal.comunder 'downloads'. The values in 'Red and Blue Color' in Excel

Worksheet, are Design Input Data which may be changed by the User, but in Tab Pages the colored values are taken from other pages and should not be changed by the user. For any query write to: techsoftinfra@gmail.com.

3.2. Design of RCC Deck Slab

3.2.1. AASHTO-LRFD Design Step 4.1 Deck Slab Design

In addition to designing the deck for dead and live loads at the strength limit state, the AASHTO-LRFD specifications require checking the deck for vehicular collision with the railing system at the extreme event limit state. The resistance factor at the extreme event limit state is taken as 1.0. This signifies that, at this level of loading, damage to the structural components is allowed and the goal is to prevent the collapse of any structural components.

The AASHTO-LRFD Specifications include two methods of deck design. The first method iscalled the approximate method of deck design (S4.6.2.1) and is typically referred to as the equivalent strip method. The second is called the Empirical Design Method (S9.7.2).

The equivalent strip method is based on the following:

- A transverse strip of the deck is assumed to support the truck's axle loads.
- The strip is assumed to be supported on rigid supports at the center of the girders. The width of the strip for different load effects is determined using the equations in S4.6.2.1.
- The truck's axle loads are moved laterally to produce the moment envelopes. Multiplepresence factors and the dynamic load allowance are included. The total moment is divided by the strip distribution width to determine the live load per unit width.
- The loads transmitted to the bridge deck during vehicular collision with the railing system are determined.
- Design factored moments are then determined using the appropriate load factors for different limit states.
- The reinforcement is designed to resist the applied loads using conventional principles of reinforced concrete design
- Shear and fatigue of the reinforcement need not be investigated.

The Empirical Design Method is based on the laboratory testing of deck slabs. This testing indicates that the loads on the deck are transmitted to the supporting components mainly through arching action in the deck, not through shears and moments as assumed by traditional design. Certain limitations on the geometry of the deck are listed in S9.7.2. Once these limitations are satisfied, the specifications give reinforcement ratios for both the longitudinal and transverse reinforcement for both layers of deck reinforcement. No other design calculations are required for the interior portions of the deck. The overhang region is then designed for vehicular collision with the railing system and for dead and live loads acting on the deck. The Empirical Design Method requires less reinforcement in the interior portions of the deck than the Approximate Method.

For this example, the Approximate Method (Strip Width Method) is used.

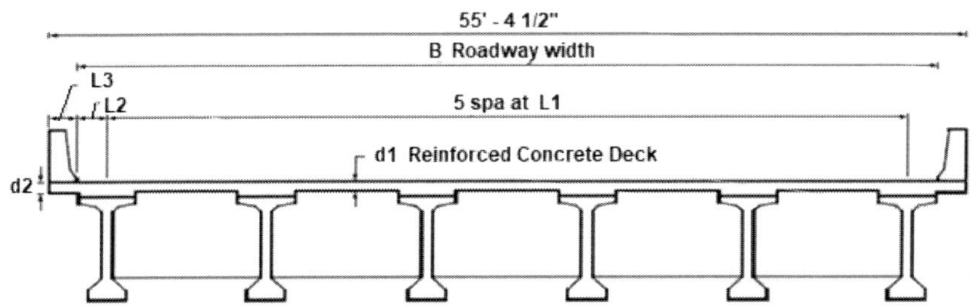

Figure 3.1. Bridge Cross-Section. Source: Figure 4-1 from AASHTO LRFD Bridge Design Specifications, FHWA Office of Bridges and Structures, 1200 New Jersey Ave., SE, Washington, DC 20590.

Required Design Input Data:
Width: = 55.375 ft
Roadway Width B = 52 ft

Distance from Outer girder CL to parapet edge L2 = 1.83 ft
Railings wide at the base L3 = 1.69 ft

Deck slab thickness d1 = 8 in.
Overhang slab thickness d2 = 9 in.
Girder spacing L1 = 9.66 ft
Top cover tc = 2.5 in. (S5.12.3)
(includes ½ in. integral wearing surface)

Bottom cover = 1 in. (S5.12.3)
Steel yield strength = 60 ksi
Slab conc. compressive strength = 4 ksi
Concrete density = 150 pcf
Future wearing surface density = 30 psf

3.2.2. AASHTO-LRFD Design Step 4.2 Deck Thickness

The specifications require that the minimum thickness of a concrete deck, excluding any provisions for grinding, grooving and sacrificial surface, should not be less than 7 in. (S9.7.1.1). Thinner decks are acceptable if approved by the bridge owner. For slabs with depths less than 1/20 of the design span, consideration should be given to pre-stressing in the direction of that span in order to control cracking.

Most jurisdictions require a minimum deck thickness of 8 in., including the 0.5 inch integral wearing surface.

In addition to the minimum deck thickness requirements of S9.7.1.1, some jurisdictions check the slab thickness using the provisions of S2.5.2.6.3. The provisions in this article are meant for slab-type bridges and their purpose is to limit deflections under live loads. Applying these provisions to the design of deck slabs is not the common practice.

For this example, a slab thickness of 8 in., including the ½ inch integral wearing surface, is assumed. The integral wearing surface is considered in the weight calculations. However, for resistance calculations, the integral wearing surface is assumed to not contribute to the section resistance, i.e., the section thickness for resistance calculations is assumed to be 7.5 in.

3.2.3. AASHTO-LRFD Design Step 4.3 Overhang Thickness

For decks supporting concrete parapets, the minimum overhang thickness is 8 in. (S13.7.3.1.2), unless a lesser thickness is proven satisfactory through crash testing of the railing system. Using a deck overhang thickness of approximately 0.75 in. to 1 in. thicker than the deck thickness has proven to be beneficial in past designs.

For this example, an overhang thickness of 9 in., including the 0.5 in. sacrificial layer is assumed in the design.

3.2.4. AASHTO-LRFD Design Step 4.4 Concrete Parapet

A Type-F concrete parapet is assumed. The dimensions of the parapet are shown in Figure 3.2. The railing crash resistance was determined using the provisions of SA13.3.1. The characteristics of the parapet and its crash resistance are summarized below.

Concrete Parapet General Values and Dimensions:
Weight per unit length = 650 lb/ft
Width at base = 1.687 ft

Moment capacity at the base calculated assuming the parapet acts as a vertical cantilever, Mc/length = 17.83 k-ft/ft

Parapet height, H " = 42 in.

Length of parapet failure mechanism, Lc = 235.2 in.

Collision load capacity, Rw = 137.22 k

Notice that each jurisdiction typically uses a limited number of railings. The properties of each parapet may be calculated once and used for all deck slabs.

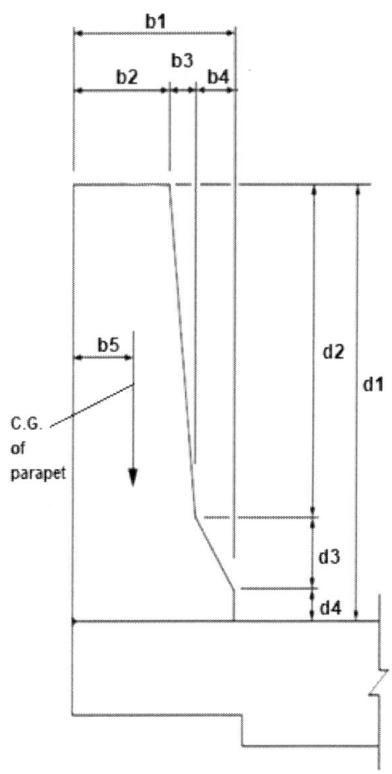

Figure 3.2. Parapet Cross-Section. Source: Figure 4-2 from AASHTO LRFD Bridge Design Specifications, FHWA Office of Bridges and Structures, 1200 New Jersey Ave., SE, Washington, DC 20590.

b1 = 1.67 ft

b2 = 1 ft

b3 = 0.28 ft

b4 = 0.41 ft

b5 = 0.63 ft

d1 = 3.5 ft

d2 = 2.67 ft

d3 = 0.58 ft

d4 = 0.25 ft

The load capacity of this parapet exceeds the minimum required by the specifications. The deck (SA13.4.2) overhang region is required to be designed to have a resistance larger than the computed resistance of the concrete parapet.

3.2.5. AASHTO-LRFD Design Step 4.5 Equivalent Strip Method (S4.6.2)

Moments are calculated for a deck transverse strip assuming rigid supports at web centerlines. The reinforcement is the same in all deck bays. The overhang is designed for cases of DL + LL at the strength I limit state and for collision with the railing system at the Extreme event II limit state.

3.2.6. AASHTO-LRFD Design Step 4.5.1 Design Dead Load Moments

Strength I load factors: (S3.4.1)

Slab and parapet:
Minimum = 0.9
Maximum = 1.25

Future wearing surface:
Minimum = 0.65
Maximum = 1.5

It is not intended to maximize the load effects by applying the maximum load factors to some bays of the deck and the minimum load factors to others. Therefore, for deck slabs the maximum load factor controls the design and the minimum load factor may be ignored.

Dead loads represent a small fraction of the deck loads. Using a simplified approach to determine the deck dead load effects will result in a negligible difference in the total (DL +LL) load effects. Traditionally, dead load positive and negative moments in the deck, except for the overhang, for a unit width strip of the deck are calculated using the following approach:

M = wl2/c

where:

M = dead load positive or negative moment in the deck for a unit width strip (k-ft/ft

w = dead load per unit area of the deck (ksf)
l = girder spacing (ft.)
c = constant, typically taken as 10 or 12

For this example, the dead load moments due to the self weight and future wearing surface are calculated assuming c = 10.

Self weight of the deck = 8 x (150)/12 = 100 psf

w = 100 psf
l = 9.66 ft
c = 10

Un-factored self weight positive or negative moment
= (100/1000)(9.66)2/10 = 0.93 k-ft/ft

Future wearing surface = 30 psf

Un-factored FWS positive or negative moment
= (30/1000) x (9.66)2/10
= 0.28 k-ft/ft

3.2.7. AASHTO-LRFD Design Step 4.6 Distance from the Center of the Girder to the Design Section for Negative Moment

For precast I-shaped and T-shaped concrete beams, the distance from the centerline of girder to the design section for negative moment in the deck should be taken equal to one-third of the flange width from the centerline of the support (S4.6.2.1.6), but not to exceed 15 in.

Girder top flange width = 42 in.
One-third of the girder top flange width = 14 in. < 15 in.

3.2.8. AASHTO-LRFD Design Step 4.7 Determining Live Load Effects

Using the approximate method of deck analysis (S4.6.2), live load effects may be determined by modeling the deck as a beam supported on the girders. One or more axles may be placed side by side on the deck (representing axles from trucks in different traffic lanes) and move them transversely across the deck to maximize the moments

(S4.6.2.1.6). To determine the live load moment per unit width of the bridge, the calculated total live load moment is divided by a strip width determined using the appropriate equation from AASHTO-LRFDTable S4.6.2.1.3-1. The following conditions have to be satisfied when determining live load effects on the deck:

Minimum distance from center of wheel to the inside
face of parapet = 1 ft. (S3.6.1.3)

Minimum distance between the wheels of two adjacent trucks = 4 ft.

Dynamic load allowance = 33% (S3.6.2.1)
Load factor (Strength I) = 1.75 (S3.4.1)
Multiple presence factor: (S3.6.1.1.2)

Single lane = 1.2
Two lanes = 1
Three lanes = 0.85

(Note: the "three lanes" situation never controls for girder spacing up to 16 ft.)

Trucks were moved laterally to determine extreme moments (S4.6.2.1.6)
Fatigue need not be investigated for concrete slabs in multi-girder bridges (S9.5.3 and S5.5.3.1)

Resistance factors, Φ, for moment:

0.9 for strength limit state (tension-controlled) (S5.5.4.2)
1.0 for extreme event limit state (S1.3.2.1)

In lieu of this procedure, the specifications allow the live load moment per unit width of the deck to be determined using AASHTO-LRFDTable SA4-1. This table lists the positive and negative moment per unit width of decks with various girder spacing and with various distances from the design section to the centerline of the girders for negative moment. This table is based on the analysis procedure outlined above and will be used for this example. In addition, some states use their own slab design charts and tables.

Table SA4-1 does not include the girder spacing of 9.67 ft. It does include girder spacing of 9.5 ft and 9.75 ft. Interpolation between the two girder spacing is allowed. However, due to the small difference between the values, the moments corresponding to the girder spacing of 9.75 ft are used which gives slightly more conservative answers than interpolating. Furthermore, the table lists results for the design section for negative moment at 12 in. and 18 in. from the center of the girder. For this example, the distance from the design section for negative moment to the centerline of the girders is 14 in. Interpolation for the values listed for 12 in. and 18 in. is allowed. However, the value corresponding to the 12 in. distance may be used without interpolation resulting in a more conservative value. The latter approach is used for this example.

3.2.9. AASHTO-LRFD Design Step 4.8 Design for Positive Moment in the Deck

The reinforcement determined in this section is based on the maximum positive moment in the deck. For interior bays of the deck, the maximum positive moment typically takes place at approximately the center of each bay. For the first deck bay, the bay adjacent to the overhang, the location of the maximum design positive moment varies depending on the overhang length and the value and distribution of the dead load. The same reinforcement is typically used for all deck bays.

3.2.9.1. Factored loads

Live Load

From AASHTO-LRFD Table SA4-1, for the girder spacing of 9.75 ft (conservative):

Un-factored live load positive moment per unit width = 6.74 k-ft/ft
Maximum factored positive moment = 1.75 k-ft/ft
Maximum factored positive moment per unit width = (1.75X6.74)
= 11.8 k-ft/ft

This moment is applicable to all positive moment regions in all bays of the deck. (S4.6.2.1.1)

Deck weight
(1.25X0.93) = 1.16 k-ft/ft
Future wearing surface
(1.5X0.28) = 0.42 k-ft/ft

Dead Load + Live Load Design Factored Positive Moment (Strength I Limit State)

MDL+LL = (11.8+1.16+0.42)

= 13.38 k-ft/ft

Notice that the total moment is dominated by the live load.

Resistance factor for flexure at the strength limit state, $\phi = 0.90$ (tension-controlled) (S5.5.4.2.1)

The flexural resistance equations in the AASHTO-LRFD bridge design specifications are applicable to reinforced concrete and pre-stressed concrete sections. Depending on the provided reinforcement, the terms related to pre-stressing, tension reinforcing steel and/or compression reinforcing steel, are set to zero. The following text is further explanation on applying these provisions to reinforced concrete sections and the possible simplifications to the equations for this case.

For rectangular section behavior, the depth of the section in compression,

c, is determined using Eq. S5.7.3.1.1-4: (S5.7.3.1.1-4)

c = (Apsfpu + Asfy - A'sf'y) / [α1f'c β1b +k Aps(fpu/dp)] (S5.7.3.1.1-4)

where:

Aps = area of pre-stressing steel (in^2)

fpu = specified tensile strength of pre-stressing steel (ksi)

fpy = yield strength of pre-stressing steel (ksi)

As = area of mild steel tension reinforcement (in^2)

A's = area of compression reinforcement (in^2)

fy = yield strength of tension reinforcement (ksi)

f'y = yield strength of compression reinforcement (ksi) b = width of rectangular section (in.)

dp = distance from the extreme compression fiber to the centroid of the pre-stressing tendons (in.)

c = distance between the neutral axis and the compressive face (in.)

α1 = stress block factor specified in S5.7.2.2 where α1 shall be taken as 0.85 for specified concrete compressive strengths not exceeding 10.0 ksi.

β1 = stress block factor specified in (S5.7.2.2)

K= 2 x (1.04-fpy / fpu)

For reinforced concrete sections (no pre-stressing) without reinforcement on the compression side of the section (or if the compression reinforcement is ignored), the above equation is reduced to:

C = Asfy / α1f'cβ1b

The depth of the compression block, a, may be calculated as:

$a = c\beta 1$

These equations for "a" and "c" are identical to those traditionally used in reinforced concrete design. Many text books use the following equation to determine the reinforcement ratio, ρ, and area of reinforcement, As:

$k' = Mu/(\phi bd2)$
$\rho = 0.85(f'c / f'y)[1.0 - \sqrt{2k'} / 0.85f'c]$ $As = \rho de$

A different method to determine the required area of steel is based on using the above equation for "a" and "c" with the Eq. S5.7.3.2.2-1 as shown below. The nominal flexural resistance, Mn, maybe taken as:

$Ma = Apsfps(dp - a/2) + Asfy(d's - a/2) - A'sf'y(d's - a/2) + \alpha 1f'c(b-bw)hf(a/2 - hf/2)$

where:

fps = average stress in pre-stressing steel at nominal bending resistance specified in Eq. S5.7.3.1.1-1 (ksi)
ds = distance from extreme compression fiber to the centroid of non-pre-stressed tensile reinforcement (in.)
d's = distance from extreme compression fiber to the centroid of compression reinforcement (in.)
b = width of the compression face of the member (in.)
bw = web width or diameter of a circular section (in.)
hf = compression flange depth of an I or T member (in.)
α1 = stress block factor specified in S5.7.2.2

For rectangular reinforced concrete sections (no pre-stressing) without reinforcement on the compression side of the section, the above equation is reduced to:

$Ma = Asfy(ds - a/2)$

From the equations for "c" and "a" above, substituting for:

$a = c\beta 1 = Asfy / \alpha 1f'cb$ in the equation for Ma above yields:

$Ma = Asfy\ (ds - a/2) = fydsAs - (f2y / 1.7\ f'cb)A2s$

Only As is unknown in this equation. By substituting for b = 12 in., the required area of reinforcement per unit width can be determined by solving the equation.

Both methods outlined above yield the same answer. The first method is used throughout the following calculations.

For the positive moment section:

Bar diameter = 0.625

Integral wearing surface = 0.5

$\Phi = 0.9$

d = 6.19

de = effective depth from the compression fiber to the centroid of the tensile force in the tensile reinforcement (in.)

= total thickness – bottom cover – ½ bar diameter – integral wearing surface

= (8-1-0.5X0.625-0.5)

= 6.19 in.

$k' = Mu/(\phi bd2)$

= [13.38/(0.9X1X6.19^2)]

= 0.388 k/in^2

$\rho = 0.85 \times (f'c / fy) \times [1.0-\sqrt{\{1.0-(2k' / 0.85f'c)\}}]$

= 0.00688

Therefore,

Required As = ρde = (0.00688X6.19) = 0.0426 in^2/in.

Required #5 bar spacing with bar area 0.31 in^2 = (0.31/0.0426) = 7.28 in.

Use #5 bars at 7 in. spacing

Check Maximum and Minimum Reinforcement

Based on past experience, maximum and minimum reinforcement requirements never control the deck slab design. The minimum reinforcement requirements are presented in S5.7.3.3.2. These provisions are illustrated later in this example.

Maximum reinforcement requirements were deleted from S5.7.3.3.1 in 2005. However, reinforced concrete sections in flexure should be checked for ductility in accordance with S5.7.2.1. The strain in the tension steel should be equal to or greater than 0.005 just as the concrete in compression reaches a strain of 0.003. This condition is called "tension-controlled".

Calculate depth of compression block:

T = tensile force in the tensile reinforcement (k)
= (0.31X60)
= 18.6 k
a = [18.6/(0.85X4X7)]
= 0.78 in.

β1 = ratio of the depth of the equivalent uniformly stressed compression zone assumed in the strength limit state to the depth of the actual compression zone

= 0.85 for f'c = 4 ksi (S5.7.2.2)

c = (0.78/0.85)
= 0.918 in.

in which:

βs = 1 + dc / 0.7(h - dc)

where:

γe = exposure factor
= 1 for Class 1 exposure condition
dc = thickness of concrete cover measured from extreme tension fiber to center of the flexural reinforcement located closest thereto (in.) = 1.3125 in.
fss = tensile stress in steel reinforcement at the service limit state (ksi)
= Maximum stress = 36 ksi
h = overall thickness or depth of the component (in.)
= 7.5 in.

By substituting dc and h,
βs [1+1.3125/(0.7X(7.5-1.3125))] 1.303

Therefore,
s ≤ 700γe / βsfs - 2dc
is required ≤ [700X1/1.303X36-(2X1.3125)] 12.3 in.
is provided = 7 in. < 12.3 in. OK

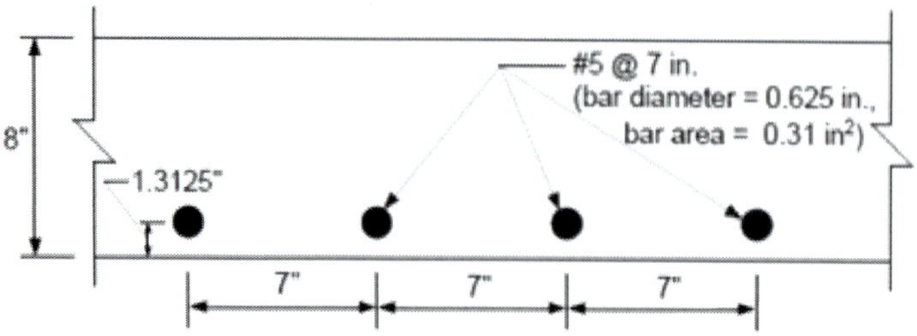

Figure 3.3. Bottom Transverse Reinforcement. Source: Figure 4-3 from AASHTO LRFD Bridge Design Specifications, FHWA Office of Bridges and Structures, 1200 New Jersey Ave., SE, Washington, DC 20590.

Stresses under Service Loads (S5.7.1)

In calculating the transformed compression steel area, the specifications require the use of two different values for the modular ratio when calculating the service load stresses caused by dead and live loads, 2n and n, respectively. For deck design, it is customary to ignore the compression steel in the calculation of service load stresses and, therefore, this provision is not applicable.

For tension steel, the transformed area is calculated using the modular ratio, n.

Modular ratio for 4 ksi concrete, n = 8

Assume stresses and strains vary linearly

Dead load service load moment = 0.93 + 0.28 = 1.21 k-ft/ft
Live load service load moment = 6.74 k-ft/ft
Dead load + live load service load positive moment = 7.95 k-ft/ft

The transformed moment of inertia is calculated assuming elastic behavior, i.e., linear stress and strain distribution. In this case, the first moment of area of the transformed steel on the tension side about the neutral axis is assumed equal to that of the concrete in compression. The process of calculating the transformed moment of inertia is illustrated in Figure 3.4 and by the calculations below.

For 4 ksi concrete, the modular ratio, n = 8 (S6.10.1.1.1b or by dividing the modulus of elasticity of the steel by that of the concrete and rounding to the nearest integer number as required byS5.7.1).

Assume the neutral axis is at a distance "y" from the compression face of the section.

Assume the section width equals the reinforcement spacing = 7 in.

The transformed steel area = (steel area)(modular ratio) = [0.31X8) = 2.48 in^2.

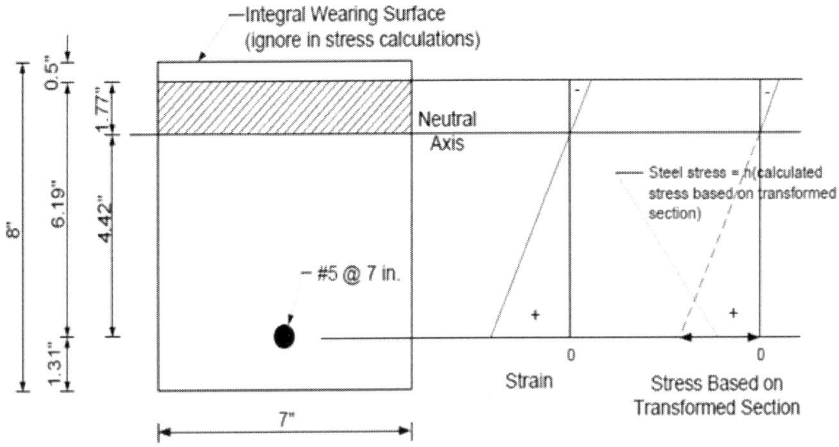

Figure 3.4. Crack Control for Positive Moment Reinforcement under Live Loads. Source: Figure 4-4 from AASHTO LRFD Bridge Design Specifications, FHWA Office of Bridges and Structures, 1200 New Jersey Ave., SE, Washington, DC 20590.

By equating the first moment of area of the transformed steel to that of the concrete, both about the neutral axis:

$$2.48(6.19 - y) = 7y(y/2)$$

Solving the equation results in y = 1.77 in.

$I_{transformed} = 2.48 \text{ x } (6.19 - 1.77)^{\wedge}2 + 7 \text{ x } (1.77)^{\wedge}3/3 = 61.4$ in4

Stress in the steel, fs = (Mc/I)n, where M is the moment acting on 7 in. width of the deck.

fs = [[(7.95 x (12/12) x 7 x (4.42)]/61.4]8 = 32.05 ksi

Allowable service load stress = 36 ksi > 32.05 ksi OK

3.2.10. AASHTO-LRFD Design Step 4.9 Design for Negative Moment at Interior Girders

3.2.10.1. Live Load

From AASHTO-LRFDTable SA4-1, for girder spacing of 9.75' and the distance from the design section for negative moment to the centerline of the girder equal to 12 in. (see Design Step 3.2.8 (AASHTO-LRFD Design Step 4.7) for explanation):

Un-factored live load negative moment per unit width of the deck = 4.21 k-ft/ft

Maximum factored negative moment per unit width at the design section for negative moment = 1.75 x 4.21 = 7.37 k-ft/ft

3.2.10.2. Dead Load

Factored dead load moments at the design section for negative moment:
Dead weight

1.25 x 0.93 = 1.16 k-ft/ft

Future wearing surface

1.5 x 0.28 = 0.42 k-ft/ft

Dead Load + live load design factored negative moment
= (1.16+0.42+7.37) = 8.95 k-ft/ft
d = distance from compression face to centroid of tension reinforcement (in.)
= total thickness – top cover – ½ bar diameter
Assume #5 bars; bar diameter = 0.625 in bar area = 0.31 in^2

d = (8-2.5-0.5X0.625) = 5.19 in.
Required area of steel = 0.0339 in^2/in.
Required spacing = (0.31/0.0339) = 9.14 in.
Use #5 at 9 in. spacing

As indicated earlier, checking the minimum and maximum reinforcement is not expected to control in deck slabs.

Check for cracking under service limit state (S5.7.3.4).

As explained earlier, service load stresses are calculated using a modular ratio, n = 8

Dead load service load moment at the design section for negative moment near the middle = -1.21 k-ft/ft.

D = 8 in
d1 = 5.19 in
d2 = 2.81 in
d3 = 1.44 in
d4 = 3.75 in
b = 9 in
girder = -4.21 k-ft/ft.

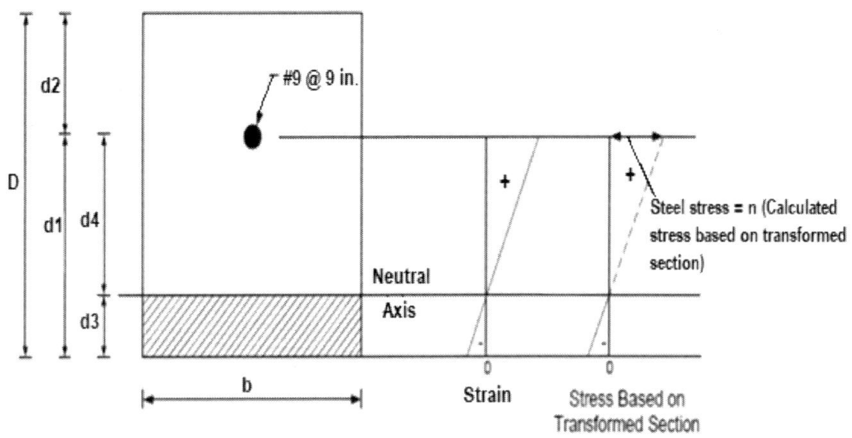

Figure 3.5. (a). Crack Control for Negative Moment Reinforcement Under Live Loads. Source: Figure 4-5a from AASHTO LRFD Bridge Design Specifications, FHWA Office of Bridges and Structures, 1200 New Jersey Ave., SE, Washington, DC 20590.

Transformed section properties may be calculated as done for the positive moment section in Design Step 3.2.9 (AASHTO-LRFD Design Step 4.8). Refer to Figure 3.5 (a) for the section dimensions and location of the neutral axis. The calculations are shown below.

Maximum dead load + live load service load moment = 5.42 k-ft/ft

$n = 8$

$I_{transformed} = 43.83\ in^4$

Total DL + LL service load stresses = [[5.42 x9 x(3.75)]/43.83] x 8

= 33.39 ksi

Maximum spacing of tension steel:

$s \leq 700\gamma e/\ \beta sfs - 2dc$

Concrete cover = 2 ½ in. – ½ in. integral wearing surface = 2 in.

where:

dc = clear cover + ½ bar diameter

= (2 +0.5 x0.625)

= 2.31 in.

fs = 33.39 ksi

$\gamma e = 1.00$

h = 7.5 in.

$\beta s = (1+2.31/0.7\ x(7.5 - 2.31)) = 1.64$

s = 9 in

s ≥ (700x1/1.64x33.39-2x2.31)= 8.16 in NG

To satisfy the crack control provisions, the most economical change is to replace the reinforcement bars by smaller bars at smaller spacing (area of reinforcement per unit width is the same). However, in this particular example, the #5 bar size cannot be reduced as this bar is customarily considered the minimum bar size for deck main reinforcement. Therefore, the bar diameter is kept the same and the spacing is reduced.

Assume reinforcement is #5 at 8 in. spacing; refer to Figure 3.5 (b).

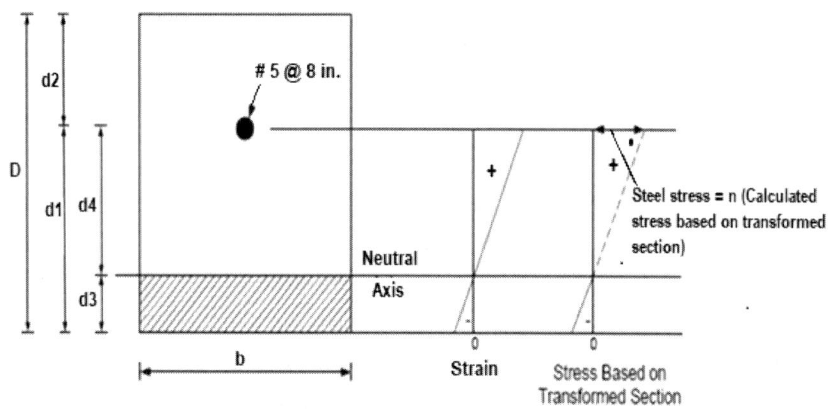

Figure 3.5. (b). Crack Control for Negative Moment Reinforcement under Live Loads. Source: Figure 4-5b from AASHTO LRFD Bridge Design Specifications, FHWA Office of Bridges and Structures, 1200 New Jersey Ave., SE, Washington, DC 20590.

$I_{transformed}$ = 42.77 in4

Total DL + LL service load stresses

= [(5.42 x 8 x3.68 /42.77)x8]

= 29.85 ksi

s = 8 in

s ≤ (700X1/1.636X29.85-2X2.31)= 9.71 in OK

Use main negative moment reinforcement #5 at 8 in. spacing

D = 8 in

d1 = 5.19 in

d2 = 2.81 in

d3 = 1.51 in

d4 = 3.68 in

b = 8 in

3.2.11. AASHTO-LRFD Design Step 4.10 Design of the Overhang

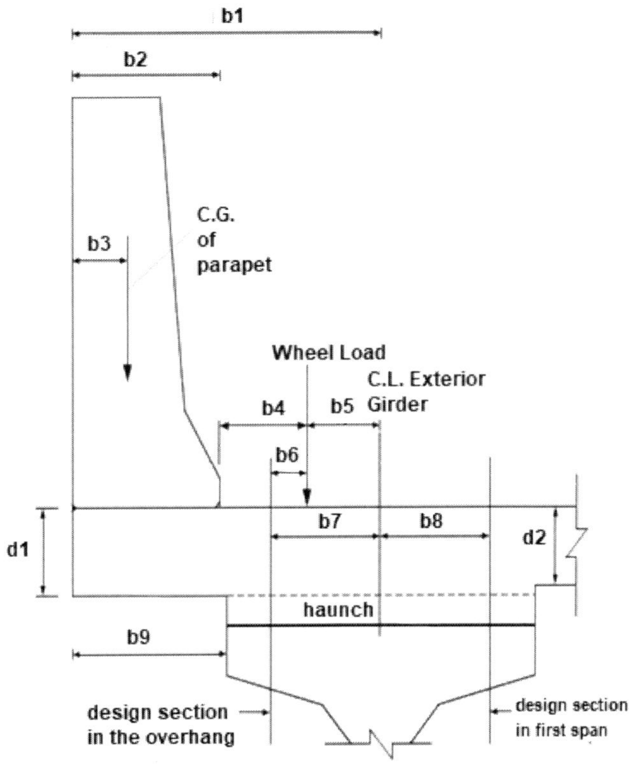

Figure 3.6. Overhang Region, Dimensions and Truck Loading. Source: Figure 4-6 from AASHTO LRFD Bridge Design Specifications, FHWA Office of Bridges and Structures, 1200 New Jersey Ave., SE, Washington, DC 20590.

b1 = 42.25 in

b2 = 20.25 in

b3 = 7.61 in

b4 = 12 in

b5 = 10 in

b6 = 4 in

b7 = 14 in

b8 = 14 in

b9 = 21.25 in

d1 = 9 in

d2 = 8 in

Assume that the bottom of the deck in the overhang region is 1 inch lower than the bottom of other bays as shown in Figure 3.6. This results in a total overhang thickness equal to 9 in. This is usually beneficial in resisting the effects of vehicular collision.

However, a section in the first bay of the deck, where the thickness is smaller than that of the overhang, must also be checked.

3.2.11.1. Assumed Loads

Self weight of the slab in the overhang area = 0.1125 k/ft2 of the deck overhang surface area.

Weight of parapet = 0.65 k/ft of length of parapet
Future wearing surface = 0.03 k/ft2 of deck surface area

As required by SA13.4.1, there are three design cases to be checked when designing the deck overhang regions.

Design Case 1: Check Overhang for Horizontal Vehicular Collision Load (SA13.4.1, Case 1)

The overhang is designed to resist an axial tension force from vehicular collision acting simultaneously with the collision + dead load moment.

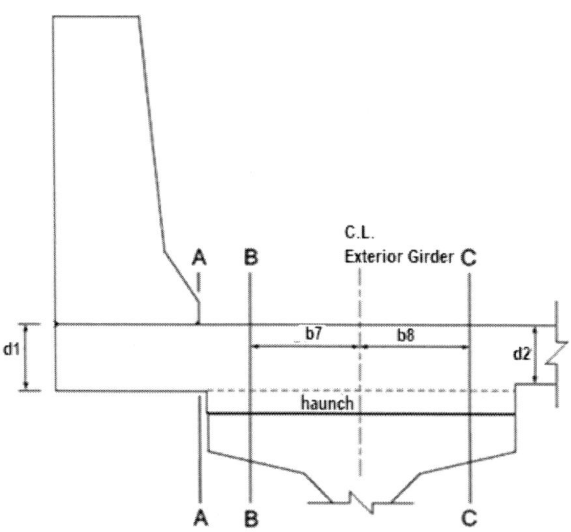

Figure 3.7. Design Sections in the Overhang Region. Source: Figure 4-7 from AASHTO LRFD Bridge Design Specifications, FHWA Office of Bridges and Structures, 1200 New Jersey Ave., SE, Washington, DC 20590.

The resistance factor, $\phi = 1.0$ for extreme event limit state (S1.3.2.1). The Specification requires that load effects in the extreme event limit state be multiplied by $\eta i \geq 1.05$ for bridges deemed important or $\eta i \geq 0.95$ for bridges deemed not important. For this example, a value of $\eta i = 1.0$ was used.

a. At inside face of parapet (Section A-A in Figure 3.7) (see Design Step 4.4 for parapet characteristics)

Mc = moment capacity of the base of the parapet given as 17.83 k-ft/ft. When this moment is transmitted to the deck overhang it subjects the deck to negative moment.

For a complete railing design example that includes sample detailed calculations of railing parameters, see Lecture 16 of the participant notebook of the National Highway Institute Course No. 13061.

$M_{DL,\, slab}$ = (0.1125 x(20.25/12)^2/2) = 0.16 k-ft/ft
$M_{DL,\, parapet}$ = (0.65X20.25-7.61/12) = 0.68 k-ft/ft
Design factored moment = (-17.83-1.25X(0.16+0.68))
= -18.88 k-ft/ft

Design axial tensile force (SA13.4.2) = RW/(Lc + 2H)
(Rw = 137.22 k,
Lc = 235.2 in., H = 42 in.)
= (137.22/[(235.2/12 + 2 x42/12]
= 5.16 k/ft

h slab = 9 in.

Assuming #5 reinforcement bars,

d = overhang slab thickness – top cover – ½ bar diameter
= (9-2.5-(0.5x0.625)) = 6.19 in.
Assume required area of steel = 0.7 in^2/ft (1)

The over-reinforced section check is not expected to control. However, due to the additional reinforcement in the overhang, it is prudent to perform this check using the provisions of S5.7.3.3.1.

Effective depth of the section, h = 6.19 in.

(Notice that the overhang has 1 inch additional thickness at its bottom)
For a section under moment and axial tension, P, the nominal resistance, Mn, may be calculated as:

Mn = T(d – a/2) – P(h/2 – a/2)
Tension in reinforcement, T = (0.7X60) = 42 k/ft
Compression in concrete, C = (42-5.16) = 36.84 k/ft
Minimum distance between the wheels of two adjacent trucks = 4 ft.
Three lanes = 0.85

a = C/bβ1f'c
= (36.84/12 x0.85x4)
= 0.9 in.
c = a/β1 = (0.9/0.85)= 1.06 in.
Mn = (42 x [6.19 - (0.9/2)]-5.16 x [(6.19/2)- (0.9/2)]
= 18.95 k-ft/ft

Notice that many designers determine the required reinforcement for sections under moment and axial tension, P, as the sum of two components:

1) the reinforcement required assuming the section is subjected to moment
2) P/fy

This approach is acceptable as it results in more conservative results, i.e., more reinforcement.

The ductility check is not expected to control. However, due to the additional reinforcement in the overhang, it is prudent to perform this check using the provisions of S5.7.2.1.

Resistance factor = 1 for extreme event limit state (S1.3.2.1)
Mr = ϕMn
(1 x18.95) = 18.95 k-ft/ft > Mu = 18.88 k-ft/ft OK
εs = 0.003 x (d/c-1) (0.003x(6.19/1.06-1))= 0.0145 > 0.005 OK

Steel has sufficiently yielded before concrete crushes, i.e., the section is ductile.

3.2.11.2. At Design Section in the Overhang (Section B-B in Figure 3.7, 4-7 AASHTO)

Assume that the minimum haunch thickness is at least equal to the difference between the thickness of the interior regions of the slab and the overhang thickness, i.e., 1 in. This means that when designing the section in the overhang at 14 in. from the center of the girder, the total thickness of the slab at this point can be assumed to be 9 in. For thinner haunches, engineering judgment should be exercised to determine the thickness to be considered at this section.

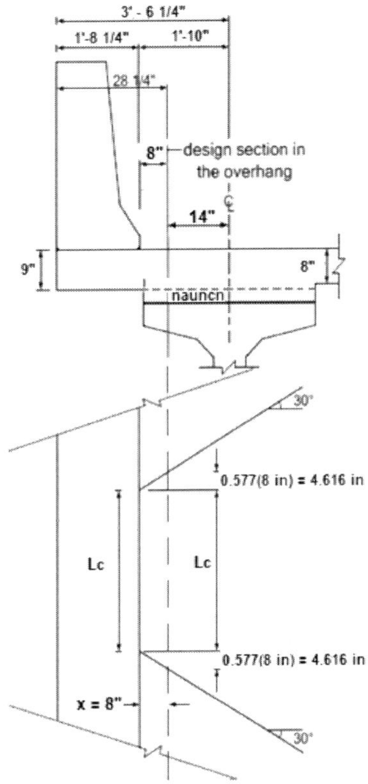

Figure 3.8. Assumed Distribution of Collision Moment Load in the Overhang. Source: Figure 4-8 from AASHTO LRFD Bridge Design Specifications, FHWA Office of Bridges and Structures, 1200 New Jersey Ave., SE, Washington, DC 20590.

At the inside face of the parapet, the collision forces are distributed over a distance Lc for the moment, where Lc is the critical length of the yield line failure pattern, and Lc + 2H for axial force. It is reasonable to assume that the distribution length will increase as the distance from the section to the parapet increases. The value of the distribution angle is not specified in the specifications and is determined using engineering judgment. In this example, the distribution length was increased using a 30° angle from the base of the parapet (see Figure 3.8). Some designers assume a distribution angle of 45°, this angle would have also been acceptable.

Collision moment at the design section = Mc x Lc/[Lc + 2(0.577)x8]
= (-17.83 x235.2/(235.2+ 2 x0.58x8))
= -17.15 k-ft/ft

Dead load moment at the design section:

$M_{DL,\,slab}$ (0.1125x (28.25/12)^2/2) = 0.31 k-ft/ft
$M_{DL,\,parapet}$ = (0.65 x (28.25 –7.61)/12) = 1.12 k-ft/ft

$M_{DL, FWS}$ = (0.03 x (8/12)^2/2) = 0.007 k-ft/ft

Factored design M = (-17.15 –1.25(0.31 +1.12) –1.5(0.007)) = -18.95 k-ft/ft

Design tensile force = Rw/[Lc + 2H + 2(0.577)x 8/12]
= (137.22/((235.2 +2 x (42) +2 x (0.577) x (8))/12))
= 5.01 k/ft

h slab = 9 in.

By inspection, for Section A-A, providing an area of steel = 0.70 in^2/ft resulted in a moment resistance of 18.95 k-ft/ft ≅ the design moment for Section B-B.

Therefore, the required area of steel for Section B-B = 0.70 in^2/ft (2)

3.2.11.3. Check Dead Load + Collision Moments at Design Section in First Span (Section C-C in Figure 3.7)

The total collision moment can be treated as an applied moment at the end of a continuous strip. The ratio of the moment M2/M1 (see Figure 3.9) can be calculated for the transverse design strip. As an approximation, the ratio M2/M1 may be taken equal to 0.4. This approximation is based on the fact that M2/M1 = 0.5 if the rotation at the first interior girder is restrained. Since this rotation is not restrained, the value of M2 will be less than 0.5M1.

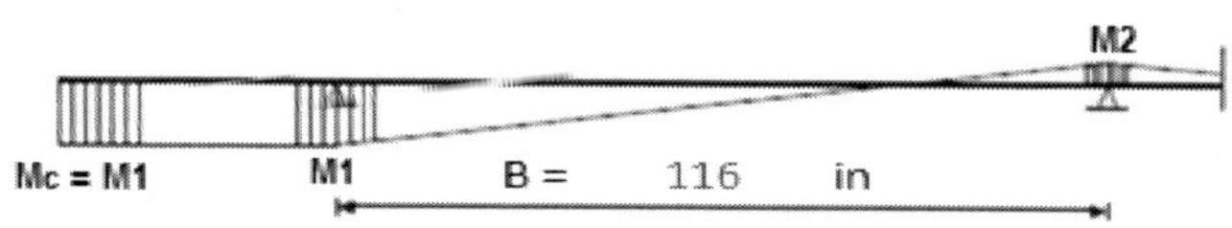

Figure 3.9. Assumed Distribution of the Collision Moment across the width of the Deck. Source: Figure 4-9 from AASHTO LRFD Bridge Design Specifications, FHWA Office of Bridges and Structures, 1200 New Jersey Ave., SE, Washington, DC 20590.

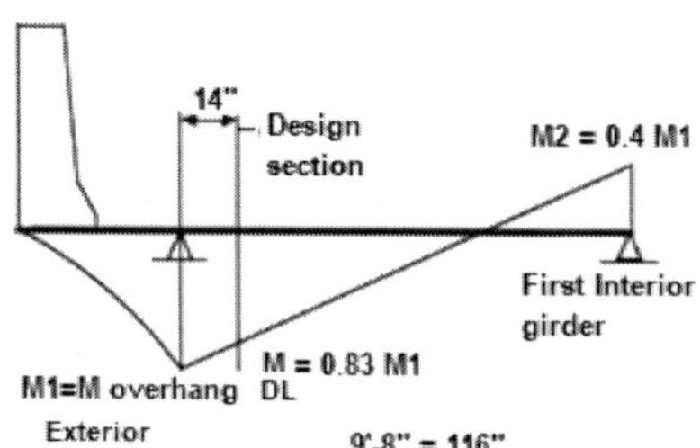

Figure 3.10. Dead Load Moment at Design Section Due to Dead Loads on the Overhang. Source: Figure 4-10 from AASHTO LRFD Bridge Design Specifications, FHWA Office of Bridges and Structures, 1200 New Jersey Ave., SE, Washington, DC 20590.

Thus, the assumption that M2/M1 = 0.4 seems to be reasonable. The collision moment per unit width at the section under consideration can then be determined by dividing the total collision moment by the distribution length. The distribution length may be determined using the 30° distribution as illustrated in Figure 3.8 except that the distance "X" will be 36 in. for Section C.

The dead load moments at the design section for negative moment to the inside of the exterior girder may be determined by superimposing two components: (1) the moments in the first deck span due to the dead loads acting on the overhang (see Figure 3.10), and (2) the effect of the dead loads acting on the first span of the deck (see Figure 3.11).

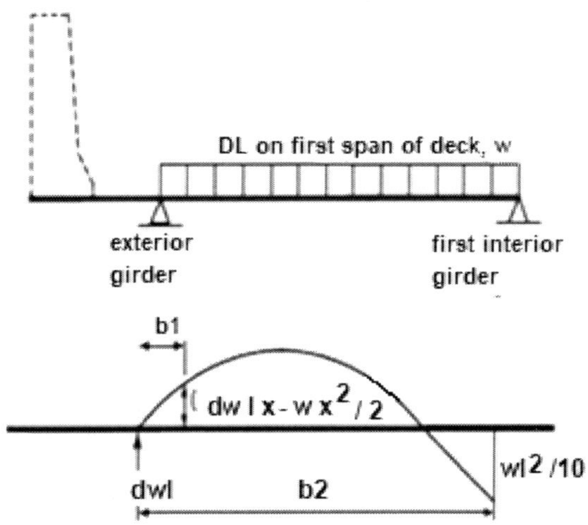

Figure 3.11. Dead Load Moment at Design Section Due to Dead Loads on the First Deck Span. Source: Figure 4-11 from AASHTO LRFD Bridge Design Specifications, FHWA Office of Bridges and Structures, 1200 New Jersey Ave., SE, Washington, DC 20590.

b1 = 14 in

b2 = 116 in

d = 0.4

Collision moment at exterior girder, M1 = -17.83 k-ft/ft

Collision moment at first interior girder, M2 = (0.4 x (17.83) = 7.13 k-ft/ft

By interpolation for a section in the first interior bay at 14 in. from the exterior girder:

Total collision moment = -17.83 + 14 x (17.83 + 7.13)/116 = -14.82 k-ft/ft

Using the 30° angle distribution, as shown in Figure 3.8:

Design collision moment = (-14.82X235.2/(235.2 + 2 x (0.577) x (22+14))) = -12.6 k-ft/ft

where Lc = 235.2 in.

Dead load moment at the centerline of the exterior girder:

MDL, Slab = (-0.1125x (42.25/12)^2/2) = -0.7 k-ft/ft
MDL, Parapet = (-0.65 x (42.25 –7.61)/12) = -1.88 k-ft/ft
MDL, FWS = (-0.03 x [(42.25 – 20.25)/12]^2/2) = -0.05 k-ft/ft

Factored dead load moment at the centerline of the exterior girder:

MFDL = (1.25 x (-0.7) +1.25(-1.88) + 1.5 x (-0.05))= -3.3 k-ft/ft

Based on Figure 3.10

The design factored dead load moment at the design section due to loads on the overhang is:

MFDL,O = (0.83 x (-3.3)) -2.74 k-ft/ft

From Figure 3.11, the dead load design factored moment due to DL on the first deck span is:

M = (1.25 x (0.1125 x (0.4x (9.66) x (14/12) – (14/12)^2/2)) +1.5(0.03 x (0.4 x (9.66) x (14/12) – (14/12)^2/2)) = 0.71 k-ft/ft

Total design dead load + collision moment:

MDL + C= (-12.6+(-2.74) +0.71) -14.63 k-ft/ft
Resistance factor = 1.0 for extreme event limit state (S1.3.2.1)

Assuming the slab thickness at this section equals 8 in. and the effective depth equals 5.19 in.;

Required area of steel = 0.62 in^2/ft (3).

Design Case 2: Vertical Collision Force (SA13.4.1, Case 2)

For concrete parapets, the case of vertical collision never controls.

Design Case 3: Check DL + LL (SA13.4.1, Case 3)

Except for decks supported on widely spaced girders (approximately 12 ft. and 14 ft. girder spacing for girders with narrow flanges and wide flanges, respectively), Case 3 does not control the design of decks supporting concrete parapets. Widely spaced girders allow the use of wider overhangs which in turn may lead to live load moments that may exceed the collision moment and, thus, control the design. The deck of this example is highly unlikely to be controlled by Case 3. However, this case is checked to illustrate the complete design process.

Resistance factor = 0.9 for strength limit state (S5.5.4.2.1).

3.2.11.4. Design Section in the Overhang (Section B-B in Figure 3.7)

The live load distribution width equations for the overhang (S4.6.2.1.3) are based on assuming that the distance from the design section in the overhang to the face of the parapet exceeds 12 in. such that the concentrated load representing the truck wheel is located closer to the face of the parapet than the design section. As shown in Figure 3.12, the concentrated load representing the wheel load on the overhang is located to the inside of the design section for negative moment in the overhang. This means that the distance "X" in the distribution width equation is negative which was not intended in developing this equation. This situation is becoming common as pre-stressed girders with wide top flanges are being used more frequently. In addition, Figure 3.6 may be wrongly interpreted as that there is no live load negative moment acting on the overhang. This would be misleading since the wheel load is distributed over the width of the wheels in the axle. Live load moment in these situations is small and is not expected to control design. For such situations, to determine the live load design moment in the overhang, either of the following two approaches may be used:

1) The design section may be conservatively assumed at the face of the girder web, or
2) The wheel load may be distributed over the width of the wheels as shown in Figure 3.12 and the moments are determined at the design section for negative moment. The distribution width may be calculated assuming "X" as the distance from the design section to the edge of the wheel load nearest the face of the parapet.

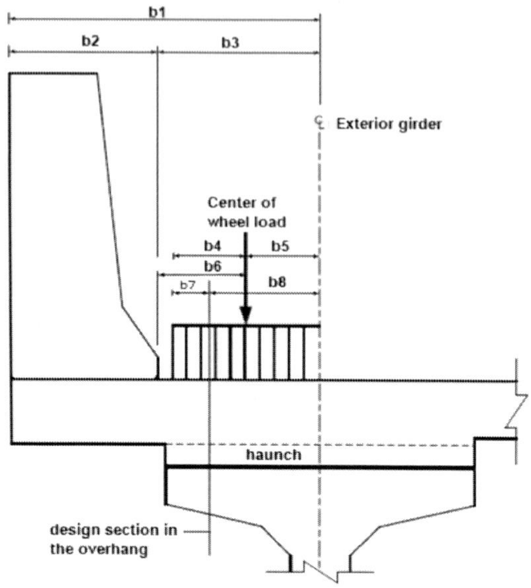

Figure 3.12. Overhang Live Load - Distributed Load. Source: Figure 4-12 from AASHTO LRFD Bridge Design Specifications, FHWA Office of Bridges and Structures, 1200 New Jersey Ave., SE, Washington, DC 20590.

The latter approach is used in this example. The wheel load is assumed to be distributed over a tire width of 20 in. as specified in S3.6.1.2.5.

b1 = 3.5208 in

b2 = 1.6875 in

b3 = 1.83 in

b4 = 10 in

b5 = 10 in

b6 = 12 in

b7 = 6 in

b8 = 14 in

Using the multiple presence factor for a single truck = 1.2 (S3.6.1.1.2) and dynamic load allowance for truck loading = 1.33 (S3.6.2.1), live load moment may be determined.

Equivalent strip width for live load = (45+10 x (6/12)) = 50 in. (S4.6.2.1.3)

Design factored moment:

Mn = (-1.25 x (0.1125) x ((42.25 – 14)/12)^2/2) -(1.25 x (0.65) x (42.25 –14 – 7.61)/12) -(1.5 x (0.03) x ((42.25 –20.25 – 14)/12)^2/2) -(1.75 x (1.33) x (1.2) x (16/(20/12) x (((6/12)^2/2)/(50/12)))

= -2.62 k-ft/ft
d = 6.19 in.

Required area of steel = 0.09 in^2/ft (4)

3.2.11.5. Check Dead Load + Live Load Moments at Design Section in First Span (Section C-C in Figure 3.7)

Assume slab thickness at this section = 8 in. (conservative to ignore the haunch)

Based on the earlier calculations for this section under collision + DL, DL factored moment at the section = -2.74 k-ft/ft.

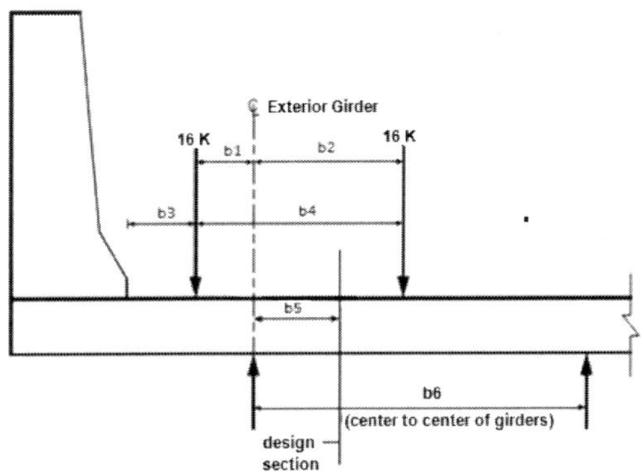

Figure 3.13. Overhang Live Load. Source: Figure 4-13 from AASHTO LRFD Bridge Design Specifications, FHWA Office of Bridges and Structures, 1200 New Jersey Ave., SE, Washington, DC 20590.

b1 = 120 in 10.000 ft
b2 = 62 in 5.167 ft
b3 = 12 in 1.000 ft
b4 = 72 in 6.000 ft
b5 = 14 in 1.167 ft
b6 = 116 in 9.667 ft

Determining live load at this section may be conducted by modeling the deck as a beam supported on the girders and by moving the design load across the width of the deck to generate the moment envelopes. However, this process implies a degree of accuracy that may not be possible to achieve due to the approximate nature of the distribution width and other assumptions involved, e.g., the girders are not infinitely rigid and the top flange is not a point support. An approximate approach suitable for hand calculations is illustrated in Figure 3.13. In this approximate approach, the first axle of the truck is applied to a simply supported beam that consists of the first span of the deck and the overhang. The negative moment at the design section is then calculated. The multiple presence factor for a single lane (1.2) and dynamic load allowance (33%) are also applied. Based on the dimensions and the critical location of the truck axle shown in Figure 3.13, the un-factored live load moment at the design section for negative moment is 3.03 k-ft.

Live load moment (including the load factor, dynamic load allowance and multiple presence factor) = 3.03 x (1.75) x (1.33) x (1.2) = 8.46 k-ft

Since the live load negative moment is produced by a load on the overhang, use the overhang strip width at the girder centerline.

Equivalent strip width = 45 + 10 x (10/12) = 53.3 in. (S4.6.2.1.3)

Design factored moment (DL + LL) = 2.74 + 8.46/(53.3/12) = 4.65 k-ft/ft

Required area of steel = 0.19 in^2/ft (5).

3.2.12. AASHTO-LRFD Design Step 4.11 Detailing of Overhang Reinforcement

From the different design cases of the overhang and the adjacent region of the deck, the required area of steel in the overhang is equal to the largest of (1), (2), (3), (4) and (5) = 0.7 in^2/ft

The provided top reinforcement in the slab in regions other than the overhang region is: #5 at 8 in. = (0.31 x (12/8)) = 0.465 in^2/ft

0.465 in^2/ft provided < 0.7 in2/ft required, therefore, additional reinforcement is required in the overhang.

Bundle one #4 bar to each top bar in the overhang region of the deck.

Provided reinforcement = (0.2 + 0.31) x (12/8) = 0.77 in^2/ft > 0.7 in^2/ft OK

Notice that many jurisdictions require a #5 minimum bar size for the top transverse reinforcement. In this case, the #4 bars used in this example would be replaced by #5 bars. Alternatively, to reduce the reinforcement area, a #5 bar may be added between the alternating main bars if the main bar spacing would allow adding bars in between without resulting in congested reinforcement.

Check the depth of the compression block:

T = 60x (0.77)

= 46.2 kips

a = 46.2/[0.85X(4) x (12)]

= 1.13 in.

$\beta 1$ = .85 for f'c = 4 ksi (S5.7.2.2)
c = 1.13/0.85
= 1.33 in.

Among Sections A, B and C of Figure 3.7, Section C has the least slab thickness. Hence, the ratio de/c is more critical at this section, making the strain in the reinforcing steel more critical.

de at Section C-C = 5.19 in.
Minimum εs = 0.003 (de/c-1) = 0.003 x (5.19/1.33-1) = 0.0087
0.0087 > 0.005 OK (S5.7.2.1)

The above calculation shows that the section is tension controlled.

Cracking under service load in the overhang needs to be checked. The reinforcement area in the overhang is 65% larger than the negative moment reinforcement in the interior portions of the deck, yet the applied service moment (2.74 + 3.03 = 5.77 k-ft/ft) is 6% larger than the service moment at interior portions of the deck (5.42 k-ft/ft from Step 4.9).

By inspection, cracking under service load does not control.

Determine the point in the first bay of the deck where the additional bars are no longer needed by determining the point where both (DL + LL) moment and (DL + collision) moments are less than or equal to the moment of resistance of the deck slab without the additional top reinforcement. By inspection, the case of (DL + LL) does not control and only the case of (DL + collision) needs to be checked.

Negative moment resistance of the deck slab reinforced with #5 bars at 8 in. spacing is 10.15 k-ft/ft for strength limit state (resistance factor = 0.9), or 11.27 k-ft/ft for the extreme event limit state (resistance factor = 1.0). By calculating the moments at different points along the deck first span in the same manner they were calculated for Section C-C for (DL + collision), it was determined that the design negative moment is less than 11.27 k-ft/ft at a point approximately 25 in. from the centerline of the exterior girder.

The theoretical termination point from the centerline of the exterior girder is 25 in.

Extend the additional bars beyond this point for a distance equal to the cut-off length. In addition, check that the provided length measured beyond the design section for moment is larger than the development length (S5.11.1.2.1).

Cut-Off Length Requirement (S5.11.1.2.1)

Checking the three requirements of S5.11.1.2.1, the cut-off length is controlled by 15 times the bar diameter.

Cut-off length = 15 x (0.625) = 9.375 in.

Required length past the centerline of the exterior girder= 25 + 9.375 = 34.375

Development Length (S5.11.2)

The basic development length, l_{db}, is taken as the larger of:

$1.25A_bf_y/\sqrt{f'c} = 1.25 \times (0.31) \times (60)/\sqrt{4} = 11.625$ in.

OR

$0.4d_bf_y = 0.4 \times (0.625) \times (60) = 15$ in.

OR

12. in

Therefore, the basic development length = 15 in.

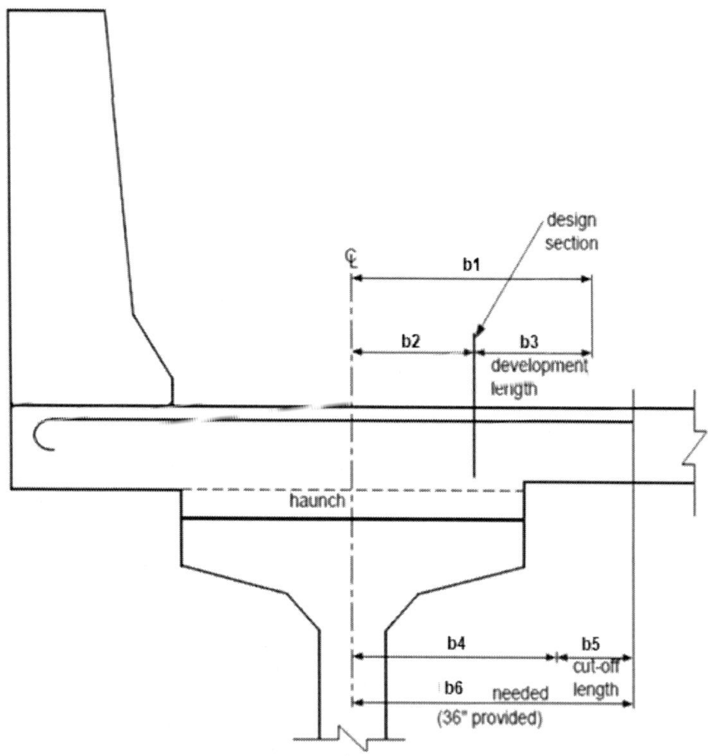

Figure 3.14. Length of the Overhang Additional Bars. Source: Figure 4-14 from AASHTO LRFD Bridge Design Specifications, FHWA Office of Bridges and Structures, 1200 New Jersey Ave., SE, Washington, DC 20590.

Correction factors:

Epoxy-coated bars = 1.2 (S5.11.2.1.2)

Two bundled bars = 1.0 (S5.11.2.3)

Spacing > 6 in. = 0.8 (S5.11.2.1.3)

Development length = 15 x (1.2)X(1) x (0.8) = 14.4 in.

Required length of the additional bars past the centerline of the exterior girder = 14 + 14.4 = 28.4 in.

28.4 < 34.375 in (needed to be satisfy cut off requirements) OK

Extend the additional bars in the overhang a minimum of 34.375 in. (say 3 ft.) beyond the centerline of the exterior girder.

b1 = 28.4 in

b2 = 14 in

b3 = 14.4 in

b4 = 26.625 in

b5 = 9.375 in

b6 = 34.375

3.2.13. AASHTO-LRFD Design Step 4.12 Longitudinal Reinforcement

Bottom Distribution Reinforcement (S9.7.3.2)

Percentage of longitudinal reinforcement = $220/\sqrt{S} \leq 67\%$

where:

S = the effective span length taken as equal to the effective length specified in S9.7.2.3 (ft.); the distance between sections for negative moment and sections at the ends of one deck span

(116 – 14– 14)/(12) = 7.33 ft.

Percentage = $220/\sqrt{7.33}$ = 81% > 67%

Use 67% of transverse reinforcement

Transverse reinforcement = #5 at 7 in. spacing = 0.53 in^2/ft

Required longitudinal reinforcement = 0.67 x (0.53)= 0.36 in^2/ft

Use #5 bars;

bar diameter = 0.625 in.

bar area = 0.31 in^2

Required spacing = 0.31/0.36 = 0.86 ft. 10.32 in.

Use #5 bars at 10 in. spacing

3.2.13.1. Top Longitudinal Reinforcement

There are no specific requirements to determine this reinforcement. Many jurisdictions use #4 bars at 12 in. spacing for the top longitudinal reinforcement.

3.2.13.2. Design Step 4.13. Deck Top Longitudinal Reinforcement in the Girder Negative Moment Region

This design step computes the deck top longitudinal reinforcement in the girder negative moment region, i.e., over the intermediate supports of the girders.

For simple span precast girders made continuous for live load: design according to S5.14.1.4.

(Notice that for continuous steel girders, this reinforcement is designed according to S6.10.1.7).

The required reinforcement area is determined during girder design. See Section 5.6 for the calculations for this reinforcement.

Use #6 bars at 4.5 in. spacing in the top layer

#6 bars at 8.5 in. spacing in the bottom layer

3.2.13.3. Design Step 4.14. Check Shrinkage and Temperature Reinforcement according to S5.10.8

Reinforcement for shrinkage and temperature stresses is provided near surfaces of concrete exposed to daily temperature changes.

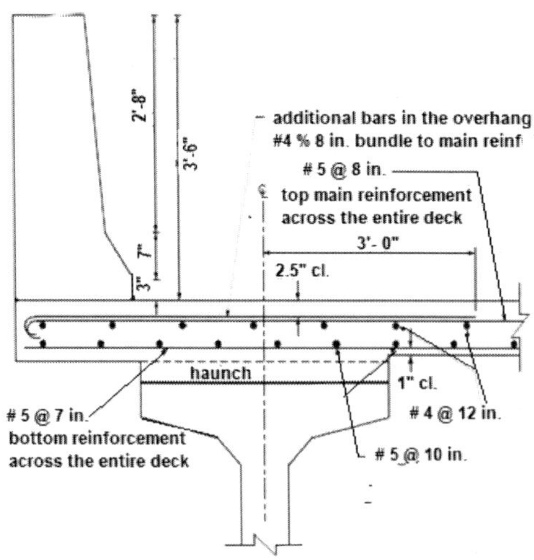

Figure 3.15. Deck Reinforcement at Mid-span of Girders. Source: Figure 4-15 from AASHTO LRFD Bridge Design Specifications, FHWA Office of Bridges and Structures, 1200 New Jersey Ave., SE, Washington, DC 20590.

Shrinkage and temperature reinforcement is added to ensure that the total reinforcement on exposed surfaces is not less than the following:

$As \geq (1.30bh)/(2 \times (b+h)fy)$ (S5.10.8-1)

$0.11 \leq As \leq 0.60$ (S5.10.8-2)

For members less than 18.0 in. thick, temperature and shrinkage reinforcement is not required.

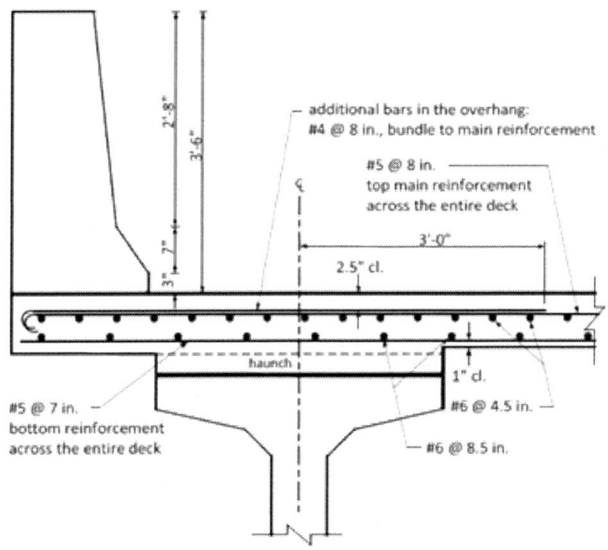

Figure 3.16. Deck Reinforcement at Intermediate Pier. Source: Figure 4-16 from AASHTO LRFD Bridge Design Specifications, FHWA Office of Bridges and Structures, 1200 New Jersey Ave., SE, Washington, DC 20590.

3.3. Design of PSC I-Girder

3.3.1. AASHTO-LRFD Design Step 5.1 - Live Load Distribution Factors (S 4.6.2.2)

The AASHTO-LRFD specifications allow the use of advanced methods of analysis to determine the live load distribution factors. However, for typical bridges, the specifications list equations to calculate the distribution factors for different types of bridge superstructures. The types of superstructures covered by these equations are described in AASHTO-LRFD Table S4.6.2.2.1-1. From this table, bridges with concrete decks supported on precast concrete I or bulb-tee girders are designated as cross-section "K". Other tables in S4.6.2.2.2 list the distribution factors for interior and exterior girders including cross-section "K". The distribution factor equations are largely based on work conducted in the NCHRP Project 12-26 and have been verified to give accurate results compared to 3-dimensional bridge analysis and field measurements. The multiple presence factors are already included in the distribution factor equations except when the tables call for the use of the lever rule. In these cases, the computations need to account for the multiple presence factors. It is to be noted that the distribution factor tables include a column with the heading "range of applicability". The ranges of applicability listed for each equation are based on the range for each parameter used in the study leading to the development of the equation. When the girder spacing exceeds the listed value in the "range of applicability" column, the specifications require the use of the lever rule (S 4.6.2.2.1). One or more of the other parameters may be outside the listed range of applicability. In this case, the equation could still remain valid, particularly when the value(s) is (are) only slightly out of the range of applicability. However, if one or more of the parameters greatly exceed the range of applicability, engineering judgment needs to be exercised.

Article S4.6.2.2.2d of the specifications states: "In beam-slab bridge cross-sections with diaphragms or cross-frames, the distribution factor for the exterior beam shall not be taken less than that which would be obtained by assuming that the cross-section deflects and rotates as a rigid cross-section." This provision was added to the specifications because the original study that developed the distribution factor equations did not consider intermediate diaphragms. Application of this provision requires the presence of a sufficient number of intermediate diaphragms whose stiffness is adequate to force the cross section to act as a rigid section. For pre-stressed girders, different jurisdictions use different types and numbers of intermediate diaphragms. Depending on the number and stiffness of the intermediate diaphragms, the provisions of S4.6.2.2.2d may not be applicable. For this example, one deep reinforced concrete diaphragm is located at the mid-span of each span. The stiffness of the diaphragm was deemed sufficient to force the cross-section to act as a rigid section, therefore, the provisions of S4.6.2.2.2d apply.

For this example, the distribution factors listed in S4.6.2.2.2 will be used.

It is to be noted that fatigue in the pre-stressing steel need not be checked for conventional pre-stressed girders (S5.5.3) when maximum stress in the concrete at Service III limit state is taken according to AASHTO-LRFD Table S5.9.4.2.2-1. This statement is valid for this example. The fatigue distribution factors are calculated in the following sections to provide the user with a complete reference for the application of the LRFD distribution factors.

Required information:

AASHTO Type I-Beam (28/72)
Non-composite beam area, Ag = 1,085 in^2
Non-composite beam moment of inertia, Ig = 733,320 in4
Deck slab thickness, ts = 8 in.
Span length, L = 110 ft.
Girder spacing, S = 9.666 ft.
Modulus of elasticity of the beam, EB = 4,696 ksi
Modulus of elasticity of the deck, ED = 3,834 ksi
C.G. to top of the basic beam = 35.62 in.
C.G. to bottom of the basic beam = 36.38 in.

3.3.1.1. AASHTO-LRFD Design Step 5.1.1

Calculate n, the modular ratio between the beam and the deck.

n = EB/ED (S 4.6.2.2.1-2)
= 4696/3834

3.3.1.2. AASHTO-LRFD Design Step 5.1.2

Calculate CG, the distance between the center of gravity of the non-composite beam and the deck. Ignore the thickness of the haunch in determining e.g., It is also possible to ignore the integral wearing surface, i.e., use ts = 7.5 in. However, the difference in the distribution factor will be minimal.

eg = NAYT + ts/2
= 35.62 + 8/2
= 39.62 in.

3.3.1.3. AASHTO-LRFD Design Step 5.1.3

Calculate 'Kg', the longitudinal stiffness parameter.

Kg = n x (I + Aeg2) (S 4.6.2.2.1-1)
= 1.225 x [733320 + 1085x(39.62)^2]
= 2,984,704 in4

3.3.1.4. AASHTO-LRFD Design Step 5.1.4: Interior Girder

Calculate the moment distribution factor for an interior beam with two or more design lanes loaded using AASHTO-LRFD Table S4.6.2.2.2b-1.

DM = 0.075 + (S/9.5)0.6 (S/L)0.2 (Kg/12.0Lts3)0.1
= 0.075+ (9.666/9.5)^0.6)x((9.666/110)^0.2)x(2984704/
(12x(110)x(8)^3))^0.1 = 0.796 lane (1)

3.3.1.5. AASHTO-LRFD Design Step 5.1.5

According to S 4.6.2.2.2e, a skew correction factor for moment may be applied for bridge skews greater than 30 degrees. The bridge in this example is skewed 20 degrees, and, therefore, no skew correction factor for moment is allowed.

Calculate the moment distribution factor for an interior beam with one design lane loaded using AASHTO-LRFD Table S 4.6.2.2.2b-1.

DM = 0.06 + (S/14) x 0.4 x (S/L) x 0.3 x (Kg/12.0Lts3)^0.1
= 0.06 + ((9.666/14)^0.4) x ((9.666/110)^0.3) x (2984704/(12x110x(8^3))^0.1) =
0.542 lane (2)

It is to be noted that the distribution factor calculated above for a single lane loaded already includes the 1.2 multiple presence factor for a single lane, therefore, this value may be used for the service and strength limit states. However, multiple presence factors should not be used for the fatigue limit state. Therefore, the multiple presence factor of 1.2 for the single lane is required to be removed from the value calculated above to determine the factor used for the fatigue limit state.

For single-lane loading to be used for fatigue design, remove the multiple presence factor of 1.2.

Multiple presence factor = 1.2
D_M = 0.542/1.2
= 0.452 lane (3)

3.3.1.6. AASHTO-LRFD Design Step 5.1.6: Skew Correction Factor for Shear

According to S 4.6.2.2.3c, a skew correction factor for support shear at the obtuse corner must be applied to the distribution factor of all skewed bridges. The value of the correction factor is calculated using AASHTO-LRFD Table S 4.6.2.2.3c-1.

SC = 1.0 + 0.20 x (12.0Lts3/Kg)0.3 x tan θ
= 1.0 + 0.20 x [[[12.0 x 110 x (8^3)]/2984704]^0.3] x tan20 = 1.047

3.3.1.7. AASHTO-LRFD Design Step 5.1.7

Calculate the shear distribution factor for an interior beam with two or more design lanes loaded using AASHTO-LRFD Table S 4.6.2.2.3a-1.

Dv = 0.2 + (S/12) – (S/35)2
= 0.2 + (9.666/12) - (9.666/35)^2
= 0.929 lane

Apply the skew correction factor:

Dv = 1.047 x (0.929)
= 0.973 lane (4)

3.3.1.8. AASHTO-LRFD Design Step 5.1.8

Calculate the shear distribution factor for an interior beam with one design lane loaded using AASHTO-LRFD Table S 4.6.2.2.3a-1.

Dv = 0.36 + (S/25.0)
= 0.36 + (9.666/25)
= 0.747 lane

Apply the skew correction factor:

Dv = 1.047 x (0.747) = 0.782 lane (5)

For single-lane loading to be used for fatigue design, remove the multiple presence factor of 1.2.

Dv = 0.782/1.2 = 0.652 lane (6)

3.3.1.9. AASHTO-LRFD Design Step 5.1.9

From (1) and (2), the service and strength limit state moment distribution factor for the interior girder is equal to the larger of 0.796 and 0.542 lane. Therefore, the moment distribution factor is 0.796 lane.

From (3): The fatigue limit state moment distribution factor is 0.452 lane.

From (4) and (5), the service and strength limit state shear distribution factor for the interior girder is equal to the larger of 0.973 and 0.782 lane. Therefore, the shear distribution factor is 0.973 lane.

From (6): The fatigue limit state shear distribution factor is 0.652 lane.

3.3.1.10. AASHTO-LRFD Design Step 5.1.10: Exterior Girder

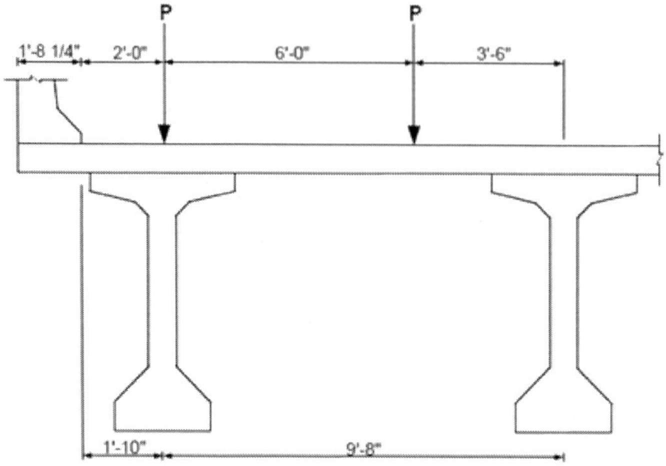

Figure 3.16. (a). Lever Rule. Source: Figure 5.1-1 from AASHTO LRFD Bridge Design Specifications, FHWA Office of Bridges and Structures, 1200 New Jersey Ave., SE, Washington, DC 20590.

b1 = 1.69 ft
b2 = 2 ft
b3 = 6 ft
b4 = 3.5 ft
b5 = 1.833 ft
b6 = 9.67 ft

3.3.1.11. AASHTO-LRFD Design Step 5.1.11

Calculate the moment distribution factor for an exterior beam with two or more design lanes using AASHTO-LRFD Table S 4.6.2.2.2d-1.

$D_M = eg_{MInterior}$
$e = 0.77 + de/9.1$

where de is the distance from the exterior web of the exterior girder to the inside face of the curb or barrier (conservatively use the centerline of web for this example).

$e = 0.77 + 1.833/9.1$
$= 0.97$
$D_M = 0.97x(0.796)$
$= 0.772$ lane (7)

3.3.1.12. AASHTO-LRFD Design Step 5.1.12

Calculate the moment distribution factor for an exterior beam with one design lane using the lever rule as per AASHTO-LRFD Table S 4.6.2.2.2d-1.

$D_M = [(3.5 + 6) + 3.5]/9.666$
$= 1.345$ wheels/2
$= 0.673$ lane (8) (Fatigue)

It is to be noted that this value does not include the multiple presence factor, therefore, it is adequate for use with the fatigue limit state. For service and strength limit states, the multiple presence factor for a single lane loaded needs to be included.

$D_M = 0.673x(1.2)$
$= 0.808$ lane (9) (Strength and Service)

3.3.1.13. AASHTO-LRFD Design Step 5.1.13

Calculate the shear distribution factor for an exterior beam with two or more design lanes loaded using AASHTO-LRFD Table S 4.6.2.2.3b-1.

$DV = eDV_{interior}$

where:

$e = 0.6 + de/11$
$= 0.6 + 1.833/10$
$= 0.783$
$DV = 0.783 \text{ x } (0.973)$
$= 0.762$ lane (10)

3.3.1.14. AASHTO-LRFD Design Step 5.1.14

Calculate the shear distribution factor for an exterior beam with one design lane loaded using the lever rule as per AASHTO-LRFD Table S 4.6.2.2.3b-1. This value will be the same as the moment distribution factor with the skew correction factor applied.

$DV = 1.047 \text{ x } 0.673$
$= 0.705$ lane (11) (Fatigue)
$DV = 1.047 \text{ x } 0.808$
$= 0.846$ lane (12) (Strength and Service)

It is to be noted that S 4.6.2.2.2d includes additional requirements for the calculation of the distribution factors for exterior girders when the girders are connected with relatively stiff cross-frames that force the cross-section to act as a rigid section. As indicated in Design Step 3.3.1 AASHTO-LRFD Design Step 5.1, these provisions are applied to this example; the calculations are shown below.

3.3.1.15. AASHTO-LRFD Design Step 5.1.15: Additional Check for Rigidly Connected Girders (S 4.6.2.2.2d)

The multiple presence factor, m, is applied to the reaction of the exterior beam (Table S 3.6.1.1.2-1)

m1 = 1.20
m2 = 1.00
m3 = 0.85

R = NL/Nb + Xext(Σe)/Σx2 (SC 4.6.2.2.2d-1)

where:

R = reaction on exterior beam in terms of lanes
N_L = number of loaded lanes under consideration
e = eccentricity of a design truck or a design lane load from the center of gravity of the pattern of girders (ft.)
x = horizontal distance from the center of gravity of the pattern of girders to each girder (ft.)
Xext = horizontal distance from the center of gravity of the pattern to the exterior girder (ft.)

See Figure 3.17 for dimensions. One lane loaded (only the leftmost lane applied):

NL = 1
Nb = 6
Xext = 24.167
Σe = 21
x1 = 24.167
x2 = 14.5
x3 = 4.833
R = 1/6 + 24.167x(21)/[2x(24.167^2 + 14.5^2 +4.833^2)]
= 0.167 + 0.31
= 0.477 (Fatigue)

Add the multiple presence factor of 1.2 for a single lane:

R = 1.2 x (0.477)
= 0.572 (Strength)

Two lanes loaded:

R = 2/6 + 24.167x(21 + 9)/[2x(24.167^2 + 14.5^2 +4.833^2)]
= 0.333 + 0.443
= 0.776

Add the multiple presence factor of 1.0 for two lanes loaded:

R = 1.0 x 0.776
= 0.776 (Strength)

Three lanes loaded:

R = 3/6 + 24.167x(21 + 9 -3)/[2x(24.167^2 + 14.5^2 +4.833^2)]^0.5 + 0.399 = 0.899

Add the multiple presence factor of 0.85 for three or more lanes loaded:

R = 0.85 x 0.899
= 0.764 (Strength)

These values do not control over the distribution factors summarized in AASHTO-LRFD Design Step 3.2.1.16 AASHTO-LRFD Design Step 5.1.16.

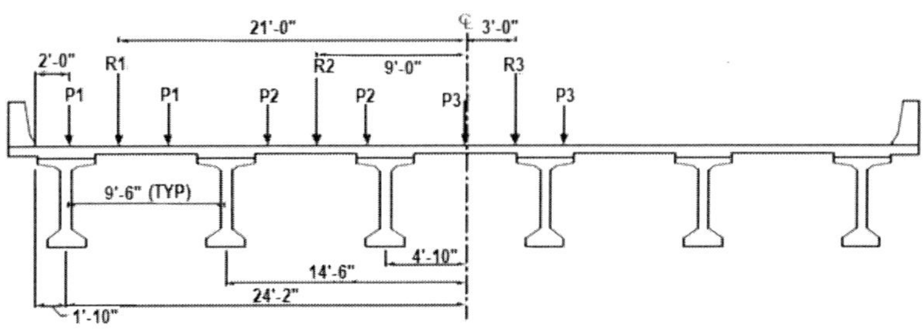

Figure 3.17. General Dimensions. Source: Figure 5.1-2 from AASHTO LRFD Bridge Design Specifications, FHWA Office of Bridges and Structures, 1200 New Jersey Ave., SE, Washington, DC 20590.

3.2.1.16. AASHTO-LRFD Design Step 5.1.16

From (7) and (9), the service and strength limit state moment distribution factor for the exterior girder is equal to the larger of 0.772 and 0.808 lane. Therefore, the moment distribution factor is 0.808 lane.

Table 3.1. (Ref. AASHTO LRFD Table 5.1-1) –
Summary of Service and Strength Limit State Distribution Factors

Distribution Factor Method	Load Case	Moment interior beams	Moment exterior beams	Shear interior beams	Shear exterior Beams
Distribution factors from Tables in S4.6.2.2.2	Multiple lanes loaded	0.796	0.772	0.973	0.762
	Single lane loaded	0.542	0.806	0.782	0.845
Additional check for rigidly connected girders	Multiple lanes loaded	NA	0.776	NA	0.776
	Single lane loaded	NA	0.572	NA	0.572
Design value		0.796	0.806	0.973	0.845

From (8): The fatigue limit state moment distribution factor is 0.673 lane.

From (10) and (12), the service and strength limit state shear distribution factor for the exterior girder is equal to the larger of 0.762 and 0.846 lane. Therefore, the shear distribution factor is 0.846 lane.

From (11): The fatigue limit state shear distribution factor is 0.705 lane.

Table 3.2. (Ref. AASHTO LRFD Table 5.1-2) –
Summary of Fatigue Limit State Distribution Factors

Distribution Factor Method	Load Case	Moment interior beams	Moment exterior beams	Shear interior beams	Shear exterior Beams
Distribution factors from Tables in S4.6.2.2.2	Multiple lanes loaded	NA	NA	NA	NA
	Single lane loaded	0.452	0.672	0.652	0.704
Additional check for rigidly connected girders	Multiple lanes loaded	NA	NA	NA	NA
	Single lane loaded	NA	0.477	NA	0.477
Design value		0.452	0.672	0.652	0.704

3.3.2. AASHTO-LRFD Design Step 5.2 - Dead Load Calculation

Calculate the dead load of the bridge superstructure components for the interior girder, which is designed in this example. Values for the exterior girder have also been included for reference. The girder, slab, haunch, and exterior diaphragm loads are applied to the non-composite section; the parapets and future wearing surface are applied to the composite section.

3.3.2.1. Interior Girder

Girder weight

DCgirder (I) = Ag(γgirder)

where:

Ag = beam cross-sectional area (in^2)

= 1,085 in^2

γ = unit weight of beam concrete (kcf)

= 0.15 kcf

DCgirder (I) = (1085/144) x (0.15)

= 1.13 k/ft/girder

Deck slab weight

The total thickness of the slab is used in calculating the weight.

Girder spacing = 9.666 ft.

Slab thickness = 8 in.

DCslab (I) = 9.666 x (8/12) x (0.15)

= 0.967 k/ft/girder

3.3.2.2. Exterior Girder

Girder weight

DCgirder (I) = 1.13 k/ft/girder

Deck slab weight

Slab width = Overhang width + ½ Girder spacing

overhang width = 3.521 ft

Slab width = 3.521 + ½ x (9.666)

= 8.35 ft.

Slab thickness = 8 in.

DCslab (E) = 8.35x(8/12) x (0.15)

= 0.835 k/ft/girder

3.3.2.3. Haunch Weight

Width = 42 in.

Thickness = 4 in.

DChaunch = [42 x (4)/144] x (0.15)

= 0.175 k/ft/girder

It is to be noted that the haunch weight in this example is assumed as a uniform load along the full length of the beam. This results in a conservative design as the haunch typically have a variable thickness that decreases toward the middle of the span length. Many jurisdictions calculate the haunch load effects assuming the haunch thickness to vary parabolically along the length of the beam. The location of the minimum thickness varies depending on the grade of the roadway surface at bridge location and the presence of a vertical curve. The use of either approach is acceptable and the difference in load effects is typically negligible. However, when analyzing existing bridges, it may be necessary to use the variable haunch thickness in the analysis to accurately represent the existing situation

3.3.2.4. Concrete Diaphragm Weight

A concrete diaphragm is placed at one-half the non-composite span length.

Location of the diaphragms:

Span 1 = 54.5 ft. from centerline of end bearing

Span 2 = 55.5 ft. from centerline of pier

Diaphragm thickness = 10 in

For this example, arbitrarily assume that the thickness of the diaphragm is 10 in. The diaphragm spans from beam to beam minus the web thickness and has a depth equal to the distance from the top of the beam to the bottom of the web. Therefore, the concentrated load to be applied at the locations above is:

DCdiaphragm = 0.15 x (10/12) x [9.666 – (8/12)] x (72 – 18)/12

= 5.0621 k/girder

The exterior girder only resists half of this loading.

3.3.2.5. Parapet Weight

According to the S 4.6.2.2.1, the parapet weight may be distributed equally to all girders in the cross section.

Parapet width = 1.6875 ft

Parapet cross-sectional area = 4.33 ft^2

DCparapet = 4.33 x (0.15)

= 0.65 k/ft

= 0.65/6 girders

= 0.108 k/ft/girder for one parapet

Therefore, the effect of two parapets yields:

DCparapet = 0.216 k/ft per girder

Note that some jurisdictions distribute parapet loads to the 2 or 3 most exterior girders on each side of the cross section.

3.3.2.6. Future Wearing Surface

Interior girder
Weight/ft2 = 0.03 k/ft2
Width = 9.666 ft.
DW_{FWS} (I) = 0.03 x (9.666)
= 0.29 k/ft/girder

Exterior Girder
Weight/ft2 = 0.03 k/ft2
Width = Slab width – Parapet width
= 8.35 – 1.6875
= 6.663 ft.

DW_{FWS} (E) = 0.03 x (6.663)
= 0.2 k/ft/girder

It is to be noted that some jurisdictions divide the weight of the future wearing surface equally between all girders (i.e., apply a uniform load of 0.26 k/ft to all girders). Article S 4.6.2.2.1 states that permanent loads of and on the deck may be distributed uniformly among the beams. This method would also be acceptable and would minimally change the moments and shears given in the tables in Design Step 3.3.3.1 (AASHTO-LRFD Design Step 5.3.1).

3.3.3. AASHTO-LRFD Design Step 5.3 - Un-Factored and Factored Load Effects

3.3.3.1. AASHTO-LRFD Design Step 5.3.1 Summary of Loads

The dead load moments and shears were calculated based on the loads shown in step 3.3.2 AASHTO-LRFD Design Step 5.2.

Table 3.3-1 (Ref. AASHTO-LRFD Table 5.3-1) -
Summary of Un-factored Moments
Interior girder, Span 1 shown, Span 2 symmetric

Location*	**Non-Composite Moments**					**Composite Moments**	**Live Load Moments**		
	Girder		Slab	Diaphragm	Total	Parapet	FWS	Positive	Negative
	**	***							
(ft.)	(k-ft)	(k-ft)	(k-ft)	(k-ft)	(k-ft)	(k-ft)	(k-ft)	(k-ft)	(k-ft)
0	47	0	0	0	0	0	0	0	0
1.0	108	61	62	3	125	9	12	92	-11
5.5	368	322	325	14	661	46	62	476	-58
11.0	656	609	615	28	1,252	85	114	886	-116
16.5	909	863	871	42	1,776	118	158	1,230	-174
22.0	1,128	1,082	1,093	56	2,230	144	193	1,509	-233
27.5	1,313	1,267	1,279	70	2,616	164	220	1,724	-291
33.0	1,464	1,417	1,432	84	2,933	177	237	1,882	-349
38.5	1,580	1,534	1,549	98	3,181	183	246	1,994	-407
44.0	1,663	1,616	1,633	111	3,360	183	246	2,047	-465
49.5	1,664	1,664	1,681	125	3,471	177	237	2,045	-523
54.5	1,725	1,679	1,696	138	3,512	165	222	2,015	-576
55.0	1,725	1,678	1,695	137	3,511	164	220	2,010	-581
60.5	1,705	1,658	1,675	123	3,456	144	194	1,927	-640
66.0	1,650	1,604	1,620	109	3,333	118	159	1,794	-698
71.5	1,562	1,515	1,531	95	3,141	86	115	1,613	-756
77.0	1,439	1,392	1,407	81	2,880	46	62	1,388	-814
82.5	1,282	1,236	1,248	67	2,551	1	1	1,124	-872
88.0	1,091	1,044	1,055	53	2,152	-52	-69	825	-1,124
93.5	865	819	827	39	1,686	-110	-148	524	-1,223
99.0	606	560	565	25	1,150	-176	-236	297	-1,371
104.5	312	265	268	11	546	-248	-332	113	-1,663
108.0	110	61	62	3	125	-297	-398	33	-1,921
109.0	47	0	0	0	0	-311	-418	15	-2,006
Spam 2 - 0	-	0	0	0	0	-326	-438	0	-2,095

* Distance from the centerline of the end bearing.

** Based on the simple span length of 110.5 ft. and supported at the ends of the girders. These values are used to calculate stresses at transfer.

*** Based on the simple span length of 109 ft. and supported at the centerline of bearings. These values are used to calculate the final stresses.

Table 3.3-2 (Ref. AASHTO-LRFD Table 5.3-2) –
Summary of Factored Moments
Interior girder, Span 1 shown, Span 2 symmetric

Location	**Strength I Moment**	**Service I Moments ****		**Service III Moments *****	
		Non-Composite	**Composite**	**Non-Composite**	**Composite**
(ft.)	(k-ft)	(k-ft)	(k-ft)	(k-ft)	(k-ft)
0	0	0	0	0	0
1.0	345	125	112	125	94
5.5	1,809	661	584	661	488
11.0	3,394	1,252	1,085	1,252	908
16.5	4,756	1,776	1,506	1,776	1,260
22.0	5,897	2,230	1,846	2,230	1,544
27.5	6,821	2,616	2,108	2,616	1,763
33.0	7,536	2,933	2,296	2,933	1,920
38.5	8,063	3,181	2,423	3,181	2,024
44.0	8,381	3,360	2,477	3,360	2,067
49.5	8,494	3,471	2,459	3,471	2,050
54.5	8,456	3,512	2,402	3,512	1,999
55.0	8,440	3,511	2,394	3,511	1,992
60.5	8,163	3,456	2,265	3,456	1,880
66.0	7,690	3,333	2,070	3,333	1,712
71.5	7,027	3,141	1,813	3,141	1,490
77.0	6,181	2,880	1,497	2,880	1,219
82.5	5,158	2,551	1,126	2,551	901
88.0	3,967	2,152	-1,245	2,152	-1,020
93.5	2,664	1,686	-1,481	1,686	-1,237
99.0	-1,535	1,150	-1,783	1,150	-1,509
104.5	-3,035	546	-2,242	546	-1,910
108.0	-4,174	125	-2,616	125	-2,232
109.0	-4,525	0	-2,734	0	-2,333
Span 2 - 0	-4,729	0	-2,858	0	-2,439

Load Factor Combinations

Strength I = 1.25 x (DC) + 1.5 x (DW) + 1.75 x (LL + IM) Service I = 1.0 x [DC + DW + (LL + IM)]

Service III = 1.0 x (DC + DW) + 0.8 x (LL + IM)

* Distance from the centerline of the end bearing.

** For service limit states, moments are applied to the section of the girder, i.e., non-composite or composite, that resists these moments. Hence, non-composite and composite moments have to be separated for service load calculations.

Table 3.3-3. (Ref. AASHTO-LRFD Table 5.3-3) - Summary of Un-factored Shear Interior girder, Span 1 shown, Span 2 symmetric

Location*	Non-composite Shears				Composite Shears	Live Load Shears*		
	Girder	Slab and Haunch	Diaphragm	Total Non-comp.	Parapet	FWS	Positive HL-93	Negative HL-93
(ft.)	(k)	(k)	(k)	(k)	(k)	(k)	(k)	(k)
0	61.6	62.2	2.5	126.4	8.9	12.0	113.3	12.9
1.0	60.5	61.1	2.5	124.1	8.7	11.7	111.7	-12.9
5.5	55.4	55.9	2.5	113.9	7.7	10.4	104.3	-13.0
11	49.2	49.7	2.5	101.4	6.5	8.8	95.5	-13.4
16.5	43	43.4	2.5	88.9	5.4	7.2	86.9	-15.9
22.0	36.7	37.1	2.5	76.4	4.2	5.6	78.7	-20.6
27.5	30.5	30.8	2.5	63.9	3.0	4.0	70.8	-26.0
33.0	24.3	24.6	2.5	51.4	1.8	2.4	63.1	-32.8
38.5	18.1	18.3	2.5	38.9	0.6	0.8	55.9	-39.8
44.0	11.9	12.0	2.5	26.4	-0.6	-0.8	48.9	-46.8
49.5	5.7	5.7	2.5	13.9	-1.8	-2.4	42.4	-54.0
54.5	0	0	2.5	-2.5	-2.9	-3.8	36.8	-60.5
55.0	-0.6	-0.6	2.5	-3.7	-3.0	-4.0	36.2	-61.2
60.5	-6.8	-6.9	2.5	-16.2	-4.2	-5.6	30.4	-68.4
66.0	-13	-13.1	2.5	-28.7	-5.3	-7.2	25.0	-75.7
71.5	-19.2	-19.4	2.5	-41.2	-6.5	-8.8	20.0	-82.9
77.0	-25.4	-25.7	2.5	-53.7	7.7	-10.4	15.4	-90.1
82.5	-31.7	-32.0	-2.5	-66.1	-8.9	-12.0	11.3	-97.3
88.0	-37.9	-38.3	-2.5	-78.6	-10.1	-13.6	8.2	-104.3
93.5	-44.1	-44.5	-2.5	-91.1	-11.3	-15.1	5.5	-111.3
99.0	-50.3	-50.8	-2.5	-103.6	-12.5	-16.7	3.2	-118.0
104.5	-56.5	-57.1	-2.5	-116.1	-13.7	-18.3	1.2	-124.7
108.0	-60.5	-61.1	-2.5	-124.1	-14.4	-19.4	0.4	-128.7
109.0	-61.6	-62.2	-2.5	-126.4	-14.6	-19.6	0.2	-129.9
Span 2-0	0	0	0	0	-14.8	-19.9	0	-131.1

* Distance from the centerline of the end bearing.

**Table 3.3-4. (Ref. AASHTO-LRFD Table 5.3-4) –
Summary of Factored Shear
Interior girder, Span 1 shown, Span 2 symmetric**

Location*	**Strength I Shear**	**Service I Shear**	**Service III Shear**
(ft.)	(k)	(k)	(k)
0	385.4	260.6	237.9
1.0	379.0	256.2	233.8
5.5	350.0	236.2	215.4
11.0	315.1	212.1	193.0
16.5	280.7	188.3	170.9
22.0	246.8	164.8	149.1
27.5	213.4	141.6	127.5
33.0	180.6	118.7	106.1
38.5	148.3	96.2	85.0
44.0	116.7	74.0	64.2
49.5	85.7	52.1	43.6
54.5	-118.4	-69.7	-57.6
55.0	-121.3	-71.8	-59.6
60.5	-153.5	-94.3	-80.6
66.0	-185.7	-116.9	-101.7
71.5	-217.9	-139.4	-122.8
77.0	-250.0	-161.8	-143.8
82.5	-282.0	-184.3	-164.8
88.0	-313.8	-206.6	-185.7
93.5	-345.4	-228.8	-206.6
99.0	-376.8	-250.9	-227.3
104.5	-407.9	-272.8	-247.8
108.5	-427.4	-286.6	-260.8
109.0	-433.0	-290.5	-264.5
Span 2-0	-277.8	-165.8	-139.6

Load Factor Combinations

Strength I = 1.25 x (DC) + 1.5 x (DW) + 1.75 x (LL + IM)

Service I = 1.0 x [DC + DW + (LL + IM)]

Service III = 1.0 x (DC + DW) + 0.8 x (LL + IM)

* Distance from the centerline of the end bearing.

Table 3.3-5. (Ref. AASHTO-LRFD Table 5.3-5) - Summary of Un-factored Moments Exterior girder, Span 1 shown, Span 2 symmetric

Location*	Noncomposite Moments					Composite Moments	Live Load Moments*		
	Girder		Slab and	Diaphra-gm	Total	Parapet (k)	FWS	Positive HL-	Negative
	**	***							
(ft.)	(k-ft)	(k-ft)	(k-ft)	(k-ft)	(k-ft)	(k-ft)	(k-ft)	(k-ft)	(k-ft)
0	47	0	0	0	0	0	0	0	0
1.0	108	61	55	1	117	9	8	93	-11
5.5	368	322	288	7	616	46	41	482	-59
11.0	656	609	545	14	1,168	85	77	897	-118
16.5	909	863	771	21	1,655	118	106	1,245	-177
22.0	1,128	1,082	967	28	2,076	144	130	1,528	-236
27.5	1,313	1,267	1,132	35	2,434	164	148	1,746	-294
33.0	1,464	1,417	1,267	42	2,726	177	160	1,906	-353
38.5	1,580	1,534	1,371	49	2,954	183	165	2,019	-412
44.0	1,663	1,616	1,445	56	3,117	183	166	2,073	-471
49.5	1,711	1,664	1,488	63	3,215	177	160	2,071	-530
54.5	1,725	1,679	1,501	69	3,248	165	149	2,041	-583
55.0	1,725	1,678	1,501	68	3,247	164	148	2,035	-589
60.5	1,705	1,658	1,482	61	3,202	144	130	1,951	-648
66.0	1,650	1,604	1,434	54	3,092	118	107	1,816	-706
71.5	1,562	1,515	1,355	48	2,917	96	77	1,633	-765
77.0	1,439	1,392	1,245	41	2,678	46	42	1,406	-824
82.5	1,282	1,236	1,105	34	2,374	1	1	1,139	-883
88.0	1,091	1,044	934	27	2,005	-52	-47	836	-1,138
93.5	865	819	732	20	1,571	-110	-100	531	-1,238
99.0	606	560	500	13	1,072	-176	-159	300	-1,389
104.5	312	266	238	6	509	-248	-224	114	-1,683
108.0	110	61	55	1	117	-297	-268	33	-1,945
109.0	47	0	0	0	0	-311	-281	15	-2,031
Span 2-0	-	0	0	0	0	-326	-294	0	-2,121

* Distance from the centerline of the end bearing

** Based on the simple span length of 110.5 ft. and supported at the ends of the girders. These values are used to calculate stresses at transfer.

*** Based on the simple span length of 109 ft. and supported at the centerline of bearings. These values are used to calculate the final stresses.

**Table 3.3-6 (Ref. AASHTO-LRFD Table 5.3-6) –
Summary of Factored Moments
Exterior girder, Span 1 shown, Span 2 symmetric**

Location*	**Strength I**	**Service I Moments****		**Service III Moments****	
		NC	Comp.	NC	Comp.
(ft.)	(k-ft)	(k-ft)	(k-ft)	(k-ft)	(k-ft)
0	0	0	0	0	0
1.0	331	117	110	117	91
5.5	1,735	616	570	616	473
11.0	3,251	1,168	1,059	1,168	879
16.5	4,554	1,655	1,469	1,655	1,220
22.0	5,644	2,076	1,801	2,076	1,496
27.5	6,524	2,434	2,057	2,434	1,708
33.0	7,203	2,726	2,242	2,726	1,861
38.5	7,702	2,954	2,368	2,954	1,964
44.0	8,001	3,117	2,422	3,117	2,007
49.5	8,103	3,215	2,407	3,215	1,993
54.5	8,061	3,248	2,355	3,248	1,947
55.0	8,047	3,247	2,347	3,247	1,940
60.5	7,793	3,202	2,226	3,202	1,836
66.0	7,351	3,092	2,041	3,092	1,678
71.5	6,727	2,917	1,796	2,917	1,469
77.0	5,928	2,678	1,494	2,678	1,213
82.5	4,961	2,374	1,140	2,374	912
88.0	3,834	2,005	-1,237	2,005	-1,009
93.5	2,605	1,571	-1,448	1,571	-1,201
99.0	-1,547	1,072	-1,723	1,072	-1,445
104.5	-2,954	509	-2,154	509	-1,818
108.0	-4,031	117	-2,510	117	-2,121
109.0	-4,364	0	-2,623	0	-2,217
Span 2-0	-4,560	0	-2,741	0	-2,317

Load Factor Combinations

Strength I = 1.25 x (DC) + 1.5 x (DW) + 1.75 x (LL + IM)

Service I = 1.0 x [DC + DW + (LL + IM)]

Service III = 1.0 x (DC + DW) + 0.8 x (LL + IM)

*Distance from the centerline of the end bearing.

**For service limit states, moments are applied to the section of the girder, i.e., non-composite or composite, that resists these moments. Hence, non-composite and composite moments have to be separated for service load calculations.

Table 3.3-7. (Ref. AASHTO-LRFD Table 5.3-7) -
Summary of Un-factored Shear
Exterior girder, Span 1 shown, Span 2 symmetric

Location*	**Non-composite Shears**				**Composite Shears**	**Live Load = IM Shears**		
	Girder	Slab and	Diaphra-gm	Total.	Parapet	FWS	Positive HL-	Negative
(ft.)	(k)	(k)	(k)	(k)	(k)	(k)	(k)	(k)
0	61.6	55.1	2.5	126.4	8.9	12.0	113.3	12.9
1.0	60.5	54.1	2.5	124.1	8.7	11.7	111.7	-12.9
5.5	55.4	49.9	2.5	113.9	7.7	10.4	104.3	-13.0
11	49.2	44	2.5	101.4	6.5	8.8	95.5	-13.4
16.5	43	38.4	2.5	88.9	5.4	7.2	86.9	-15.9
22.0	36.7	32.8	2.5	76.4	4.2	5.6	78.7	-20.6
27.5	30.5	27.3	2.5	63.9	3.0	4.0	70.8	-26.0
33.0	24.3	21.7	2.5	51.4	1.8	2.4	63.1	-32.8
38.5	18.1	16.2	2.5	38.9	0.6	0.8	55.9	-39.8
44.0	11.9	10.6	2.5	26.4	-0.6	-0.8	48.9	-46.8
49.5	5.7	5.1	2.5	13.9	-1.8	-2.4	42.4	-54.0
54.5	0	0	2.5	-2.5	-2.9	-3.8	36.8	-60.5
55.0	-0.6	-0.5	2.5	-3.7	-3.0	-4.0	36.2	-61.2
60.5	-6.8	-6.1	2.5	-16.2	-4.2	-5.6	30.4	-68.4
66.0	-13	-11.6	2.5	-28.7	-5.3	-7.2	25.0	-75.7
71.5	-19.2	-17.2	2.5	-41.2	-6.5	-8.8	20.0	-82.9
77.0	-25.4	-22.7	2.5	-53.7	-7.7	-10.4	15.4	-90.1
82.5	-31.7	-28.3	-2.5	-66.1	-8.9	-12.0	11.3	-97.3
88.0	-37.9	-33.9	-2.5	-78.6	-10.1	-13.6	8.2	-104.3
93.5	-44.1	-39.4	-2.5	-91.1	-11.3	-15.1	5.5	-111.3
99.0	-50.3	-45.0	-2.5	-103.6	-12.5	-16.7	3.2	-118.0
104.5	-56.5	-50.5	-2.5	-116.1	-13.7	-18.3	1.2	-124.7
108.0	-60.5	-54.1	-2.5	-124.1	-14.4	-19.4	0.4	-128.7
109.0	-61.6	-55.1	-2.5	-126.4	-14.6	-19.6	0.2	-129.9
Span 2-0	0	0	0	0	-14.8	-19.9	0	-131.1

* Distance from the centerline of the end bearing.

Table 3.3-8. (Ref. AASHTO-LRFD Table 5.3-8) – Summary of Factored Shear Exterior girder, Span 1 shown, Span 2 symmetric

Location*	**Strength I Shear**	**Service I Shear**	**Service III Shear**
(ft.)	(k)	(k)	(k)
0	342.9	233.3	213.7
1.0	337.2	229.4	210.0
5.5	311.3	211.4	193.3
11.0	280.1	189.7	173.2
16.5	249.3	168.3	153.2
22.0	219.0	147.1	133.4
27.5	189.1	126.2	113.9
33.0	159.7	105.5	94.6
38.5	130.9	85.2	75.5
44.0	102.5	65.1	56.6
49.5	74.8	45.4	38.0
54.5	-101	-59.3	-48.7
55.0	-103.6	-34.1	-50.5
60.5	-132.4	-81.4	-69.5
66.0	-161.3	-101.8	-88.6
71.5	-190.1	-122.1	-107.7
77.0	-218.8	-142.4	-126.7
82.5	-247.5	-162.6	-145.7
88.0	-276	-182.8	-164.7
93.5	-304.4	-202.8	-183.5
99.0	-332.5	-222.8	-202.2
104.5	-360.4	-242.5	-220.9
108.5	-377.9	-255.0	-232.7
109.0	-382.9	-258.6	-236.0
Span 2-0	-237.8	-142.1	-119.3

Load Factor Combinations
Strength I = 1.25(DC) + 1.5(DW) + 1.75(LL + IM)
Service I = 1.0[DC + DW + (LL + IM)]
Service III = 1.0(DC + DW) + 0.8(LL + IM)
* Distance from the centerline of the end bearing.

The live load moments and shears were calculated using a generic live load analysis computer program. The live load distribution factors from Design Step 2.2.1 are applied to these values.

Based on the analysis results, the interior girder controls the design. The remaining sections covering the superstructure design are based on the interior girder analysis. The exterior girder calculations would be identical.

3.3.3.2. AASHTO-LRFD Design Step 5.3.2 Analysis of Creep and Shrinkage Effects

AASHTO-LRFD Design Step5.3.2.1 Creep Effects

The compressive stress in the beams due to pre-stressing causes the pre-stressed beams to creep. For simple span pre-tensioned beams under dead loads, the highest compression in the beams is typically at the bottom, therefore, creep causes the camber to increase, i.e., causes the upward deflection of the beam to increase. This increased upward deflection of the simple span beam is not accompanied by stresses in the beam since there is no rotational restraint of the beam ends. When simple span beams are made continuous through a connection at the intermediate support, the rotation at the ends of the beam due to creep taking place after the connection is established are restrained by the continuity connection. This results in the development of fixed end moments (FEM) that maintain the ends of the beams as flat. As shown schematically in Figure 3.18 for a two-span bridge, the initial deformation is due to creep that takes place before the continuity connection is established. If the beams were left as simple spans, the creep deformations would increase; the deflected shape would appear as shown in part "b" of the figure. However, due to the continuity connection, fixed end moments at the ends of the beam will be required to restrain the end rotations after the continuity connection is established as shown in part "c" of the figure. The beam is analyzed under the effects of the fixed end moments to determine the final creep effects.

Similar effects, albeit in the opposite direction, take place under permanent loads. For ease of application, the effect of the dead load creep and the pre-stressing creep are analyzed separately. Figures 3.19 and 3.20 show the creep moment for a two-span bridge with straight strands. Notice that the creep due to pre-stressing and the creep due to dead load result in restrained moments of opposite sign. The creep from pre-stressing typically has a larger magnitude than the creep from dead loads.

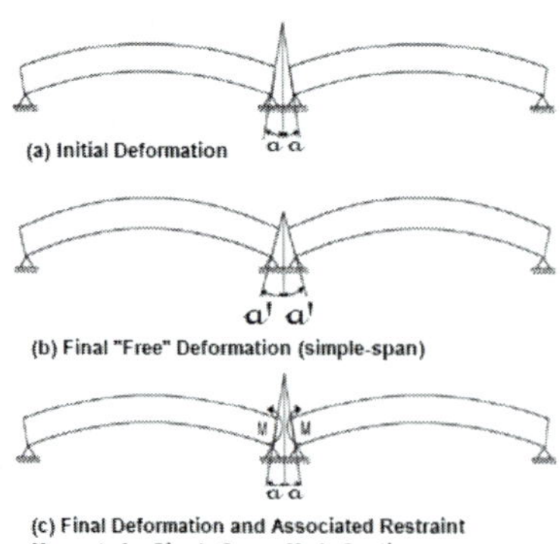

Figure 3.18. Pre-stressed Creep Deformations and Restraint Moments. Source: Figure 5.3-1 from AASHTO LRFD Bridge Design Specifications, FHWA Office of Bridges and Structures, 1200 New Jersey Ave., SE, Washington, DC 20590.

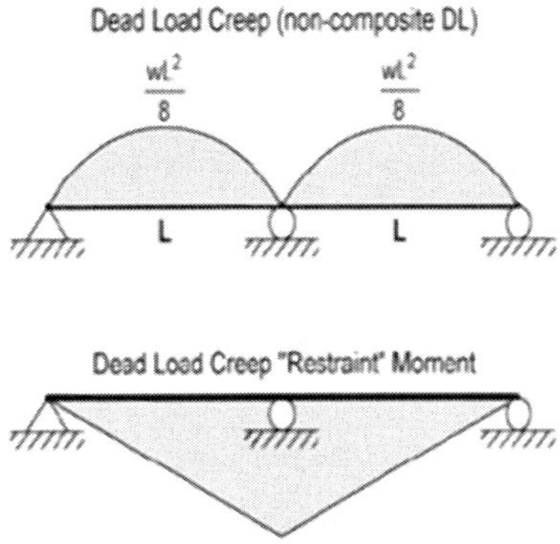

Figure 3.19. Dead Load Creep Moment. Source: Figure 5.3-2 from AASHTO LRFD Bridge Design Specifications, FHWA Office of Bridges and Structures, 1200 New Jersey Ave., SE, Washington, DC 20590.

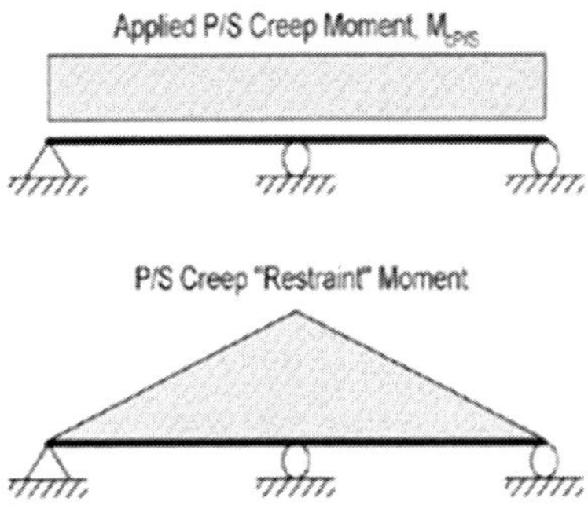

Figure 3.20. Pre-stressed Creep Moment. Source: Figure 5.3-3 from AASHTO LRFD Bridge Design Specifications, FHWA Office of Bridges and Structures, 1200 New Jersey Ave., SE, Washington, DC 20590.

AASHTO-LRFD Design Step 5.3.2.2 Shrinkage Effects

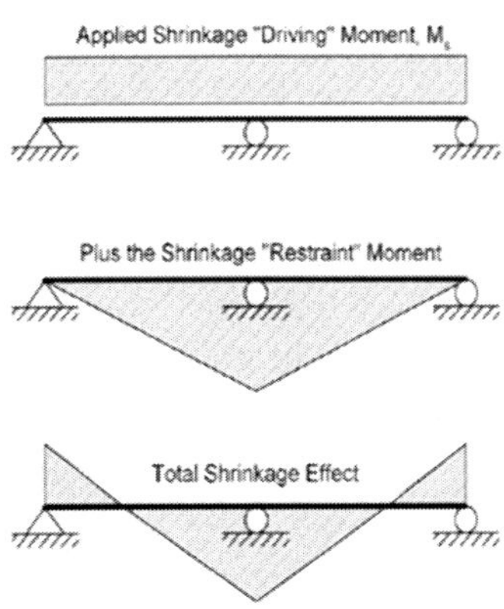

Figure 3.21. Shrinkage Moment. Source: Figure 5.3-4 from AASHTO LRFD Bridge Design Specifications, FHWA Office of Bridges and Structures, 1200 New Jersey Ave., SE, Washington, DC 20590.

The shrinkage of the pre-tensioned beams is different from the shrinkage of the deck slab. This is due to the difference in the age, concrete strength, and method of curing of the two concretes. Unlike creep, differential shrinkage induces stresses in all pre-stressed composite beams, including simple spans. The larger shrinkage of the deck causes the composite beams to sag as shown in Figure 3.21. The restraint and final moments are also shown schematically in the figure.

Calculations of Creep and Shrinkage Effects

The effect of creep and shrinkage should be determined using the methods outlined AASHTO LRFD 5.14.1.4 and NCHRP Report 519, 2004.

AASHTO-LRFD Design Step 5.3.2.3 Effect of Beam Age at the Time of the Continuity Connection Application

The age of the beam at the time of application of the continuity connection has a great effect on the final creep and shrinkage moments. AASHTO LRFD 5.14.1.4.4 through 5.14.1.4.10 provides further discussion on this subject.

For this example, creep and shrinkage effects were ignored.

3.3.4. AASHTO-LRFD Design Step 5.4 - Loss of Pre-Stress (S 5.9.5)

3.3.4.1. Design Step 5.4.1 General

Loss of pre-stress can be characterized as that due to instantaneous loss and time-dependent loss. Losses due to anchorage set, friction and elastic shortening are instantaneous. Losses due to creep, shrinkage and relaxation are time-dependent.

For pre-tensioned members, pre-stress loss is due to elastic shortening, shrinkage, creep of concrete and relaxation of steel. For members constructed and pre-stressed in a single stage, relative to the stress immediately before transfer, the loss may be taken as:

$\Delta f_{pT} = \Delta f_{pES} + \Delta f_{pLT}$ (S5.9.5.1-1)

where:

Δf_{pES} = sum of all losses or gains due to elastic shortening or extension at the time of application of pre-stress and/or external loads (ksi)

Δf_{pLT} = losses due to long-term shrinkage and creep of concrete, and relaxation of the steel (ksi)

Notice that an additional loss occurs during the time between jacking of the strands and transfer. This component is the loss due to the relaxation of steel at transfer.

The stress limit for pre-stressing strands of pre-tensioned members given in S5.9.3 is for the stress immediately prior to transfer referred to as fpbt in the LRFD specifications. To determine the jacking stress, the loss due to relaxation at transfer needs to be added to the stress limits in S5.9.3.Practices differ from state to state as what strand stress is to be shown on the contract drawings. The specifications assume that the designer will determine the stress in the strands immediately before transfer. The fabricator is responsible for determining the jacking force by adding the relaxation loss at transfer, jacking losses and seating losses to the engineer-determined stress immediately prior to transfer. The magnitude of the jacking and seating losses depends on the jacking equipment and anchorage hardware used in the pre-casting yard. It is recommended that the engineer conduct preliminary calculations to determine the anticipated jacking stress.

An accurate estimation of the total pre-stress loss requires recognition that the time-dependent losses resulting from creep and relaxation are interdependent. If required, rigorous calculation of the pre-stress losses should be made in accordance with a method supported by research data. However, for conventional construction, such a refinement is seldom warranted or even possible at the design stage, since many of the factors are either unknown or beyond the designer's control. Thus, three methods of estimating time-dependent losses are provided in the LRFD specifications: (1) the approximate estimate, (2) a refined estimate, and (3) the background necessary to perform a rigorous time-step analysis.

The Approximate Method for calculating the time-dependent losses is presented in S5.9.5.3. The values obtained from this method include the loss due to relaxation at transfer, ΔfpR1. To determine the time-dependent loss after transfer for pre-tensioned members, ΔfpR1 needs to be estimated and deducted from the total time-dependent losses calculated using S5.9.5.3. The refined method of calculating time-dependent losses is presented in S5.9.5.4. The method described above is used in this example.

3.3.4.2. AASHTO-LRFD Design Step 5.4.2 Calculate the Initial Stress in the Tendons Immediately Prior to Transfer (S5.9.3)

fpt + Δ fpES = 0.75fpu

fpt = 270

fpt + Δ fpES = 0.75 x (270)

= 202.5 ksi

3.3.4.3. AASHTO-LRFD Design Step 5.4.3 Determine the Instantaneous Losses (S5.9.5.2)

Friction (S5.9.5.2.2)

The only friction loss possible in a pre-tensioned member is at hold-down devices for draping or harping tendons. The LRFD specifications only specify the consideration of these losses.

For this example, all strands are straight strands and hold-down devices are not used.

Elastic Shortening, ΔfpES (S5.9.5.2.3)

The pre-stress loss due to elastic shortening in pre-tensioned members is taken as the concrete stress at the centroid of the pre-stressing steel at transfer, fcgp, multiplied by the ratio of the modulus of elasticities of the pre-stressing steel and the concrete at transfer. This is presented in Eq. S5.9.5.2.3a-1.

ΔfpES = (Ep/Ect)fcgp (S5.9.5.2.3a-1)

where:

fcgp = sum of concrete stresses at the center of gravity of pre-stressing tendons due to the pre-stressing force immediately after transfer and the self-weight of the member at the sections of maximum moment (ksi)

Ep = modulus of elasticity of the pre-stressing steel (ksi)

Ect = modulus of elasticity of the concrete at transfer (ksi)

Applying this equation requires estimating the stress in the strands after transfer. Proposed estimates for pre-tensioned members are given in S5.9.5.2.3a. For information on additional elastic shortening gains/losses due to strand elongation/shortening caused by external loading and discussion on non-transformed vs. transformed beam sections, see SC5.9.5.2.3a.

Alternatively, the loss due to elastic shortening may be calculated using Eq. C5.9.5.2.3a-1:

where:

e54.5' = average pre-stressing steel eccentricity at mid-span (in.)

fpbt = stress in pre-stressing steel immediately prior to transfer; 0.75fpu (ksi)

Mg = mid-span moment due to member self-weight (k-in)

The alternative approach is used for this example.

Nstrands = 44

Aps = 0.153

Nstrands.Aps = 6.732
fpbt = 202.5
Ig = 733320
e54.5' = 31.38
Ag = 1085
Mg = 20142
Eci = 4200
Ep = 28500
ΔfpES = 6.732[(202.5)[733320
+ 31.38^2(1085)]-31.38(20142) (1085)/(6.732)[733320
+ 31.38^2(1085)]+(1085(733320)(4200) /28500)
ΔfpES = 13.7 ksi

3.3.4.4. AASHTO-LRFD Design Step 5.4.4 Calculate the Pre-Stressing Stress Immediately after Transfer

fpi , = Stress immediately prior to transfer – ΔfpES
= fpbt – ΔfpES
= 202.5 – 13.7
= 188.8 ksi

3.2.4.5. AASHTO-LRFD Design Step 5.4.5 Calculate the Pre-Stressing Force at Transfer

Pi = Nstrands(Aps)(fpi)
= 44 x (0.153) x (188.8)
= 1,271 kips (initial loss = 6.77%)

3.3.4.6. AASHTO-LRFD Design Step 5.4.6 Time-Dependent Losses after Transfer, Refined Method (S5.9.5.4)

Refined estimated time-dependent losses are specified in S5.9.5.4. The refined method can provide a better estimate of total losses than the Approximate Method of S5.9.5.3.

The variables used in these calculations and their values are repeated or calculated here for convenience:

The average annual ambient relative humidity may be obtained from local weather statistics or taken from the map of Figure S5.4.2.3.3-1 shown below.

f'c = 6.0 ksi = concrete compressive strength of beam at 28 days
f'ci = 4.8 ksi = concrete compressive strength of beam at transfer
Ag = 1,085 in^2 = gross area of beam
Ig = 733,320 in^4 = moment of inertia of beam
e54.5 = 31.38 in. = pre-stressing eccentricity at 54 ft.
Ac = 1,795 in^2 = area of composite section
Ic = 1,402,024 in^4 = moment of inertia of composite section
epc = 46.96 in. = pre-stressing eccentricity in composite section
(V/S)b = Ag/perimeter = 1085 / 246.2 = 4.406 in. = volume-to-surface ratio of beam
(V/S)d = 4.0 in. = volume-to-surface ratio of deck
ti = 1 day = equivalent age at transfer (steam cure)
td = 90 days = age at deck placement
tf = 10,000 days = final time
ed = 23.79 in. = eccentricity of deck with respect to gross composite section, positive in typical construction where deck is above girder (in.)
Mg = 20,142 k-in. = mid-span moment due to member self-weight
44-0.5 in. = Diameter, low realization, Grade 270 strands
Aps = 0.153 in^2 = area per strand (based on 8" slab (includes 1/2" wearing surface used for shrinkage calculation later)
fpy = 243 ksi = pre-stressing steel yield strength
fpu = 270 ksi = pre-stressing steel ultimate strength
Ep = 28,500 ksi = pre-stressing steel modulus
Ad = 928 in^2 = area of deck concrete
f'cs = 4.0 ksi = slab concrete strength at 28 days
f'cis = 3.2 ksi = slab concrete strength at loading (set at 80% of f'cs)
Ec = 4696 ksi = girder final elastic modulus
Eci = 4.2 ksi = girder elastic modulus at transfer
Ecd = 3,834 ksi = slab elastic modulus

The average annual ambient relative humidity may be obtained from local weather statistics or taken from the map of Figure 3.22 (Ref. AASHTO-LRFD Figure S5.4.2.3.3-1) shown below.

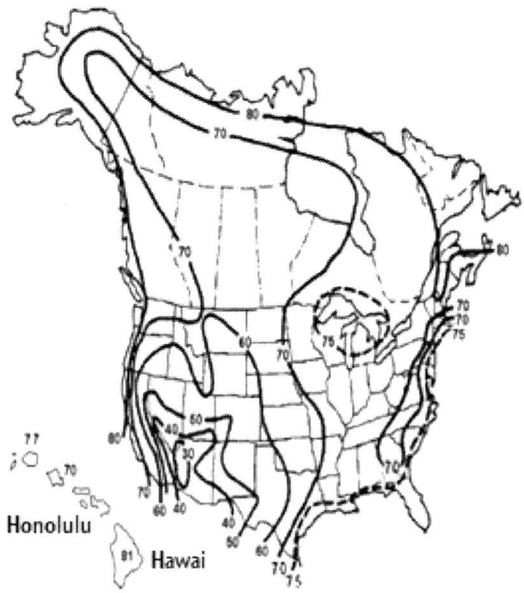

Figure 3.22 – Annual Average Ambient Relative Humidity in Percent. Source: Figure 5.4.2.3.3-1 from AASHTO, AASHTO LRFD Bridge Design Specifications.555 12th Street NW Suite 1000, Washington, DC 20004.

AASHTO-LRFD Design Step 5.4.6.1 Creep Coefficients (S5.4.2.3.2)

The creep coefficients are used in the expression for creep and shrinkage losses in S5.9.5.4 and are calculated from Eq. S5.4.2.3.2-1:

$$\Psi(t, ti) = 1.9kskhckfktdti\text{-}0.118 \qquad \text{(S5.4.2.3.2-1)}$$

in which:

$$ks = 1.45 - 0.13 \times (V/S) \geq 1.0 \qquad \text{(S5.4.2.3.2-2)}$$

$$Khc = 1.56 - 0.008H \qquad \text{(S5.4.2.3.2-3)}$$

$$kf = 5/1 + f'ci \qquad \text{(S5.4.2.3.2-4)}$$

$$Ktd = (t / 12((100\text{-}4f'ci)/(f'ci + 20)) + t) \qquad \text{(S5.4.2.3.2-4)}$$

where:

ks = factor for the effect of the volume-to-surface ratio of the component
khc = humidity factor for creep
kf = factor for the effect of concrete strength
ktd = time development factor

For the girder concrete,

ks = 1.45-0.13 x (4.406) = 0.877 (Must be $\geq$ 1.0; therefore use ks = 1.0)
Khc = 1.56 - 0.008 x (70) = 1
kf = 5/1+4.8 = 0.862
For final age, t = tf - ti = 10,000 days - 1 day = 9,999 days,
Ktd = 9999/[12x ((100-4x (4.8))/(4.8+20))+9999] = 1
For age at deck placement, t = td - ti = 90 - 1 = 89 days,
Ktd = 89/[12 x ((100-4 x (4.8))/(4.8+20))+89] = 0.695

Substituting the above factors into the creep coefficient equation results in the following values:

Between age at transfer, ti= 1 day and final age, tf = 10,000 days,

Ψ(tf, ti) = 1.9X1 x 1 x0.862 x 1X1^(-0.118) = 1.638

Between age at transfer, ti = 1 day and age at deck placement, td = 90 days,

Ψ(td, ti) = 1.9 x 1 x 1 x 0.862 x 0.695 x 1^(-0.118)= 1.138

Between age at deck placement, td = 90 days and final age, tf = 10,000 days,

Ψ(tf, td) = 1.9 x 1 x 1 x 0.862 x 1 x 90^(-0.118) = 0.963

For the deck concrete,

Ks = 1.45-0.13 x (4) = 0.93 (Must be $\geq$1.0; therefore, use ks = 1.0)
khc = 1.56-0.008 x (70) = 1
kf = 5/1+3.2 = 1.19
For final age, tfs = tfb - td = 10,000 - 90 = 9910 days,
ktd = 9910 /[12x ((100-4 x (3.2)) / (3.2+20))+9910]= 1

Between age of deck at deck placement, td = 1 day and final age, tfs = 9,910 days,

Ψ(tfs, td) = 1.9 x 1 x 1 x 1.19 x 1X1^(-0.118) = 2.261

AASHTO-LRFD Design Step 5.4.6.2 Shrinkage Strain (S5.4.2.3.3)

The shrinkage strains are used in the expressions for shrinkage losses in S5.9.5.4 and are calculated from Eq. S5.4.2.3.3-1:

$\varepsilon sh = ks \times khs \times kf \times ktd \times 0.48\times10-3$ (S5.4.2.3.3-1)

in which the equations for kvs, kf and ktd are the same as those given in the previous step for creep coefficient calculations, and:

$khs = 2.00 - 0.014H$ (S5.4.2.3.3-2)

where:

khs = humidity factor for shrinkage

For the girder concrete,

H = 70

khs = 2-0.014x (70)= 1.02

Substituting the above factor and the values for kvs, kf and ktd previously calculated (kvs = 1.00,kf = 0.862 , ktd = 1.0 for tf = 10,000 days and ktd =for 0.695 for td = 90 days) results in the following values for shrinkage strain:

kvs = 1

kf = 0.862

ktd = 1

tf = 10000

ktd = 0.695

td = 90 days

Between age at transfer,

ti = 1 day and final age, tf = 10,000 days, εbif = 1 x1.02 x0.862 x1 x 0.48×10^(-3) = 0.000422

Between age at transfer,

ti = 1 day and age at deck placement, td = 90 days, εbid = 1 x 1.02 x 0.862 x 0.695 x0.48×10^(-3) = 0.000293

Between age at deck placement,

td 90 days and final age, tf = 10,000 days, εbif = 0.000422 - 0.000293 = 0.000129

For the deck concrete, between age of deck at deck placement, td = 1 day and final age, tfs = 9,910 days, the previously calculated values for the factors may be used (kvs = 1.00, khs = 1.02, kf = 1.19, and ktd = 1.0 for tf = 10,000 days). This results in the following shrinkage strain:

td = 1 day
khs = 1.02
tfs = 9,910 days
kf = 1.19
ktd = 1
kvs = 1
tf = 10000 days
$\varepsilon ddf = 1 \times 1.02 \times 1.19 \times 1 \times (0.48 \times 10^{(-3)}) = 0.000583$

AASHTO-LRFD Design Step 5.4.6.3 Shrinkage Losses (S5.9.5.4.2a, S5.9.5.4.3a, and S5.9.5.4.3d)

The pre-stress loss due to shrinkage of the girder concrete between time of transfer and deck placement is given as Eq. S5.9.5.4.2a-1:

$$\Delta fpSR = \varepsilon\, bid \times Ep \times Kid \qquad (S5.9.5.4.2a\text{-}1)$$

where:

Kid = transformed section coefficient that accounts for time-dependent interaction between concrete and bonded steel in the section being considered for time period between transfer and deck placement;
Ep = 28500
Aps = 0.153
Eci = 4200
Ag = 1085
epg = 31.38
$\Psi b = 0.7$
tf , ti = 1.638
Ig = 733320

Substituting the values given previously in this section,

$$K_{id} = \frac{1}{1 + \frac{28500}{4200} \times \frac{(44X0.153)}{1085} \times \left[1+ \frac{1085X(31.38)^2}{733320}\right] X \left[1+ (0.7\,X1.638)\right]}$$

= 0.818

The loss due to shrinkage of girder concrete between transfer and deck placement is, therefore:

ΔfpSR = (0.000293) x (28500) x (0.818) = 6.831 ksi

The pre-stress loss due to shrinkage of the girder concrete between time of deck placement and final time is given as Eq. S5.9.5.4.3a-1:

ΔfpSD = ε bdf EpKdf (S5.9.5.4.3a-1)

in which:

$$K_{df} = \frac{1}{1+\frac{E_p}{E_{ci}}\frac{A_{ps}}{A_c}\left(1+\frac{A_c e_{pc}^2}{I_c}\right)\left[1=0.7\Psi_b\left(t_f, t_i\right)\right]} \quad (S5.9.5.4.4a\text{-}2)$$

where:

Kdf = transformed section coefficient that accounts for time-dependent interaction between concrete and bonded steel in the section being considered for time period between deck placement and final time.

Substituting the values given previously in this section,

Ep = 28500
Eci = 4200
Aps = 0.153
Ac = 1795
Epc = 46.96
Ic = 1402024
Ψb(tf, ti) = 1.638
Kdf = 1/(1+(28500/4200)x(44x0.153/1795)x(1+(1795x46.96^2)/ 1402024)x(1+(0.7x1.638)))
= 0.827

The loss due to shrinkage of girder concrete between transfer and deck placement is, therefore:

ΔfpSD = (0.000129) x (28500) x (0.827) = 3.04 ksi

The pre-stress loss due to shrinkage of the deck composite section is given as Eq. S5.9.5.4.3d-1:

$$\Delta f_{pSS} = \frac{E_p}{E_c} \Delta f_{cdf} K_{df} \left[1 + 0.7\Psi_b(t_f, t_d)\right]$$

in which:

$$\Delta f_{cdf} = \frac{\varepsilon_{ddf} A_d E_{cd}}{\left[1 + 0.7\Psi_d(t_f, t_d)\right]} \left(\frac{1}{A_c} - \frac{e_{pc} e_d}{I_c} \right)$$

where:

Δfcdf= change in concrete stress at centroid of pre-stressing strands due to shrinkage of deck concrete.

Substituting the values given previously in this section,

εddf = 0.000583
Ad = 928
Ecd = 3834
Ψd(tf, ti) = 2.261
Ac = 1795
Epc = 46.96
Ed = 23.79
Ic = 1402024
Δfcdf = ((0.000583x928x3834)/(1+0.7x2.261))x((1/1795)-(46.96x 23.79/1402024))
= -0.193 ksi

The loss due to shrinkage of the deck is, therefore:

Ep = 28500
Ec = 4696
Δfcdf = -0.193
Kdf = 0.827
Ψb(tf, td) = 0.963

Δf_{pSS}= ((28500/4696)x(-0.193)x0.827x(1+0.7x0.963)
= -1.622 ksi

The negative sign implies a gain in tension in the strands, offsetting the losses but at the same time causing bottom fiber tensile stress increases.

AASHTO-LRFD Design Step 5.4.6.4 Creep Losses (S5.9.5.4.2b and S5.9.5.4.3b)

The pre-stress loss due to creep of girder concrete between time of transfer and deck placement is given as Eq. S5.9.5.4.2b-1:

$$\Delta f_{pCR} = E_p/E_{ci}.f_{cgp}\psi_b(t_d,t_i)K_{id} \quad (S5.9.5.4.2b\text{-}1)$$

in which:

fcpg = sum of concrete stresses at the center of gravity of pre-stressing tendons due to the pre-stressing force after jacking and the self-weight of the member at the sections of maximum moment:

$$f_{cpg} = P_i/A_g + P_i e^2_{54.5}/I_g - M_g e_{54.5}/I_g$$

Pi = 1271
Ag = 1085
e54.5 = 31.38
Ig = 733320
Mg = 20142
fcpg = 1271/1085 + 1271x(31.38)^2 / 733320 - 20142x(31.38)/733320= 2.016 ksi

The loss due to creep of the girder between transfer and deck placement is, therefore:

$$\Delta f_{pCR} = (28500/4200)\text{x}(2.016)\text{x}(1.138)\text{x}(0.818) = 12.734 \text{ ksi}$$

The pre-stress loss due to creep of girder concrete between time of deck placement and final time is given as Eq. S5.9.5.4.3b-1:

$$\Delta f_{pcD} = E_p/E_{ci} f_{cgp}[\Psi_b(t_f,t_i)-\Psi_b(t_d,t_i)]K_{df}+E_p/E_c\Delta f_{cd}\Psi_b(t_f,t_d)K_{df} \quad (S5.9.5.4.3b\text{-}1)$$

in which:

Δ_{fcd} = change in concrete stress at centroid of pre-stressing strands due to long-term losses between transfer and deck placement, combined with deck weight and superimposed loads.

Assuming relaxation loss, $\Delta fpRI = 1.2$ ksi as allowed by S5.9.5.4.2c, the total long-term losses between transfer and deck placement are:

$\Delta f_p(t_d,t_i) = \Delta f_{pSR} + \Delta f_{pCR} + \Delta f_{pRI}$
$\Delta f_{pSR} = 6.831$
$\Delta f_{pCR} = 12.734$
$\Delta f_{pRI} = 1.2$
$\Delta f_p(t_d,t_i) = 6.831+12.734+1.2 = 20.765$ ksi

The loss of force in the strands is, therefore:

$P_{loss}\ (t_d,t_i) = A_{ps}\Delta f_p(t_d,t_i)$
$A_{ps} = 0.153$
$\Delta f_p(t_d,t_i) = 20.765$
$P_{loss}\ (t_d,t_i) = 44(0.153)(20.765) = 139.8$ kips

As previously calculated,

$M_{deck} = M_{slab} + M_{diaph}$
$M_{slab} = 1696$ k-ft
$M_{diaph} = 138$ k-ft
$M_{deck} = 1696+138 = 1{,}834$ k-ft
$M_{SDL} = M_{parapet} + M_{FWS} = 165+222 = 387$ k-ft
$M_{parapet} = 165$
$M_{FWS} = 222$
$M_{SDL} = 165 + 222 = 387$ k-ft

Therefore, the change in concrete stress is:

$f_{cd} = (-P_{loss}/A_g)-(P_{loss}e^2_{54.5}/I_g)-(M_{deck}e_{54.5}/I_g)-(M_{SDL}e_{pc}/I_c)$
$P_{loss} = 139.8$
$A_g = 1085$
$e_{54.5} = 31.38$
$I_g = 733320$
$M_{deck} = 1834$
$M_{SDL} = 387$
$e_{pc} = 46.96$
$I_c = 1402024$

f_{cd} = -(139.8/1085)-(139.8 x (31.38)^2/733320)-(1834 x 12 x (31.38)/733320)-((387) x 12 x (46.96)/1402024)
= -1.414 ksi

The loss due to creep of the girder between deck placement and final time is, therefore:

Δf_{pCD} = 28500/4200(2.016)(1.638-1.138)(0.827)+28500/4696(-1.414) (0.963)(0.827)
= -1.178 ksi

AASHTO-LRFD Design Step 5.4.6.5 Relaxation Losses (S5.9.5.4.2c and S5.9.5.4.3c)

The pre-stress loss due to relaxation of pre-stressing strands between time of transfer and deck placement is given as Eq. S5.9.5.4.2c-1:

$\Delta f_{PR1} = (f_{pi}/K_L)((f_{pi}/f_{py})-0.55)$ (S5.9.5.4.2c-1)

in which:
KL = 30for low-relaxation strands

The loss due to relaxation between transfer and deck placement is, therefore

Fpi= 188.8
KL= 30
Fpi= 243
Δf_{PR1} = (188.8/30)x((188.8/243)-0.55) = 1.428 ksi

The pre-stress loss due to relaxation of pre-stressing strands in the composite section between time of deck placement and final time is given as Eq. S5.9.5.4.3c-1:

$\Delta f_{PR2} = \Delta f_{PR1}$ = 1.428 ksi (S5.9.5.4.3c-1)

3.3.4.7. AASHTO-LRFD Design Step 5.4.7 Calculate Total Time Dependent Loss after Transfer

$\Delta f_{PLT} = (\Delta f_{pSR} + \Delta f_{pCR} + \Delta f_{PR1})id + (\Delta f_{pSD} + \Delta f_{PCD} + \Delta f_{pR2} - \Delta f_{PSS})d_f$
Δf_{pCR} = 6.831
Δf_{pCR} = 12.734
Δf_{PR1} = 1.428
Δf_{pSD} = 3.04

$\Delta f_{PCD} = -1.178$
$\Delta f_{pR2} = 1.428$
$\Delta f_{PSS} = -1.622$
$\Delta f_{PLT} = [(6.831+12.734+1.428)]+[(3.04)+(-1.178)+(1.428)+(-1.622)]$

$= 22.66$ ksi (S5.9.5.4.1-1)

$\Delta f_{pT} = \Delta f_{PES} + \Delta f_{PLT}$

$= 13.7 + 22.66 = 36.36$ (S5.9.5.1-1)

If the approximate method is used, Δf_{pT} is approximately equal to 37.37 ksi. The approximate method is simpler and generally validates the more refined method. The refined method is likely to be used in software due to its ability to rapidly calculate the various quantities needed for loss evaluation.

3.3.4.8. AASHTO-LRFD Design Step 5.4.8 Calculate the Final Effective Pre-Stress Responses

Max f_{pe} = 0.80fpy (State after all losses) = 0.8 x (243)
= 194.4 ksi

Calculate the actual effective pre-stress stress after all losses

fpe = $0.75f_{pu} - \Delta f_{pT}$
fpu = 270
fpe = 0.75 x (270)-36.36 = 166.14 ksi < 194.4 ksi OK

Calculate the actual effective pre-stress force after all losses

Pe = Nstrands(Aps)(fpe)
Nstrands = 44
Aps = 0.153
Fpe = 166.14
Pe = 44 x (0.153) x (166.14) = 1118.5 kips (total loss = 18%)

3.3.4.9. AASHTO-LRFD Design Step 5.4.9 Calculate Jacking Stress, fpj

As indicated earlier, the fabricator is responsible for calculation of the jacking force. The calculations presented below are for reference purposes.

As shown earlier, the stress in the pre-stressing strands immediately prior to transfer is 202.5 ksi.

The jacking stress, fpj = stress immediately prior to transfer + relaxation loss at transfer.

Relaxation at transfer was calculated in design step 3.3.4.6.5 design Step 5.4.6.5and is as follows:

$\Delta f_{pR1} = 1.428$ ksi

Therefore, the jacking stress is:

fpj = 202.5 + 1.428 = 203.9 ksi

3.3.5. AASHTO-LRFD Design Step 5.5 - Stress in Pre-Stressing Strands

3.3.5.1. Design Step 5.5.1 Stress in Pre-Stressing Strands at Nominal Flexural Resistance

The stress in pre-stressing steel at nominal flexural resistance may be determined using stress compatibility analysis. In lieu of such analysis a simplified method is presented in S5.7.3.1.1. This method is applicable to rectangular or flanged sections subjected to flexure about one axis where the Whitney stress block stress distribution specified in S5.7.2.2 is used and for which fpe, the effective pre-stressing steel stress after losses, is not less than 0.5fpu. The average stress in pre-stressing steel, fps, may be taken as:

$$f_{ps} = f_{pu}[1 - k(c/d_p)] \qquad \text{(S5.7.3.1.1-1)}$$

where:

$$k = 2(1.04 - f_{py} / f_{pu}) \qquad \text{(S5.7.3.1.1-2)}$$

The value of "k" may be calculated using the above equation based on the type and properties of pre-stressing steel used or it may be obtained from AASHTO-LRFD Table SC5.7.3.1.1-1.

The distance from the neutral axis to the compression face of the member may be determined as follows:

for T-section behavior (Eq. S5.7.3.1.1-3):

$$c = \frac{A_{ps} f_{pu} + A_s f_y - A'_s f'_y - \alpha_1 f'_c (b - b_w) h_f}{\alpha_1 f'_c \beta_1 b_w + k A_{ps} \dfrac{f_{pu}}{d_p}}$$

for rectangular section behavior (Eq. S5.7.3.1.1-4):

$$c = \frac{A_{ps} f_{pu} + A_s f_y - A'_s f'_y}{\alpha_1 f'_c \beta_1 b + k A_{ps} \dfrac{f_{pu}}{d_p}}$$

T-sections where the neutral axis lies in the flange, i.e., “c” is less than the slab thickness, are considered rectangular sections.

From AASHTO-LRFD Table SC5.7.3.1.1-1:

$k = 0.28$ for low relaxation strands

Assuming rectangular section behavior with no compression steel or mild tension reinforcement:

$c = A_{ps}f_{pu}/[\alpha_1 f'_c\beta_1 b + kA_{ps}(f_{pu}/d_p)]$

For the midspan section:

Girder depth = 72 in

Structural slab thickness = 7.5 in

Total section depth, h = girder depth + structural slab thickness = 72 + 7.5 = 79.5 in.

dp = h – (distance from bottom of beam to location of P/S steel force)

(distance from bottom of beam to location of P/S steel force = 5 in) = 79.5 – 5 = 74.5 in.

$\beta 1 = 0.85$ for 4 ksi slab concrete (S5.7.2.2)

b = effective flange width (calculated in Section 2 of this example)

= 116 in

c = 6.732 x (270)/[0.85 x 4 x 0.85 x 116 + 0.29 x (6.732) x (270/74.5)]= 5.31 in < 7.5 in (structural slab thickness)

The assumption of the section behaving as a rectangular section is correct.

Notice that if "c" from the calculations above was greater than the structural slab thickness (the integral wearing surface is ignored), the calculations for "c" would have to be repeated assuming a T-section behavior following the steps below:

1) Assume the neutral axis lies within the precast girder flange thickness and calculate "c". For this calculation, the girder flange width and area should be converted to their equivalent in slab concrete by multiplying the girder flange width by the modular ratio between the precast girder concrete and the slab concrete. The web width in the equation for "c" will be substituted for using the effective converted girder flange width. If the calculated value of "c" exceeds the sum of the deck thickness and the precast girder flange thickness, proceed to the next step. Otherwise, use the calculated value of "c".
2) Assume the neutral axis is below the flange of the precast girder and calculate "c". The term "$0.85\ f'c(b - b_w)$" in the calculations should be broken into two terms, one refers to the contribution of the deck to the composite section flange and the second refers to the contribution of the precast girder flange to the composite girder flange.

$f_{ps} = f_{pu}[1 - k(c/d_p)]$ (S5.7.3.1.1-1)
$f_{pu} = 270$
$k = 0.28$
$c = 5.31$
$d_p = 74.5$
$f_{ps} = 270 \times [1 - 0.28 \times (5.31/74.5)] = 264.6$ ksi

3.3.5.2. Design Step 5.5.2 Transfer and Development Length

Transfer Length = 60(Strand diameter) (S5.11.4.1)
Strand diameter = 0.5 in = 60 x (0.5 in.) = 30 in.
Development Length = $ld \geq \kappa[fps - (2/3)fpe]db$ (S5.11.4.2-1)

From earlier calculations:

fps = 264.6 ksi (Design Step 3.3.5.1 Design Step 5.5.1)
fpe = 166.14 ksi (Design Step 3.3.4.8 Design Step 5.4.8)
From S5.11.4.2, $\kappa = 1.6$ for fully bonded strands
From S5.11.4.3, $\kappa = 2$ for partially de-bonded strands

For fully bonded strands (32 strands):

$ld \geq 1.6x[264.6 - (2/3)x166.14]x(0.5) = 123.07$ in. (10.26 ft.)

For partially de-bonded strands (two groups of 6-strands each):

ld ≥ 2x[264.6 – (2/3)x166.14](0.5) = 153.84 in (12.82 ft.)

3.3.5.3. Design Step 5.5.3 Variation in Stress in Pre-Stressing Steel along the Length of the Girders

According to S5.11.4.1, the pre-stressing force, fpe, may be assumed to vary linearly from 0.0 at the point where bonding commences to the effective stress after losses, fpe, at the transfer length. Between the transfer length and the development length, the strand force may be assumed to increase linearly, reaching the stress at nominal resistance, fps at the development length.

As shown in AASHTO-LRFD Figures 2-5 and 2-6, each beam contains three groups of strands:

Group 1: 32 strands fully bonded, i.e., bonded length starts 9 in. outside the centerline of bearings of the non-composite beam (note that the bond length starts at the end of the beam)

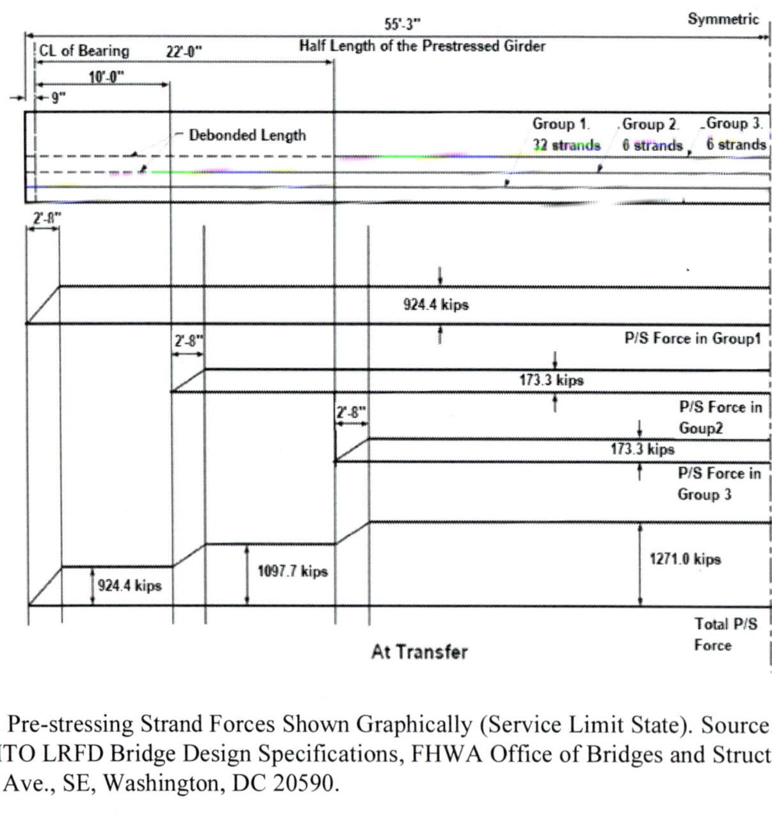

Figure 3.23. Pre-stressing Strand Forces Shown Graphically (Service Limit State). Source: Figure 5.5-1 from AASHTO LRFD Bridge Design Specifications, FHWA Office of Bridges and Structures, 1200 New Jersey Ave., SE, Washington, DC 20590.

Table 3.4. (Ref. AASHTO-LRFD Table 5.5-1) – Pre-stressing Strand Forces

Dist. From Girder end	**Dist. from CL of Brg**	**Initial Pre-stressing Force at Transfer**			
		Group 1	Group 2	Group3	Total
(ft.)	(ft.)	(k)	(k)	(k)	(k)
0*	-0.75*	0.0			0.0
0.75	0.00	277.3			277.3
2.50	1.75	924.4			924.4
7.75	7.00	924.4			924.4
10.39	9.64	924.4			924.4
10.75**	10.00**	924.4	0.0		924.4
11.75	11.00	924.4	69.3		993.7
13.25	12.50	924.4	173.3		1,097.70
17.25	16.50	924.4	173.3		1,097.70
22.75***	22.00***	924.4	173.3	0.0	1,097.70
23.73	22.98	924.4	173.3	67.90	1,165.60
25.25	24.50	924.4	173.3	173.3	1,271.00
28.25	27.50	924.4	173.3	173.3	1,271.00
33.75	33.00	924.4	173.3	173.3	1,271.00
35.73	34.98	924.4	173.3	173.3	1,271.00
39.25	38.50	924.4	173.3	173.3	1,271.00
44.75	44.00	924.4	173.3	173.3	1,271.00
50.25	49.50	924.4	173.3	173.3	1,271.00
55.25	54.50	924.4	173.3	173.3	1,271.00
55.75	55.00	924.4	173.3	173.3	1,271.00
61.25	60.50	924.4	173.3	173.3	1,271.00
66.75	66.00	924.4	173.3	173.3	1,271.00
72.25	71.50	924.4	173.3	173.3	1,271.00
74.77	74.02	924.4	173.3	173.3	1,271.00
77.75	77.00	924.4	173.3	173.3	1,271.00
83.25	82.50	924.4	173.3	173.3	1,271.00
85.25	84.50	924.4	173.3	173.3	1,271.00
86.77	86.02	924.4	173.3	67.9	1,165.60
87.75+++	87.00+++	924.4	173.3	0.0	1,097.70
87.75+++	87.00+++	924.4	173.3	0.0	1,097.70
88.75	88.00	924.4	173.3		1,097.70
94.25	93.50	924.4	173.3		1,097.70
97.25	96.50	924.4	173.3		1,097.70
99.75++	99.00++	924.4	0.0		924.4
100.11	99.36	924.4			924.4
103.25	102.5	924.4			924.4
108.00	107.25	924.4			924.4
109.75	109.00	277.3			277.3
110.5+	109.75+	0.0			0.0

*, **, *** - Point where bonding commences for strand Groups 1, 2, and 3, respectively.

+, ++, +++ - Point where bonding ends for strand Groups 1, 2, and 3, respectively.

Table 3.5. (Ref. AASHTO-LRFD Table 5.5-1) – Pre-stressing Strand Forces

Dist. from Girder End	Dist. from CL of Brg	Pre-stressing Force After Loses				Force at the Nominal Flexural Resistance			
		Group 1	Group 2	Group3	Total	Group 1	Group 2	Group 3	Total
0*	0.75*	0.0			0.0	0.0			0.0
0.75	0.00	244.0			244.0	244.0			244.0
2.50	1.75	813.4			813.4	813.4			813.4
7.75	7.00	813.4			813.4	1132.1			1132.1
10.44	9.69	813.4			813.4	1295.5			1295.5
10.75**	10.00**	813.4	0.0		813.4	1295.5	0.0		1295.5
11.75	11.00	813.4	61.0		874.4	1295.5	61.0		1356.5
13.25	12.50	813.4	152.5		965.9	1295.5	152.5		1448.0
17.25	16.50	813.4	152.5		965.9	1295.5	186.8		1482.3
22.75***	22.00***	813.4	152.5	0.0	965.9	1295.5	233.9	0.0	1529.4
23.80	23.05	813.4	152.5	64.1	10.29	1295.5	242.9	64.1	1602.4
25.25	24.5	813.4	152.5	152.5	11 18.4	1295.5	242.9	152.5	1690.9
28.25	27.5	813.4	152.5	152.5	11 18.4	1295.5	242.9	178.2	1716.6
33.75	33.00	813.4	152.5	152.5	11 18.4	1295.5	242.9	225.3	1763.7
35.8	35.05	813.4	152.5	152.5	11 18.4	1295.5	242.9	242.9	1781.3
39.25	38.50	813.4	152.5	152.5	11 18.4	1295.5	242.9	242.9	1781.3
44.75	44.00	813.4	152.5	152.5	11 18.4	1295.5	242.9	242.9	1781.3
50.25	49.50	813.4	152.5	152.5	11 18.4	1295.5	242.9	242.9	1781.3
55.25	54.50	813.4	152.5	152.5	11 18.4	1295.5	242.9	242.9	1781.3
55.75	55.00	813.4	152.5	152.5	11 18.4	1295.5	242.9	242.9	1781.3
61.25	60.50	813.4	152.5	152.5	11 18.4	1295.5	242.9	242.9	1781.3
66.75	66.00	813.4	152.5	152.5	11 18,4	1295.5	242.9	242.9	1781.3
72.25	71.50	813.4	152.5	152.5	11 18.4	1295.5	242.9	242.9	1781.3
74.70	73.95	813.4	152.5	152.5	11 18.4	1295.5	242.9	242.9	1781.3
77.75	77.00	813.4	152.5	152.5	11 18.4	1295.5	242.9	216.8	1755.2
83.25	82.50	813.4	152.5	152.5	11 18.4	1295.5	242.9	169.6	1708.0
85.25	84.5	813.4	152.5	152.5	11 18.4	1295.5	242.9	152.5	1690.9
86.70	85.95	813.4	152.5	64.1	1029.9	1295.5	242.9	64.1	1602.4
87.75+++	87.00+++	813.4	152.5	0.0	965.9	1295.5	233.9	0.0	1529.1
88.75	88.00	813.4	152.5		965.9	1295.5	225.3		1520.8
94.25	93.50	813.4	152.5		965.9	1295.5	178.2		1473.7
97.25	96.50	813.4	152.5		965.9	1295.5	152.5		1448.0
99.75++	99.00++	813.4	0.0		813.4	1295.5	0.0		1295.5
100.06	99.31	813.4			813.4	1295.5			1295.5
103.25	102.50	813.4			813.4	1101.8			1101.8
108.00	107.25	813.4			813.4	813.4			813.4
109.75	109.00	244.0			244.0	244.0			244.0
110.5+	109.75+	0.0			0.0	0.0			0.0

*, **, *** Point where bonding commences for strand Groups 1, 2, and 3, respectively.
+, ++, +++ Point where bonding ends for strand Groups 1, 2, and 3, respectively.

Group 2: 6 strands. Bonded length starts 10 ft. from the centerline of bearings of the non-composite beam, i.e., 10’-9” from the end of the beam.

Group 3: 6 strands. Bonded length starts 22 ft. from the centerline of bearings of the non-composite i.e., 22'-9" from the end of the beam.

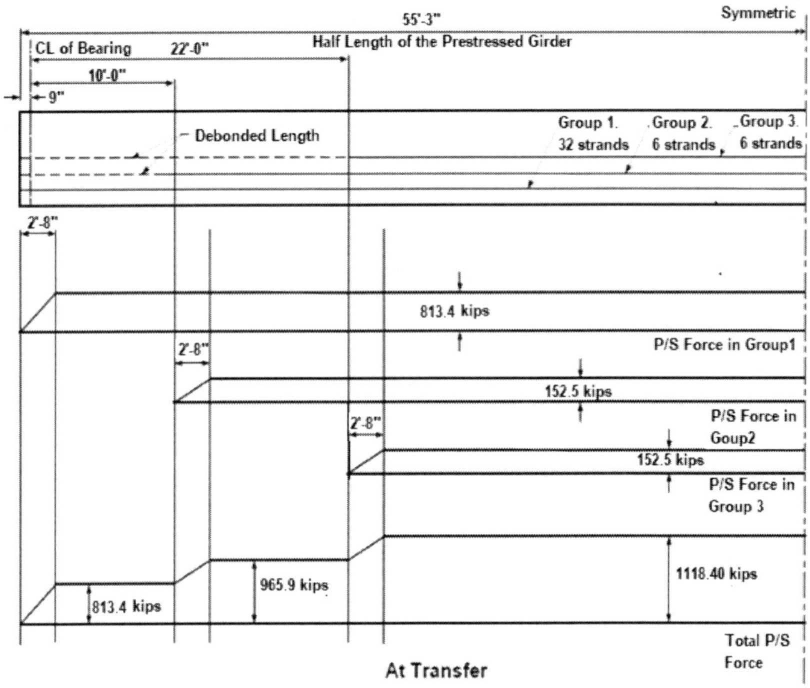

Figure 3.24. Pre-stressing Strand Forces Shown Graphically (Service Limit State). Source: Figure 5.5-1 (Continued) from AASHTO LRFD Bridge Design Specifications, FHWA Office of Bridges and Structures, 1200 New Jersey Ave., SE, Washington, DC 20590.

For each group, the stress in the pre-stressing strands is assumed to increase linearly from 0.0 at the point where bonding commences to fpe, over the transfer length, i.e., over 30 inches. The stress is also assumed to increase linearly from fpe at the end of the transfer length to fps at the end of the development length. Table 3.1 (Ref. AASHTO LRFD Table 5.1-1) shows the strand forces at the service limit state (maximum strand stress = fpe) and at the strength limit state (maximum strand stress = fps) at different sections along the length of the beams. To facilitate the calculations, the forces are calculated for each of the three groups of strands separately and sections at the points where bonding commences, end of transfer length and end of development length for each group are included in the tabulated values. Figures 3.23 and 3.24 are graphical representations of Table 3.1 (Ref. AASHTO LRFD Table 5.1-1).

3.3.5.4. Design Step 5.5.4 Sample Strand Stress Calculations

Pre-Stress Force at Centerline of End Bearing after Losses under Service or Strength

Only Group 1 strands are bonded at this section. Ignore Group 2 and 3 strands.

Transfer length = 2.5

Distance from the point bonding commences for Group 1 strands = 0.75 ft < transfer length

Percent of pre-stressing force developed in Group 1 strands = 0.75/transfer length = (0.75/2.5) x (100) = 30% = 0.3

Stress in strands = 0.3 x 166.14 = 49.84 ksi

Force in strands at the section = 32x0.153 x 49.84 = 244 kips

Pre-Stress Force at a Section 11 ft. from the Centerline of End Bearing after Losses under Service Conditions

Only strands in Group 1 and 2 are bonded at this section. Ignore Group 3 strands.

The bonded length of Group 1 strands before this section is greater than the transfer length. Therefore, the full pre-stressing force exists in Group 1 strands.

Force in Group 1 strands = 32x 0.153 x 166.14 = 813.4 kips

Distance from the point bonding commences for Group 2 strands = 1ft. < transfer length

Percent of pre-stressing force developed in Group 2 strands = 1/transfer length = (1/2.5) x (100) = 40% = 0.4

Stress in Group 2 strands = 0.4 x166.14 = 66.5 ksi

Force in Group 2 strands at the section = 6x 0.153 x66.5 = 61 kips

Total pre-stressing force at this section = force in Group 1 + force in Group 2 = 813.4 + 61 = 874.4 kips

Strands Maximum Resistance at Nominal Flexural Capacity at a Section 7.0 ft. from the Centerline of End Bearing

Only Group 1 strands are bonded at this section. Ignore Group 2 and 3 strands.

Distance from the point bonding commences for Group 1 strands, i.e., distance from end of beam = 7.75 ft. (7'- 9")

This distance is greater than the transfer length (2.5 ft.) but less than the development length of the fully bonded strands (10.26 ft.). Therefore, the stress in the strand is assumed to reach fpe, 166.13 ksi, at the transfer length then increases linearly from fpe to fps, 264.6 ksi, between the transfer length and the development length.

Stress in Group 1 strands = 166.14+ (264.6– 166.14)x[(7.75 – 2.5)/(10.26 –2.5)]= 232.8 ksi

Force in Group 1 strands = 32x0.153x232.8 = 1139.79 kips

Strands Maximum Resistance at Nominal Flexural Capacity at a Section 22 ft. from Centerline of End Bearing

Only strands in Group 1 and 2 are bonded at this section. Ignore Group 3 strands.

The bonded length of Group 1 strands before this section is greater than the development length for Group 1 (fully bonded) strands. Therefore, the full force exists in Group 1 strands.

Force in Group 1 strands = 32x0.153 x 264.6 = 1295.5 kips

The bonded length of Group 2 at this section = 22 – 10= 12 ft.

Stress in Group 2 strands= 166.14 + (264.6 – 166.14) x [(12– 2.5)/(12.82 –2.5)]= 256.8 ksi

Force in Group 2 strands = 6x0.153x256.8 = 235.7 kips

Total pre-stressing force at this section = force in Group 1 + force in Group 2

= 1295.5 + 235.7 = 1,531.20 kips

3.3.6. AASHTO-LRFD Design Step 5.6 - Flexure Design

3.3.6.1. AASHTO-LRFD Design Step 5.6.1 Flexural Stress at Transfer

AASHTO-LRFD Design Step 5.6.1.1

Stress Limits at Transfer

Compression stress:

The allowable compression stress limit for pre-tensioned concrete components is calculated according to S5.9.4.1.1.

fci = 4.8 ksi

fCompression= -0.60 (f'ci)

= (-0.6 x 4.8)

= -2.88 ksi

Tension stress:

From AASHTO LRFD Table S5.9.4.1.2-1, the stress limit in areas with bonded reinforcement sufficient to resist the tension force in the cracked concrete computed on the basis of an un-cracked section is calculated as:

fTension = 24√f'ci = 0.24X √4.8 = 0.53 ksi

AASHTO-LRFD Design Step 5.6.1.2 Stress Calculations at Transfer

Table 3.6. (Ref. AASHTO-LRFD Table 5.6-1) – Stresses at Top and Bottom of Beam at Transfer

Location	Girder Self weight Moment	F_{ps} at Transfer	Stress at Transfer	
			Top of Beam	Bottom of Beam
(ft.)(1)	(k-ft.)(2)	(kips)(3)	(ksi)	(ksi)
0	47	277.3	0.135	-0.654
1.75	153	924.4	0.451	-2.183
5.5	368	924.4	0.326	-2.055
11.0	656	993.7	0.209	-2.065
16.5	909	1,097.70	0.123	-2.171
22.0	1128	1,097.70	-0.005	-2.04
27.5	1313	1,271.00	-0.009	-2.358
33.0	1464	1 1,271.00	-0.097	-2.269
38.5	1580	1,271.00	-0.155	-2.209
44.0	1663	1 1,271.00	-0.203	-2.16
49.5	1711	1,271.00	-0.231	-2.132
54.5	1725	1 1,271.00	-0.24	-2.12
55.0	1725	1,271.00	-0.24	-2.123
60.5	1705	1 1,271.00	-0.228	-2.135
66.0	1650	1,271.00	-0.196	-2.168
71.5	1562	1 1,271.00	-0.144	-2.22
77.0	1439	1,271.00	-0.083	-2.284
82.5	1282	1 1,271.00	0.009	-2.377
88.0	1091	1,097.70	0.017	-2.063
93.5	865	1 1,097.70	0.149	-2.197
99.0	606	924.4	0.197	-1.923
104.5	312	924.4	0.358	-2.105
107.25	153	924.4	0.451	-2.200
109.0	47	1 277.3	0.135	-0.66

Notes:

1 - Distance measured from the centerline of the bearing of the simple span girder.

2 - See Section 5.3, based on 110.5 ft. length

3 - See Section 5.5 for pre-stressing forces.

Sample Calculations for Flexural Stresses at Transfer

Definitions:

Pi = Initial pre-stressing force from Table 3.1 (Ref. AASHTO LRFD Table 5.1-1) (kips)

Ag = Gross area of the basic beam (in^2)

e = Distance between the neutral axis of the non-composite girder and the center of gravity of the pre-stressing steel (in.)
St = Section moduli, top of non-composite beam (in3)
Sb = Section moduli, bottom of non-composite beam (in3)
Mg = Moment due to the girder self weight only (k-ft)

See Section 3.2 for section properties.

Sample Calculations at 1 ft. – 9 in. from CL of Bearing (2 ft. – 6 in. from Girder End)

Girder top stress:

$ftop = -P_i/A_g + P_i e_{0'}/S_t - M_g/S_t$
$P_i = 924.4$
$A_g = 1085$
$e_{0'} = 31.005$
$S_t = 20588$
$M_g = 153$
ftop = -924.4/1085 + 924.4x(31.005)/20588 -153x(12)/20588
= 0.451 ksi < 0.53 ksi OK

Girder bottom stress:

$fbottom = -P_i/A_g - P_i e_{0'}/S_b + M_g/S_b$
Sb = 20157
= -924.4/1085-924.4x(31.005)/20157+ 153x(12)/20157
= -2.183 ksi < -2.88 ksi OK

Girder top stress:

$ftop = -P_i/A_g + P_i e_{11'}/S_t - M_g/S_t$
Pi = 993.7
e11' = 31.222
Mg = 656
= -993.7/1085 +993.7x31.222/20588 -656x12/20588
= 0.209 ksi < 0.53 ksi OK

Girder bottom stress:

$Fbottom = -P_i/A_g - P_ie_{11'}/S_b + M_g/S_b$
$= -993.7/1085 - 993.7x31.222/20157+656x12/20157$
$= -2.065$ ksi < -2.88 ksi OK

Sample Calculations at 54 ft. – 6 in. From the CL of Bearing (55 ft. – 3 in. From Girder End) – Mid-span of Non-composite Beam

Girder top stress:

ftop = -Pi/Ag + Pie54.5'/St – Mg/St
= -1271/1085 + 1271x31.38/20588 -1725x12/20588
= -0.24 ksi < -2.88 ksi OK

Girder bottom stress:

Fbottom = -Pi /Ag – Pie54.5'/Sb + Mg/Sb
= -1271/1085- 1271x31.38/20157+ 1725x12/20157
= -2.123 ksi < -2.88 ksi OK

3.3.6.2. AASHTO-LRFD Design Step 5.6.2 Final Flexural Stress under Service Limit State

Maximum compression is checked under Service I limit state and maximum tension is checked under Service III limit state for jurisdictions using gross section analysis. The difference between Service I and Service III limit states is that Service I has a load factor of 1.0 for live load while Service III has a load factor of 0.8.

As indicated in Section 5.3, many jurisdictions do not include creep and shrinkage restraint moment effects in designing a pre-tensioned girder bridge. The calculations presented herein do not include creep and shrinkage moments. If creep and shrinkage are required by a specific jurisdiction, then their effects should be included. See Section 5.3 and Appendix C for calculations and values of creep and shrinkage effects for the example bridge.

AASHTO-LRFD Design Step 5.6.2.1

Stress Limits

Compression stress:

From AASHTO LRFD Table S5.9.4.2.1-1, the stress limit due to the sum of the effective pre-stress, permanent loads, and transient loads and during shipping and handling is taken as $0.6\phi wf'c$ (where ϕw equals 1.0 for solid sections such as pre-tensioned I-beams).

For pre-stressed concrete beams, f'c = 6 ksi
fComp, beam1 = -0.6x6 =-3.6 ksi
For deck slab, f'c = 4 ksi
fComp, slab = -0.6X4 = -2.4 ksi

Table 3.7. (Ref. AASHTO-LRFD Table 5.6-2) – Stresses in the Pre-Stressed Beam

Location	**Girder Non-composite Moment**	**F_{ps} After Transfer**	**Composite Dead Load Moment**	**Live load Positive Moment**
(ft.)(1)	(k-ft.)(2)	(kips)(3)	(ksi)	(k-ft.)(2)
0	0	244	0	0
1.75	217	813.4	36	170
5.5	661	813.4	108	476
11.0	1,252	874.4	199	886
16.5	1,776	965.9	276	1,230
22.0	2,230	965.9	337	1,509
27.5	2,616	1118.4	384	1,724
33.0	2,933	1118.4	414	1,882
38.5	3,181	1118.4	429	1,994
44.0	3,360	1118.4	429	2,047
49.5	3,471	1118.4	414	2,045
54.5	3,512	1118.4	387	2,015
55.0	3,511	1118.4	384	2,010
60.5	3,456	1118.4	338	1,927
66.0	3,333	1118.4	277	1,794
71.5	3,141	1118.4	201	1,613
77.0	2,880	1118.4	108	1,388
82.5	2,551	1118.4	2.0	1,124
88.0	2,152	965.9	-121	825
93.5	1,686	965.9	-258	524
99.0	1,150	813.4	-452	297
104.5	546	813.4	-580	113
107.25	217	813.4	-670	58
109.0	0	244.0	-729	15

Notes:
1 - Distance measured from the centerline of the bearing of the end abutment
2 - See Section 5.3 for load effects based on 110.5 ft. girder length
3 - See Section 5.5 for pre-stressing forces based on 110.5 ft. girder length
4 - Service I limit state for compression
5 - Service III limit state for tension

From AASHTO LRFD Table S5.9.4.2.1-1, the stress limit in pre-stressed concrete at the service limit state after losses for fully pre-stressed components in bridges other than segmentally constructed due to the sum of effective pre-stress and permanent loads shall be taken as:

fComp, beam2 = -0.45(f'c) = -0.45x6 =-2.7ksi

From Section 5.5.3.1, for pre-stressed components in other than segmentally constructed bridges, the compressive stress due to the Fatigue I load combination and one-half the sum of the effective pre-stress and permanent loads after losses shall not exceed:

fComp, beam3 = -0.40(f'c) = -0.4X6 = -2.4 ksi

Table 3.8. (Ref. AASHTO-LRFD Table 5.6-2 (cont.) – Stresses in the Pre-Stressed Beam

Location	**Final Stress under PS & DL**		**Stress Under ½ (DL+P/S) + Live Load**	**Final Stress Under All Loads**		
	Top of Beam	Bottom of Beam		Top of Beam	Bottom of Beam	Top of Slab
(ft.)(1)	(k-ft.)(4)	(kips)(4)	(ksi)(4)	(k-ft.)(4)	(k-ft.)(5)	(k-ft.)(4)
0	0.143	-0.6	0.071	0.143	-0.6	0
1.75	0.343	-1.856	0.142	0.313	-1.795	-0.04
5.5	0.071	-1.559	-0.046	-0.01	-1.39	-0.112
11.0	-0.244	-1.326	-0.274	-0.396	-1.011	-0.209
16.5	-0.508	-1.206	-0.465	-0.719	-0.769	-0.29
22.0	-0.783	-0.909	-0.65	-1.042	-0.372	-0.355
27.5	-0.917	-1.044	-0.754	-1.213	-0.43	-0.406
33.0	-1.107	-0.842	-0.876	-1.43	-0.172	-0.442
38.5	-1.254	-0.687	-0.969	-1.596	0.022	-0.466
44.0	-1.358	-0.581	-1.03	-1.709	0.147	-0.476
49.5	-1.42	-0.521	-1.061	-1.771	0.206	-0.473
54.5	-1.44	-0.509	-1.065	-1.785	0.208	-0.462
55.0	-1.438	-0.511	-1.064	-1.783	0.204	-0.461
60.5	-1.399	-0.564	-1.03	-1.729	0.121	-0.436
66.0	-1.316	-0.664	-0.966	-1.624	-0.026	-0.399
71.5	-1.191	-0.813	-0.872	-1.468	-0.239	-0.349
77.0	-1.023	-1.009	-0.75	-1.261	-0.515	-0.288
82.5	-0.813	-1.252	-0.6	-1.006	-0.852	-0.217
88.0	-0.659	-1.159	-0.471	-0.801	-0.865	-0.135
93.5	-0.364	-1.497	-0.272	-0.454	-0.311	-0.051
99.0	-0.117	-1.517	-0.11	-0.168	-0.411	003
104.5	0.257	-1.934	-0.109	0.237	-1.893	009
107.25	0.464	-2.169	-0.222	0.454	-2.149	0.118
109.0	0.268	-0.924	-0.131	0.265	-0.919	0.137

Notes:

1 - Distance measured from the centerline of the bearing of the end abutment.

2 - See Section 5.3 for load effects based on 110.5 ft. girder length.

3 - See Section 5.5 for pre-stressing forces based on 110.5 ft. girder length.

4 - Service I limit state for compression.

5 - Service III limit state for tension.

Tension stress:

From AASHTO LRFD Table S5.9.4.2.2-1, the stress limit in pre-stressed concrete at the service limit state after losses for fully pre-stressed components in bridges other than segmentally constructed, which include bonded pre-stressing tendons and are subjected to not worse than moderate corrosion conditions shall be taken as the following:

$f_{Tensile} = 0.19\sqrt{f'c} = 0.19 x \sqrt{6} = 0.47$ ksi

Definitions:

Pe = Final pre-stressing force (kips) from Design Step 3.3.4 (AASHTO-LRFD Design Step 5.4)

Stc = Section moduli, top of the beam of the composite section – gross section (in3)

Sbc = Section moduli, bottom of the beam of the composite section – gross section (in3)

Stsc = Section moduli, top of slab of the composite beam (in3)

MDNC = Moment due to the girder, slab, haunch and interior diaphragm (k-ft)

MDC = Total composite dead load moment, includes parapets and future wearing surface (k-ft)

MLLC = Live load moment (k-ft)

All tension stresses and allowable use positive sign convention. All compression stresses and allowable use negative sign convention. All loads are factored according to Table 3.4.1-1 in the AASHTO LRFD specifications for Service I and Service III limit states as applicable.

AASHTO-LRFD Design Step 5.6.2.2

Sample Calculations at 11 ft. From the CL of Bearing
(11 ft. – 9 in. From Girder End)

Girder top stress after losses under sum of all loads (Service I):

$f_{top} = -P_e/A_g + P_e e_{11'}/S_t - M_{DNC}/S_t - M_{DC}/S_{tc} - M_{LLC}/S_{tc}$

$P_e = 874.4$

$A_g = 1085$

$e_{11'} = 31.222$

$S_t = 20588$

$S_b = 20157$

$M_{DNC} = 1252$

$M_{DC} = 199$

$S_{tc} = 69949$
$S_{bc} = 26985$
$M_{LLC} = 886$
ftop = (-874.4/1085)+(874.4x31.222/20588)-(1252x12/20588)-(199x12/69949)-(886x12/69949)
= -0.396 ksi < -3.6 ksi OK

Stress limit for compression under full load
Girder top stress under pre-stressing and dead load after losses:

$f_{top} = -P_e/A_g + P_e e_{11'}/S_t - M_{DNC}/S_t - M_{DC}/S_{tc}$
= (-874.4/1085)+(874.4(31.222)/20588)-(1252(12)/20588)-(199(12)/ 69949) = -0.806 + 1.326 - 0.73 - 0.034
= -0.244 ksi < -2.7 ksi OK

Stress limit for compression under permanent load.

Girder top stress under LL + ½(PS + DL) after losses (Fatigue I with 1.5 Load Factor):

$ftop = -P_e/A_g + P_e e_{11'}/S_t - M_{DNC}/S_t - M_{DC}/S_{tc} - M_{LL}/S_{tc}$
= -874.4/(1085x2)+874.4(31.222)/(20588x2)-1252(12)/(20588x 2)-199(12)/(69949x2)- 886(12)/69949
= -0.403 + 0.663 - 0.365 - 0.017 - 0.152
= -0.274 ksi x1.5 = -0.411 < -2.4 ksi OK

Stress limit for compression under LL + ½(DL + PS).
Girder bottom stress under all loads (Service III):

fbottom = -Pe/Ag – Pee11'/Sb + MDNC/Sb + MDC/Sbc + MLLC/Sbc
= (-874.4/1085)+(874.4x(31.222)/20157)-(1252x(12)/ 20157)-199x(12)/26985)+(0.8x(886)x(12)/26985)
= -0.806-1.354+0.745+0.088+0.315
= -1.011 ksi < -2.7 ksi OK

Stress limit for compression under full load.

Note that the gross concrete composite section properties are typically used for the stress calculations due to all load components. However, some jurisdictions use the transformed section properties in calculating the stress due to live load. The transformed section properties are listed in Section 2. In this example, the gross section properties are used for this calculation.

Girder bottom stress under pre-stressing and dead load after losses:

fbottom = -Pe/Ag – Pee11'/Sb + MDNC/Sb + MDC/Sbc
= (-874.4/1085) - (874.4x(31.222)/20157)+(1252x(12)/ 20157)+ (199x(12)/26985)
= -0.81 - 1.35 + 0.745 + 0.088
= -1.326 ksi < -2.7 ksi OK

Stress limit for compression under pre-stress and permanent loads

Sample Calculations at 54 ft. – 6 in. From the CL of Bearing (55 ft. – 3 in. From Girder End) – Midspan of Non-Composite Girder.

Girder top stress after losses under sum of all loads (Service I):

$f_{top} = -P_e/A_g + P_{e e54.5'}/S_t - M_{DNC}/S_t - M_{DC}/S_{tc} - M_{LLC}/S_{tc}$
Pe = 1118.4
Ag = 1085
e54.5' = 31.38
St = 20588
MDNC = 3512
MDC = 387
MLLC = 2015
Stc = 69949
ftop = (-1118.4/1085)+(1118.4x(31.38)/20588)-(3512x(12)/20588)-(387x(12)/69949)-(2015x(12)/69949)
= -1.031+1.705-2.047-0.066-0.346
= -1.785 ksi < -3.6 ksi OK

Stress limit for compression under full load.

Girder top stress after losses under pre-stress and permanent loads:

ftop = -Pe/Ag + Pee54.5'/St – MDNC/St – MDC/Stc
= -1118.4/1085+1118.4(31.38)/20588-3512(12)/20588-387(12)/ 69949
= -1.031+1.705-2.047-0.066
= -1.44 ksi < -2.7 ksi OK

Stress limit for compression under pre-stress and permanent loads.

Girder top stress under LL + ½(PS + DL) after losses (Fatigue I with 1.5 Load Factor):

ftop = -Pe/Ag + Pee54.5'/St – MDNC/St – MDC/Stc – MLLC/Stc = -1118.4/(1085x2)+1118.4(31.38)/(20588x2)-(3512(12)/(20588x2)-387(12)/69949(2)-2015(12)/69949

= -0.515+0.852-1.024-0.033-0.346

= -1.066 ksi < -2.4 ksi OK

Stress limit for compression under LL + ½(DL + PS) load.

Girder bottom stress (Service III):

fbottom = -Pe/Ag – Pee54.5'/Sb + MDNC/Sb + MDC/Sbc+MLLC/Sbc = [(-1118.4/1085)] - [1118.4x (31.38)/20157] + [3512x(12)/20157] + [387x(12)/26985] + [0.8x (2015) x (12)/26985]

= -1.031-1.741+2.091+0.172+0.717

= 0.208 ksi < 0.465 ksi Stress limit for tension

Note that the stresses are calculated without including creep and shrinkage. Jurisdictions that do not include creep and shrinkage typically design the girders for a reduced tensile stress limit or for zero tension at final condition. Including creep and shrinkage would normally result in additional tensile stress at the bottom of the beam at the mid-span section.

Girder bottom stress after losses under pre-stress and dead load:

fbottom = -Pe/Ag – Pee54.5'/Sb + MDNC/Sb + MDC/Sbc

= (-1118.4/1085) - (1118.4x31.38/20157) + (3512x12/20157) + (387x12/26985) = -1.031-1.741+2.091+0.172

= -0.509 ksi < -2.7 ksi OK

Deck slab top stress under full load:

MDC = 387

Stsc = 50902

modular ratio ok beam = 4696

modular ratio of slab = 3834

ftop slab = (-MDC/Stsc - MLLC/Stsc)/modular ratio between beam and slab

= -387x12/50902-1x2015x12/50902/(4696/3834)

= (-0.091-0.475)/1.225

= -0.462 ksi < -2.4 ksi OK

Stresses at Service Limit State for Sections in the Negative Moment Region

Sections in the negative moment region may crack under service limit state loading due to high negative composite dead and live loads. The cracking starts in the deck and as the loads increase the cracks extend downward into the beam. The location of the neutral axis for a section subject to external moments causing compressive stress at the side where the pre-stressing force is located may be determined using a trial and error approach as follows:

1. Assume the location of the neutral axis.
2. Assume a value for the compressive strain at the extreme compression fiber (bottom of the beam). Calculate the tensile strain in the longitudinal reinforcement of the deck assuming the strain varies linearly along the height of the section and zero strain at the assumed location of the neutral axis.
3. Calculate the corresponding tension in the deck reinforcement based on the assumed strain.
4. Calculate the compressive force in the concrete.
5. Check the equilibrium of the forces on the section (pre-stressing, tension in deck steel and compression in the concrete). Change the assumed strain at the bottom of beam until the force equilibrium is achieved.
6. After the forces are in equilibrium, check the equilibrium of moments on the section(moment from pre-stressing, external moment and moment from internal compression and tension).
7. If moment equilibrium is achieved, the assumed location of the neutral axis and strains are correct. If the moments are not in equilibrium, change the assumed location of the neutral axis and go to Step 2 above.
8. After both force and moment equilibriums are achieved, calculate the maximum stress in the concrete as the product of the maximum concrete strain and the concrete modulus of elasticity.

Notice that when additional compression is introduced into the concrete due to external applied forces, the instantaneous stress in the pre-stressing steel is decreased by the modular ratio multiplied by the additional compressive stress in the surrounding concrete. The change in the pre-stressing steel force is typically small and was ignored in the following calculations.

Sample Calculations for a Section in the Negative Moment Region Under Service Limit State, Section at 107 ft. – 3 in. From the CL of End Bearing (108 ft. From Girder End)

Table 3.9. (Ref. AASHTO-LRFD Table 5.3-1)

Location	Noncomposite Moments				Composite Moments		Live Load + IM Moment	
	Girder*	Slab and Hanuch	Exterior Diaphragm	Total Noncomp.	Parapet	FWS	Positive HL-93	Negative HL-93
(ft.)	(k-ft)	(k-ft)	(k-ft)	(k-ft)	(k-ft)	(k-ft)	(k-ft)	(k-ft)
104.5	266	268	11	546	-248	-332	113	-1,663
108	61	62	3	125	-297	-398	33	-1,921

* Based on the simple span length of 109 ft.

Maximum negative moment at the section at 104.5 ft= 546+(-248)+(-332)+(-1663)= -1,697 k-ft

Maximum negative moment at the section at 108.0 ft = 125+(-297)+(-398)+(-1921) = -2,491 k-ft

By interpolation, the maximum Service I negative moment at the section under consideration is:

M_{neg} = (-2491) - (-2491 + 1697) x [(108 – 107.25) / (108 – 104.5)] = -2321 k-ft

Trial and error approach (see above) was applied to determine the location of the neutral axis. The calculations of a cycle of the process are shown below.

Referring to Figure 3.25 (Ref. AASHTO-LRFD Figure 5.6-1):

Assume neutral axis at 32.5 inches from the bottom of beam

Assume maximum concrete compressive strain = 0.00079 in./in.

Tensile strain in deck reinforcement = 0.00079 x (75.52 –32.5) / 32.5
= 0.001046 in./in.

Modulus of elasticity of concrete beam = 4,696ksi (see Section 2)

Concrete stress at bottom of beam = 0.00079 x 4696 = 3.71 ksi

Area of deck longitudinal reinforcement= 15.31 in^2 (see AASHTO-LRFD Section 5.6.5.1 for calculation)

E = 29000

Force in deck steel= 15.31x0.001046x29000 = 464.4 k

Force in pre-stressing steel = 813.4 k (see Table 1.2.4 & 1.2.5 Table 3.4 and 3.5 (Ref. AASHTO-LRFD Table 5.5-1)

Compressive Forces in the Concrete

Considering Figure 3.25 (Ref. AASHTO-LRFD Figure 5.6-1), by calculating the forces acting on different areas as the volume of the stress blocks for areas A1, A2 and A3 as the volume of a wedge, prism or pyramid, as appropriate, the forces in Table 3.9

(Ref. AASHTO-LRFD Table 5.6-3) may be calculated. Recall that the centers of gravity of a wedge, a prism with all rectangular faces, a prism with a triangular vertical face and a pyramid are at one-third, one-half, one-third and one-quarter the height, respectively. The location of the centers of gravity shown in the figure may also be calculated. The moment from internal compressive concrete forces shown in Table 3.9 (Ref. AASHTO-LRFD Table 5.6-3) is equal to the force multiplied by the distance from the neutral axis to the location of the force.

Sample force calculations for area A3.

Two components of stress act on area A3. The first component is a rectangular stress distribution with an intensity of 1.66 ksi. The second component is a triangular stress distribution with an intensity of 1.14 ksi.

Force due to the rectangular stress distribution:

Table 3.10. (Ref. AASHTO-LRFD Table 5.6-3) – Forces in Concrete under Service Load in Negative Moment Regions (Section at 107'-3" from the end bearing)

Area Designation	Force Designation	Area	Stress	Force	Distance from bot.of beam	Distance toN/A	Moment at N/A
		(in^2)	(ksi)	(Kips)	(in)	(in)	(k-ft)
A1	P1	260	3.71	482.3	10.83	21.67	871.0
A2	P2 a	160	2.80	448.0	4.00	28.50	1,064.0
A2	P2 b	160	0.91	72.8	2.67	29.83	181.0
A3 *	P3 a	100	1.66	166.0	11.33	21.17	292.9
A3 *	P3 b	100	1.14	76.0	10.50	22.00	139.3
Total				1,245.1			2,548.20

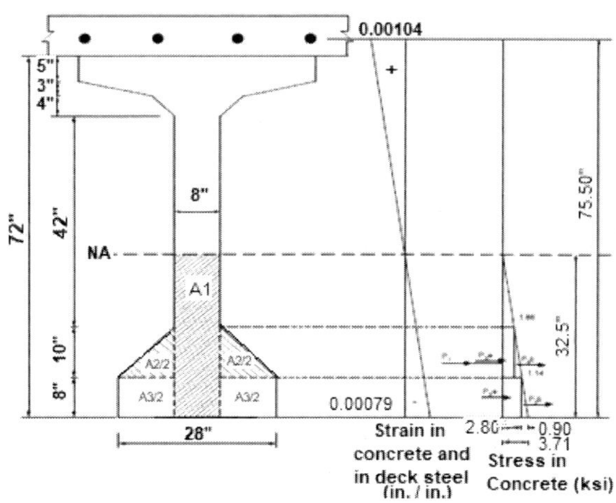

Figure 3.25. Compressive Force in the Concrete. Source: Figure 5.6-1 from AASHTO LRFD Bridge Design Specifications, FHWA Office of Bridges and Structures, 1200 New Jersey Ave., SE, Washington, DC 20590.

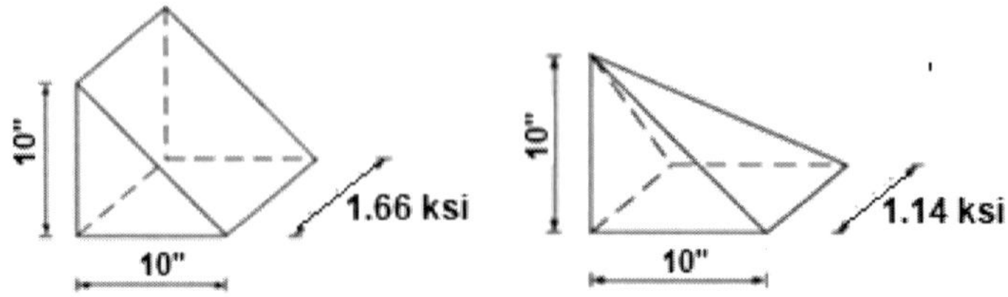

(a) Rectangular Stress Distribution (b) Triangular Stress Distribution

Figure 3.26. (a). Shapes Used in Determining Forces for A3. Source: Figure 5.6-1 (a) from AASHTO LRFD Bridge Design Specifications, FHWA Office of Bridges and Structures, 1200 New Jersey Ave., SE, Washington, DC 20590.

F rectangular = 2[0.5(10)(10)](1.66) = 166 k

The volume used to determine the effect of the triangular stress distribution is calculated using geometry of a pyramid.

Pyramid base = 10
Pyramid height = 1.14

F triangular = 2 triangles(1/3 pyramid base)(pyramid height) = 2x (1/3) x (10) x (1.14) x (10) = 76 k

Check Force Equilibrium

Net force on section = P/S steel force + concrete compressive force + deck steel force
= 813.4 + (-1245.1) +464.4
= 32.7 kips OK

(Sightly unbalanced but based on the concrete compressive force of 1245 kips, the engineer decided not to do another iteration and proceeded.)

Check Moment Equilibrium

Net M on the section = external moment + pre-stressing force moment + deck slab force moment + concrete compression moment:

= 2321 +813.4x (32.5 –5.375)/12 –464.4x(75.52 – 32.5)/12 – 2548.2
= -53.5 k-ft ≈ 0 OK

From Table 3.9 (Ref. AASHTO-LRFD Table 5.6-3), the maximum stress in the concrete is 3.71 ksi. The stress limit for compression under all loads (Table S5.9.4.2.1-1)

under service condition is 0.6f'c (where f'c is the compressive strength of the girder concrete). For this example, the stress limit equals 3.6 ksi.

The calculated stress equals 3.71 ksi or is 3% overstressed. However, as explained above, the stress in the pre-stressing steel should decrease due to compressive strains in the concrete caused by external loads, i.e., pre-stressing steel force less than 813.4 k and the actual stress is expected to be lower than the calculated stress, and the above difference (3%) is considered within the acceptable tolerance.

Notice that the above calculations may be repeated for other cases of loading in AASHTO LRFD Table S5.9.4.2.1-1 and the resulting applied stress is compared to the respective stress limit. However, the case of all loads applied typically controls. Strain compatibility software could be used to perform the above calculations which would simplify the task but the entire procedure was shown here for completeness.

3.3.6.3. AASHTO-LRFD Design Step 5.6.3 Longitudinal Steel at Top of Girder

The tensile stress limit at transfer used in this example requires the use of steel at the tension side of the beam to resist the tensile force in the concrete calculated based on an un-cracked section (Table S5.9.4.1.2-1) where reinforcement is proportioned using a stress of 0.5fy, not to exceed 30 ksi. The sample calculations are shown for the section in Table 3.6 (Ref. AASHTO-LRFD Table 5.6-1) with the highest tensile stress at transfer, i.e., the section at 1.75 ft. from the centerline of the end bearing.

By integrating the tensile stress in Figure 3.27 (Ref. AASHTO-LRFD Figure 5.6-2) over the corresponding area of the beam, the tensile force may be calculated as:

b1 = 42 in 3.5 ft
b2 = 4 in 0.33 ft
b3 = 13 in 1.08 ft
b4 = 8 in 0.67 ft
b5 = 28 in 2.33 ft
D = 72 in 6 ft
d1 = 5 in 0.42 ft
d2 = 3 in 0.25 ft
d3 = 4 in 0.33 ft
d4 = 42 in 3.5 ft
d5 = 10 in 0.83 ft
d6 = 8 in 0.67 ft
d7 = 12.33 in 1.03 ft

Tensile force = [(5x42)x(0.451 +0.268)/2] + [(7.33x8)x(0.268 + 0)/2] + [(2x4x3) x (0.268 + 0.158)/2] + [(2x4x4)/2] x [0.012 + (0.158 –0.012)x(2/3)] + [(2x3x13)/2] x [0.158 + (0.268 –0.158)x(2/3)]

= 88.46476+10.7713333

= 99.2 k

fy = 60 ksi

Required area of steel = 99.2/0.5fy

= 99.2/0.5 x 60

= 3.31 in^2

Required number of #5 bars = 3.31/0.31

= 10.68 bars

Use 10 #5 bars as shown in Figure 3.28 (Ref. AASHTO-LRFD Figure 5.6-3), for a stress in the reinforcement of ksi, a slight overstress. 99.2/(10) x (0.31) = 32

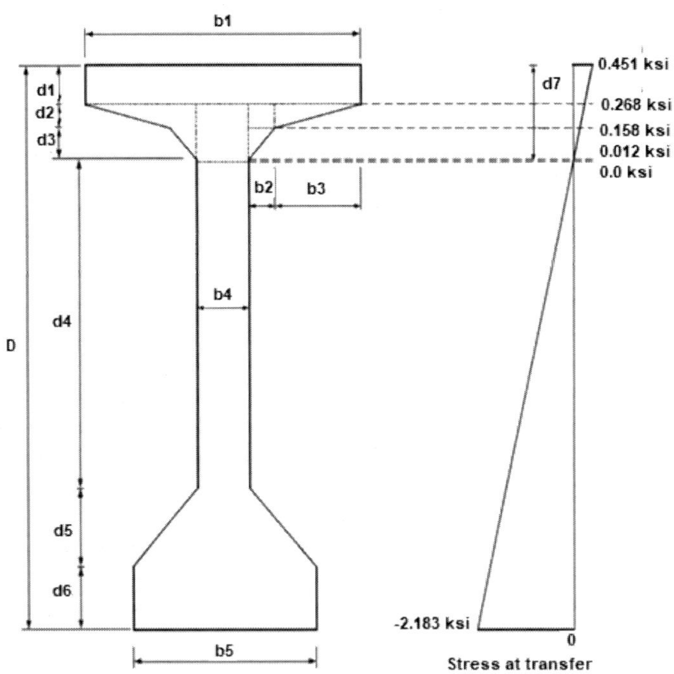

Figure 3.27. Stress at Location of Maximum Tensile Stress at Transfer. Source: Figure 5.6-2 from AASHTO LRFD Bridge Design Specifications, FHWA Office of Bridges and Structures, 1200 New Jersey Ave., SE, Washington, DC 20590.

b1 = 42 in

b2 = 4 in

b3 = 13 in
b4 = 8 in
b5 = 28 in
D = 72 in
d1 = 5 in
d2 = 3 in
d3 = 4 in
d4 = 42 in
d5 = 10 in
d6 = 8 in
d7 = 12.33 in

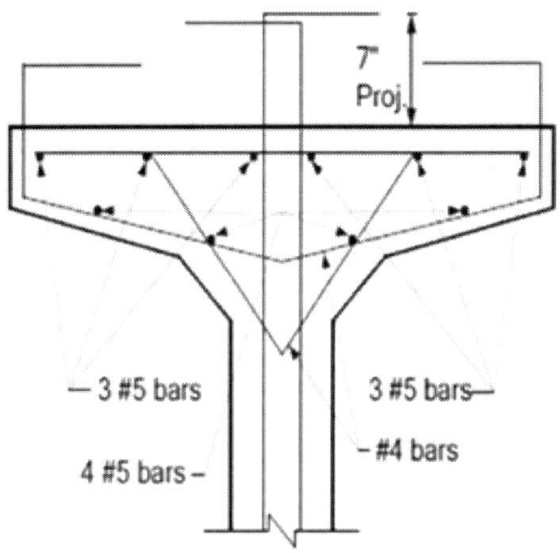

Figure 3.28. Longitudinal Reinforcement of Girder Top Flange. Source: Figure 5.6-3 from AASHTO LRFD Bridge Design Specifications, FHWA Office of Bridges and Structures, 1200 New Jersey Ave., SE, Washington, DC 20590.

3.3.6.4. AASHTO-LRFD Design Step 5.6.4 Flexural Resistance at the Strength Limit State in Positive Moment Region (S5.7.3.1)

Sample Calculations at Mid-Span

c = distance between the neutral axis and the compressive face at the nominal flexural resistance (in.)

c = 5.31 in., which is less than the slab thickness, therefore, the neutral axis is in the slab and section is treated as a rectangular section (see design Step 3.4 and 3.5 (Ref. AASHTO-LRFD Table 5.5-1) for commentary explaining how to proceed if "c" is greater than the deck thickness).

fps = stress in the pre-stressing steel at the nominal flexural resistance (ksi).

fps = 264.6 ksi

The factored flexural resistance, Mr, shall be taken as ϕMa, where Ma is determined using Eq. S5.7.3.2.2-1.

Factored flexural resistance in flanged sections (S5.7.3.2.2)

Ma = Aps x fps(dp – a/2) + As x fy(ds – a/2) – A′s x f′y(d′s – a/2) + α1 f′c(b – bw)hf(a/2 – hf/2) (S5.7.3.2.2-1)

The definition of the variables in the above equation and their values for this example are as follows:

Aps = area of pre-stressing steel (in^2)
= 6.73 in^2
fps = average stress in pre-stressing steel at nominal bending resistance specified in Eq. S5.7.3.1.1-1 (ksi)
= 264.6 ksi
dp = distance from extreme compression fiber to the centroid of pre-stressing tendons (in.)
= 74.5 in.
As = area of non-pre-stressed tension reinforcement (in^2)
= 0 in^2
fy = specified yield strength of reinforcing bars (ksi)
= 60 ksi
ds = distance from extreme compression fiber to the centroid of non-pre-stressed tensile reinforcement (in.), NA
A′s = area of compression reinforcement (in^2)
= 0 in^2
f′y = specified yield strength of compression reinforcement (ksi), NA
d′s = distance from the extreme compression fiber to the centroid of compression reinforcement (in.), NA
f′c = specified compressive strength of concrete at 28 days, unless another age is specified (ksi)
= 4 ksi (slab)
b = width of the effective compression block of the member (in.)
= width of the effective flange
= 116 in.

(See design Step 3.4 and 3.5 (Ref. AASHTO-LRFD Table 5.5-1) for commentary for the determination of the effective width, b, when the calculations indicate that the compression block depth is larger than the flange thickness.)

bw = web width taken equal to the section width "b" for a rectangular section.

$\alpha 1$ = stress block factor specified in S5.7.2.2 where $\alpha 1$ shall be taken as 0.85 for specified concrete compressive strengths not exceeding 10.0 ksi.

$\beta 1$ = stress block factor specified in S5.7.2.2, NA

hf = compression flange depth of an I or T member (in.), NA

a = $\beta 1$c; depth of the equivalent stress block (in.)

= 0.85X5.31 = 4.51 in.

The second, third and fourth terms in Eq. S5.7.3.2.2-1 are equal to zero for this example.

Substituting,

Ma = (6.73X264.6)[74.5 – (4.51/2)]

= 128651/12 = 10721 k-ft

Therefore, the factored flexural resistance, Mr, shall be taken as:

Mr = ϕMa (S5.7.3.2.1-1)

where:

ϕ = resistance factor as specified in S5.5.4.2 for flexure in pre-stressed concrete, tension-controlled sections = 1

Mr = 1X10721 k-ft = 10,721 k-ft

The maximum factored applied moment for Strength I limit state is 8,456 k-ft Table 3.3-2 (Ref. AASHTO-LRFD Table 5.3-2)

Mr = 10,721 k-ft > Mu = 8,456 k-ft OK

AASHTO-LRFD Design Step 5.6.4.1

Check if Section Is Tension-Controlled

The section is tension-controlled if the strain in the reinforcement is greater than or equal to 0.005 when the concrete in compression reaches a strain of 0.003.

The strain in the pre-stressed reinforcement, found by proportioning the linear strain diagram is:

εs = 0.003(de/c-1)

where:

c = 5.31 in. (see Section 5.5.1)

de = 74.5 in. (Conservatively use dp although S5.7.2.1 states that the distance can be to the extreme tension steel)

$\varepsilon s = 0.003X(74.5/5.31-1) = 0.039$

$0.039 > 0.005$ OK

AASHTO-LRFD Design Step 5.6.4.2

Check Minimum Required Reinforcement (S5.7.3.3.2)

Critical location is at the mid-span of the continuous span = 54.5 ft. from the end bearing. All strands are fully bonded at this location.

According to S5.7.3.3.2, unless otherwise specified, at any section of a flexural component, the amount of pre-stressed and non-pre-stressed tensile reinforcement shall be adequate to develop a factored flexural resistance, Mr, at least equal to the lesser of:

1.33 times the factored moment required by the applicable strength load combinations specified in AASHTO LRFD Table 3.4.1-1.

AND

The cracking moment strength as determined by S5.7.3.3.2-1.

Mcr = y3[y1fr+y2fcpe)Sc-Mdnc (Sc/Snc-1)]

where:

M_{DNC} = factored using Service I limit state, see Table 3.3-1 (Ref. AASHTO-LRFD Table 5.3-1).

= 3,511 k-ft

Sb = 20,157 in3

Sbc = 26,985 in3

Fcpe = 0.545 ksi (from Design Step 3.3.6.2.2 AASHTO-LRFD Design Step 5.6.2.2)

fc = 6

fr = 0.37 √f'c (S5.4.2.6)

= 0.37x√6 ksi

= 0.906 ksi

y1 = 1.6 for all concrete structures

y2 = 1.1 for bonded tendons

y2 = 1 for pre-stressed concrete structures

Mcr = 1x[1.6x0.906+1.1x0.545)x26985-3511x12x(26985/20157-1)]

Mcr = 3419 k-ft

The applied factored moment, Mu, taken from Table 3.3-2 (Ref. AASHTO-LRFD Table 5.3-2) is 8,456 k-ft (Strength I) = 1.33X8456 = 11246 k-ft

Mr has to be greater than the lesser of Mcr and 1.33Mu, i.e., 3,418 k-ft.

Mr also has to be greater than the applied factored load Mu = 8,456 k-ft (strength requirement).

Mr = 10,697 k-ft, therefore, both provisions are OK.

3.3.6.5. AASHTO-LRFD Design Step 5.6.5 Continuity Connection at Intermediate Support

AASHTO-LRFD Design Step 5.6.5.1

Negative Moment Connection at the Strength Limit State

Determine the deck steel at the intermediate pier.

Based on preliminary calculations, the top and bottom longitudinal reinforcement of the deck are assumed to be #6 bars at 5.5 in. spacing and #6 bars at 8.5 in. spacing, respectively.

Abar = 0.44

Calculate the total area of steel per unit width of slab:

As = Abar/spacing (in^2/in.)
For top row of bars: As top = 0.44/5.5 = 0.08 in^2/in
For bottom row of bars: As bot = 0.44/8.5 = 0.052 in^2/in.
Therefore, As = 0.08+ 0.052 = 0.132 in^2/in.

Calculate the center of gravity of the slab steel from the top of the slab. Calculations are made from the top of the total thickness and include the integral wearing surface in the total thickness of slab (see Figure 4-16).

Top cover = 2.5
Dia # 5 main rein. = 0.625
Dia # 6 = 0.75
Slab thickness = 8
Bottom cover = 1
As top = 0.08
CGStop = 3.5

As bot = 0.052
CGSbot = 6
As = 0.132

Top mat (B1):

CGStop = Covertop + Dia#5 main rein. + ½ Dia#6
= 2.5+ 0.625 + ½ (0.75) = 3.5 in.

Bot. mat (B2):

CGSbot = tslab – Coverbot – Dia#5 main rein. – ½ Dia#6
= 8 –1–0.625 – ½ (0.75) = 6 in.

Center of gravity of the deck longitudinal reinforcement from the top of the deck:

CGS = [As top(CGStop) + As bot(CGSbot)]/As
= [0.08 x3.5 + 0.052 x6]/0.132
= 4.48 in. from the top of slab (3.98 in. from the top of the structural thickness)

Calculate the depth to the slab steel from the bottom of the beam. The haunch depth is ignored in the following calculations:

ds = girder + slab – CGS
= 72+ 8 – 4.48
= 75.52 in.

The specification is silent about the strength of the concrete in the connection zone. Many jurisdictions use the girder concrete strength for these calculations. This reflects observations made during girder tests in the past. In these tests, the failure always occurred in the girder. This behavior is due to the confinement of the diaphragm concrete in the connection zone provided by the surrounding concrete. This confinement increases the apparent strength of the diaphragm concrete compared to the unconfined strength measured during typical testing of concrete cylinders.

Assume the neutral axis is in the bottom flange (rectangular behavior), therefore,

f'c = f'c, beam = 6 ksi
$\beta 1 = \beta 1$, beam = 0.75 (corresponds to the 6.0 ksi concrete, S5.7.2.2)
$\alpha 1 = \alpha 1$, beam = 0.85 (corresponds to the 6.0 ksi concrete, S5.7.2.2)
b = width of section = width of girder bottom flange = 28 in.

Calculate c,

$c = Asfy/\alpha 1\beta 1f'cb$ (S5.7.3.1.1-4, modified)

where:

As = area of reinforcement within the effective flange width of 116 in. (in^2)

= Asbslab

= 0.132X116

= 15.31 in^2

fy = 60 ksi

f'c = 6.0 ksi

$\beta 1 = 0.75$

$\alpha 1 = 0.85$

b = 28 in.

c = (15.31X60)/(0.85X0.75X6X28)

= 8.58 in., which is approximately equal to the thickness of the bottom flange of the beam (8 in.), therefore, the section is checked as a rectangular section. If "c" was significantly larger than the thickness of the bottom flange, a reduction in the section width should be considered.

Calculate the nominal flexural resistance according to S5.7.3.2.1 and the provisions for a rectangular section.

Mn = Asfy(ds – a/2)

where:

$a = \beta 1c$

= 0.75x 8.58

= 6.44 in.

ds = 75.52 in.

Mn = (15.31X60)X[75.52 – (6.44/2)]/12

= 5535 k-ft

The factored flexural resistance, Mr, is:

$Mr = \phi fMn$ (S5.7.3.2.1-1)

where:

$\phi f = 0.9$ for flexure in reinforced concrete (S5.5.4.2.1)

$Mr = 0.9 \times 5535$
$= 4982$ k-ft

Check moment capacity versus the maximum applied factored moment at the critical location.

Critical location is at the centerline of pier.

Strength I limit state controls.

$|Mu| = 4{,}729$ k-ft, see Table 3.3-2 (Ref. AASHTO-LRFD Table 5.3-2)

$4{,}729 < Mr = 4{,}982$ k-ft OK

Check service crack control (S5.5.2)

Actions to be considered at the service limit state are cracking, deformations, and concrete stresses, as specified in Articles S5.7.3.4, S5.7.3.6, and S5.9.4, respectively. The cracking stress is taken as the modulus of rupture specified in S5.4.2.6.

The maximum spacing of the mild steel reinforcement for control of cracking at the service limit state shall satisfy:

$S \leq 700\gamma e/\beta s f s - 2dc$

in which:

$\beta s = 1 + dc/0.7(h - dc)$

where:

γe = exposure factor = 1 for Class 1 exposure condition

dc = thickness of concrete cover measured from extreme tension fiber to center of the flexural reinforcement located closest thereto (in.) = 2 + 0.625+ ½(0.75) = 3 in.

(see Figure 3.29 (Ref. AASHTO-LRFD Figure 5.6-4).

h = overall thickness or depth of the component (in.)

= 72 + 7.5 = 79.5 in.

By substituting dc and h,

$\beta s = 1 + 3/0.7 \times (79.5 - 3) = 1.06$

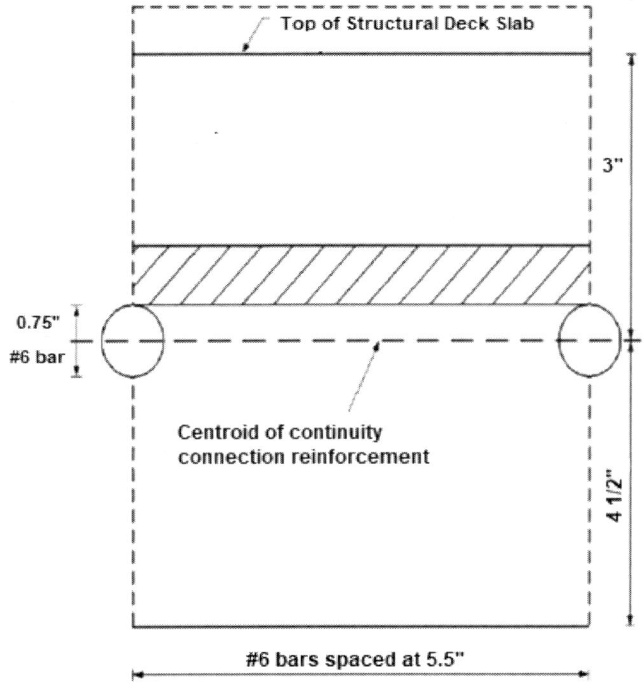

Figure 3.29. Dimensions for calculation of the area, A. Source: Figure 5.6-4 from AASHTO LRFD Bridge Design Specifications, FHWA Office of Bridges and Structures, 1200 New Jersey Ave., SE, Washington, DC 20590.

Connection moment at Service I limit state is 2,858 k-ft, see Table 3.3-2 (Ref. AASHTO-LRFD Table 5.3-2).

Assuming: Section width is equal to beam bottom flange width = 28
Modular ratio = 6 for 6 ksi concrete
Area of steel = 15.31 in^2

At the service limit state, the depth of the neutral axis and the transformed moment of inertia under service loads may be calculated using the same procedure used earlier in the example (Section 4). The neutral axis is 18.86 in. from the bottom of the beam.

Maximum service stress in the steel = fs = 33.74 ksi.
Therefore:

Srequired ≤ (700x1) / (1.06x33.74) - (2x3)= 13.6 in
Sprovided = 5.5 in. < 13.6 in. OK

So, use #6 bars @ 5.5 in. to satisfy crack control.
Total area of steel per unit width of slab:

As top = 0.44/5.5 = 0.08 $in^2/in.$
Asbot = 0.44/8.5 = 0.052 $in^2/in.$

Therefore,

As = 0.08 +0.052 = 0.132 $in^2/in.$
As = area of reinforcement within the effective flange width of 116 in. (in2) = Asbslab = 0.132 x116 = 15.31 in^2

Center of gravity of the deck longitudinal reinforcement from the top of the deck:

CGS = [As top(CGStop) + As bot(CGSbot)]/As
= [(0.08 x 3.5) + (0.052 x 6)]/0.132
= 4.48 in. (3.87 in. from the top of the structural thickness)

AASHTO-LRFD Design Step 5.6.5.2

Positive Moment Connection

For jurisdictions that consider creep and shrinkage in the design, it is likely that positive moment will develop at intermediate piers under the effect of pre-stressing, permanent loads and creep and shrinkage. These jurisdictions provide reinforcement at the bottom of the beams at intermediate diaphragms to resist the factored positive moment at these locations.

For jurisdictions that do not consider creep and shrinkage in the design, it is unlikely that live load positive moments at intermediate supports will exceed the negative moments from composite permanent loads at these locations. This suggests that there is no need for the positive moment connection. However, in recognition of the presence of creep and shrinkage effects, most jurisdictions specify some reinforcement to resist positive moments.

Two forms of the connection have been in use:

1) Figure 3.30 (Ref. AASHTO-LRFD Figure 5.6-5) shows one alternative that requires extending some of the pre-stressing strands at the end of the girder into the intermediate diaphragm. Due to the small space between girders, these strands are bent upwards into the diaphragm to provide adequate anchorage. Only strands that are fully bonded are used for the positive moment connection.
2) The second alternative requires adding mild reinforcement bars as shown in Figure 3.31 (Ref. AASHTO-LRFD Figure 5.6-6). This alternative may lead to

congestion at the end of the beam due to the presence of the pre-stressing strands at these locations.

Typical details of the top of the pier cap for expansion and fixed bearings are shown schematically in Figures 3.32 (Ref. AASHTO-LRFD Figure 5.6-7) and Figure 3.33 (Ref. AASHTO-LRFD Figure 5.6-8).

3.3.6.6. AASHTO-LRFD Design Step 5.6.6 Fatigue in Pre-Stressed Steel (S5.5.3)

Article S5.5.3 states that fatigue need not be checked when the maximum tensile stress in the concrete under Service III limit state is taken according to the stress limits of AASHTO LRFD Table S5.9.4.2.2-1. The stress limit in this table was used in this example and, therefore, fatigue of the pre-stressing steel need not be checked.

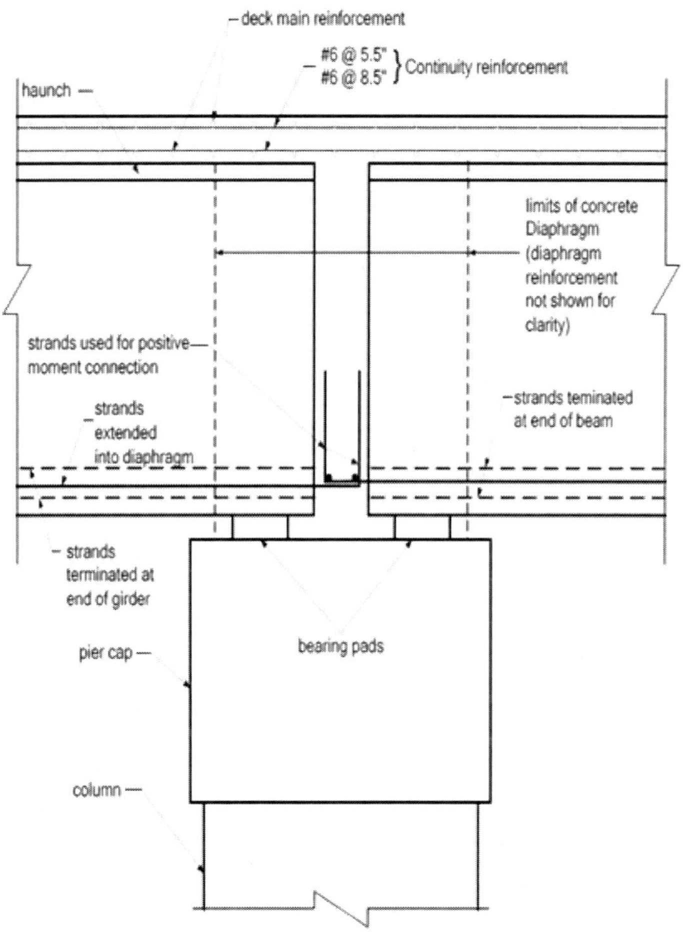

Figure 3.30. Continuity connection. Alternative 1: Strands used for positive moment connection. Source: Figure 5.6-5 from AASHTO LRFD Bridge Design Specifications, FHWA Office of Bridges and Structures, 1200 New Jersey Ave., SE, Washington, DC 20590.

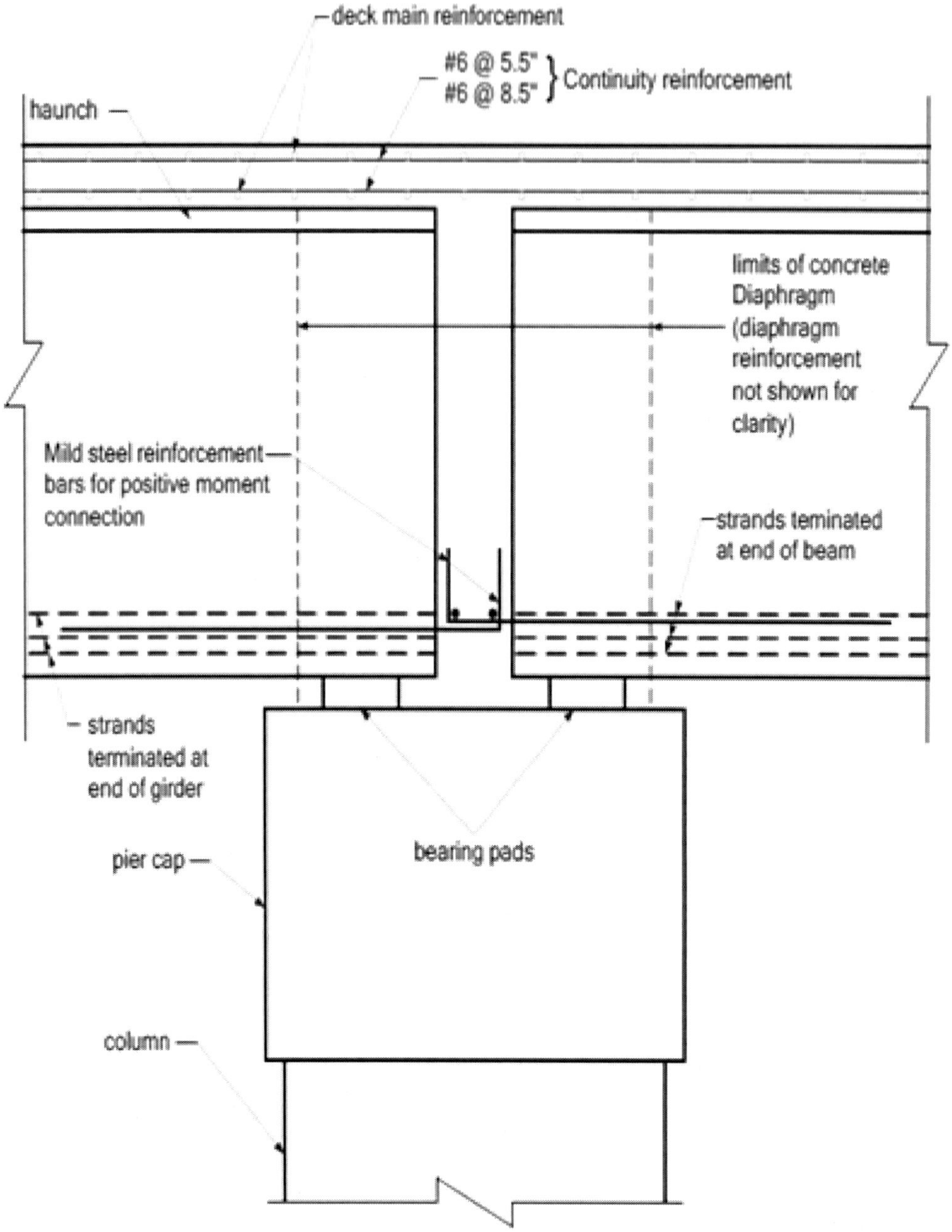

Figure 3.31. Continuity connection. Alternative 2: Reinforcement bars used for positive moment connection. Source: Figure 5.6-6 from AASHTO LRFD Bridge Design Specifications, FHWA Office of Bridges and Structures, 1200 New Jersey Ave., SE, Washington, DC 20590.

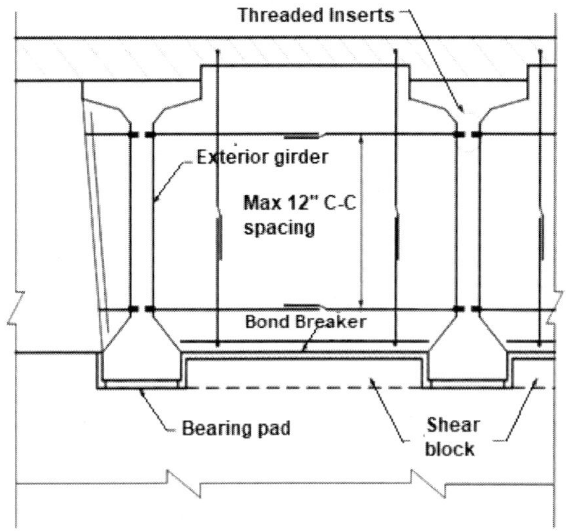

Figure 3.32. Typical diaphragm at intermediate pier (Expansion bearing). Source: Figure 5.6-7 from AASHTO LRFD Bridge Design Specifications, FHWA Office of Bridges and Structures, 1200 New Jersey Ave., SE, Washington, DC 20590.

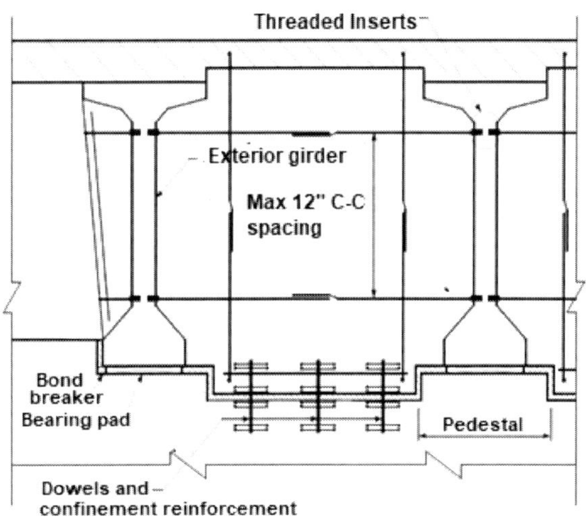

Figure 3.33. Typical diaphragm at intermediate pier (Fixed bearing). Source: Figure 5.6-8 from AASHTO LRFD Bridge Design Specifications, FHWA Office of Bridges and Structures, 1200 New Jersey Ave., SE, Washington, DC 20590.

3.3.6.7. AASHTO-LRFD Design Step 5.6.7 Camber (S5.7.3.6)

The provisions of S2.5.2.6 shall be considered.

Deflection and camber calculations shall consider dead load, live load, pre-stressing, erection loads, concrete creep and shrinkage, and steel relaxation. For determining deflection and camber, the provisions of Articles S4.5.2.1, S4.5.2.2, and S5.9.5.5 shall apply.

Instantaneous deflections are computed using the modulus of elasticity for concrete as specified in S5.4.2.4 and taking the gross moment of inertia, Ig, as allowed by S5.7.3.6.2.

Deflection values are computed based on pre-stressing, girder self-weight, slab, formwork, exterior diaphragm weight, and superimposed dead load weight. Camber values are computed based on initial camber, initial camber adjusted for creep, and final camber. Typically, these calculations are conducted using a computer program. Detailed calculations are presented below.

Deflection due to initial pre-stressing is computed as:

ΔP/S = -(PiesL2)/(8EciIg) (for straight bonded strands)
ΔP/S = -Pies[L2 – (Lt + 2Lx)2]/(8EciIg) (for de-bonded strands)

where:

Pi = applied load acting on the section (kips)
es = eccentricity of the pre-stressing force with respect to the centroid of the cross section at the mid-span of the beam (in.)
L = span length (ft.)
Lt = transfer length of the strands (in.)
Lx = distance from end of beam to point where bonding commences (in.)
Eci = modulus of elasticity of concrete at transfer (ksi)
Ig = moment of inertia (in4)

The negative sign indicates upward deflection.
Computer software is generally used to determine the deflections due to each loading.
However, sample calculations are provided for this example.
See Table 3.4 and 3.5 (Ref. AASHTO-LRFD Table 5.5-1) for pre-stressing forces.
Group 1 strands: 32 fully bonded strands:

Initial pre-stressing force = 924.4 k
Distance from bottom of the beam to the neutral axis = 36.38 in.
Distance from the bottom of the beam to the centroid of Group 1 strands = 5.375 in.

Deflection due to Group 1 strands:

L = 109
Eci = 4200
Ig = 733320

ΔP/S 1 = -(PiesL2)/(8EciIg)
= -[924.4x(36.38 – 5.375)x[109x(12)]^2]/[8x(4200)x(733320)
= -1.99 in. (upward deflection)

Group 2 strands: 6 strands de-bonded for 10 ft. from centerline of bearings:

Transfer length = 30 in.
Initial pre-stressing force = 173.3 k

From AASHTO-LRFD Figures 2-5 and 2-6, the distance from the bottom of the beam to the centroid of Group 2 is 4.0 in.

Deflection due to Group 2 strands:

Lx = 10
ΔP/S 2 = -Pies[L2 – (Lt + 2Lx)2]/(8EciIg)
= -173.3X(36.38–4)X[(109X12)^2] – [302X10X12)]^2]/
[8X(4200)X(733320)] = -0.37 in.

Group 3 strands: 6 strands de-bonded for 22 ft. from centerline of bearings:

Transfer length = 30 in.
Initial pre-stressing force = 173.3 k

From AASHTO-LRFD Figures 2-5 and 2-6, the distance from the bottom of the beam to the centroid of Group 3 is 4.0 in.

Deflection due to Group 3 strands:

ΔP/S3 = -Pies[L2 – (Lt + 2 x)2]/(8EciIg)
=-173.3 x (36.38 –4) x [[109X(12)]^2 – [30 + 2X(22) x (12)]^2]/[8X(4200) x (733320)] = -0.32 in. (upward deflection)

Total initial deflection due to pre-stressing:

ΔP/S Tot = -1.99-(-0.37)-(-0.32)
= -2.68 in. (upward deflection)

Notice that for camber calculations, some jurisdictions assume that some of the pre-stressing force is lost and only considers a percentage of the value calculated above (e.g., Pennsylvania uses 90% of the above value to compute beam seat elevations and 85% to compute probable beam sag). In the following calculations the full value is used. The user may revise these values to match any reduction required by the bridge owner's specification. The reader may also refer to the PCI Bridge Design Manual for additional methods for estimating long term deflections.

Using conventional beam theory to determine deflection of simple span beams under uniform load or concentrated loads and using the loads calculated in Section 5.2, using non-composite and composite girder properties for loads applied before and after the slab is hardened, respectively, the following deflections may be calculated:

Δsw = deflection due to the girder self-weight = 1.16 in.
Δs = deflection due to the slab, formwork, and exterior diaphragm weight = 1.12 in.
ΔSDL = deflection due to the superimposed dead load weight = 0.104 in.

All deflection from dead load is positive (downward).

AASHTO-LRFD Design Step 5.6.7.1

Camber to Determine Bridge Seat Elevations

Initial camber, Ci:

Ci = ΔP/S Tot + Δsw
= -2.68 +1.16
= -1.52 in. (upward deflection)

Initial camber adjusted for creep, CiA:

CiA = CiCr

where:

Cr = constant to account for creep in camber (S5.4.2.3.2)
= 1.9kvskhckfktdti-0.118 (S5.4.2.3.2-1)
Cr = 1.436 from Section 5.4.6.1

Therefore, the initial camber, CiA is:

CiA = -1.52X1.436
= -2.18 in. (upward deflection)

Final camber, CF:

CF = CiA + Δs + ΔSDL

= -2.18 + 1.12 + 0.104
= -0.96 in. (upward deflection)

This camber is used to determine bridge seat elevation.

AASHTO-LRFD Design Step 5.6.7.2

Haunch Thickness

The haunch thickness is varied along the length of the girders to provide the required roadway elevation. For this example, the roadway grade is assumed to be 0.0. Therefore, the difference between the maximum haunch thickness at the support and the minimum haunch thickness at the center of the beam should equal the final camber, i.e., 0.96 in. in this example. Minimum haunch thickness is not included in the specifications and is typically specified by the bridge owner. Figure 3.34 (Ref. AASHTO-LRFD Figure 5.6-9) shows schematically the variation in haunch thickness. Haunch thickness at intermediate points is typically calculated using a computer program.

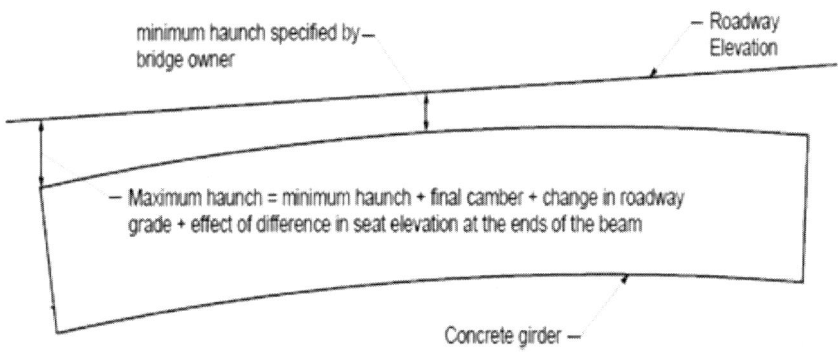

Figure 3.34. Schematic view of haunch. Source: Figure 5.6-9 from AASHTO LRFD Bridge Design Specifications, FHWA Office of Bridges and Structures, 1200 New Jersey Ave., SE, Washington, DC 20590.

AASHTO-LRFD Design Step 5.6.7.3

Camber to Determine Probable Sag in Bridge

To eliminate the possibility of sag in the bridge under permanent loads, some jurisdictions require that the above calculations for CF be repeated assuming a further reduction in the initial P/S camber. The final CF value after this reduction should show upward deflection.

3.3.6.8. AASHTO-LRFD Design Step 5.6.8 Optional Live Load Deflection Check

Service load deformations may cause deterioration of wearing surfaces and local cracking in concrete slabs and in metal bridges which could impair serviceability and durability, even if self limiting and not a potential source of collapse.

As early as 1905, attempts were made to avoid these effects by limiting the depth-to-span ratios of trusses and girders, and starting in the 1930's, live load deflection limits were prescribed for the same purpose. In a study of deflection limitations of bridges ASCE (1958), an ASCE committee, found numerous shortcomings in these traditional approaches and noted them. For example:

"The limited survey conducted by the committee revealed no evidence of serious structural damage that could be attributed to excessive deflection. The few examples of damaged stringer connections or cracked concrete floors could probably be corrected more effectively by changes in design than by more restrictive limitations on deflection. On the other hand, both the historical study and the results from the survey indicate clearly that unfavorable psychological reaction to bridge deflection is probably the most frequent and important source of concern regarding the flexibility of bridges. However, those characteristics of bridge vibration which are considered objectionable by pedestrians or passengers in vehicles cannot yet be defined."

Since that time, there has been extensive research on human response to motion, and it is now generally agreed that the primary factor affecting human sensitivity is acceleration as opposed to deflection, velocity, or the rate of change of acceleration for bridge structures, but the problem is a difficult and subjective one. Thus, to this point in history there are no simple definitive guidelines for the limits of tolerable static deflection or dynamic motion. Among current specifications, the Ontario Highway Bridge Design Code of 1983 contains the most comprehensive provisions regarding vibrations tolerable to humans.

The deflection criteria in S2.5.2.6.2 is considered optional. The bridge owner may select to invoke these criteria if desired. If an owner chooses to invoke deflection control, the following principles may apply:

- When investigating the maximum absolute deflection for straight girder systems, all design lanes should be loaded, and all supporting components should be assumed to deflect equally.
- For curved steel box and I-girder systems, the deflection of each girder should be determined individually based on its response as part of a system.
- fFr composite design, the stiffness of design cross-section used for determination of deflection should include the entire width of the roadway and the structurally continuous portions of the railings, sidewalks and median barriers.

- For straight girder systems, the composite bending stiffness of an individual girder may be taken as the stiffness determined as specified above, divided by the number of girders.
- When investigating maximum relative displacements, the number and position of loaded lanes should be selected to provide the worst differential effect.
- The live load portion of load combination Service I of AASHTO LRFD Table S3.4.1-1 should be used, including the dynamic load allowance IM.
- The live load shall be taken from S3.6.1.3.2.
- The provisions of S3.6.1.1.2 should apply.
- For skewed bridges, a right cross-section may be used; for curved and curved skewed bridges a radial cross-section may be used.

If the owner invokes the optional live load deflection criteria, the deflection should be taken as the larger of:

- That resulting from the design truck alone, or
- That resulting from 25 percent of the design truck taken together with the design lane load.

According to S2.5.2.6.2, the deflection criteria for vehicular live load limits deflection to L/800 (not including pedestrian loads).

110 x (12)/800 = 1.65 in.

The calculated live load deflection determined by using computer software is 0.324 in.

0.324 in. < 1.65 in. OK

3.3.7. AASHTO-LRFD Design Step 5.7 - Shear Design (S 5.8)

Shear design in the AASHTO-LRFD Specifications is based on the modified compression field theory. This method takes into account the effect of the axial force on the shear behavior of the section. The angle of the shear cracking, θ, and the shear constant, β, are both functions of the level of applied shear stress and the axial strain of the section. Figure 3.35 (Ref. AASHTO-LRFD Figure S5.8.3.4.2-1 (reproduced below) illustrates the shear parameters.

The transverse reinforcement (stirrups) along the beam is shown in AASHTO-LRFD Figure 3.36 (Ref. AASHTO-LRFD Figure 5.7-1). Table 3.11 (Ref. AASHTO-LRFD Table 5.7-1) lists the variables required to be calculated at several sections along the beam for shear analysis.

A sample calculation for shear at several sections follows the table.

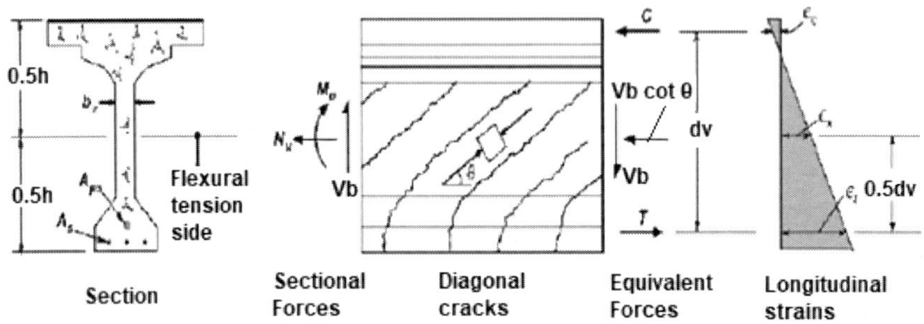

Figure 3.35. Illustration of shear parameters for section containing at least the minimum amount of transverse reinforcement, Vp = 0. Source: Figure S5.8.3.4.2-1 from AASHTO LRFD Bridge Design Specifications, FHWA Office of Bridges and Structures, 1200 New Jersey Ave., SE, Washington, DC 20590.

Notice that many equations contain the term Vp, the vertical component of the prestressing force. Since draped strands do not exist in the example beams, the value of Vp is taken as zero.

3.3.7.1. AASHTO-LRFD Design Step 5.7.1 Critical Section for Shear near the End Support

According to S5.8.3.2, where the reaction force in the direction of the applied shear introduces compression into the end region of a member, the location of the critical section for shear shall be taken as dv from the internal face of the support. Otherwise, the design section shall be taken at the internal face of the support where the beam-type element extends on both sides of the reaction area, the design section on each side of the reaction shall be determined separately based upon the loads on each side of the reaction and whether their respective contribution to the total reaction introduces tension or compression into the end region.

Therefore, the critical section near the end support is taken as dv.

3.3.7.2. AASHTO-LRFD Design Step 5.7.2 Shear Analysis for a Section in the Positive Moment Region. Sample Calculations

Section dv from the centerline of the end bearing

Table 3.11. (Ref. AASHTO-LRFD Table 5.7-1) - Shear Analysis at Different Sections

Dist [(1)]	A_{ps}[(2)]	A_{sl}[(3)]	CGS [(4)]	d_e[(5)]	C (Rectangular behavior) [(6)]	C (T-section behavior)[(7)]	de-β1c/2	0.9de	dv [(8)]	Vu [(9)]	Vp [(9,10)]	vu [(11)]	Mu [(9,12)]	Mu/dv
(ft.)	(in^2)	(in^2)	(in.)	(in.)	(in.)	(in.)	(in.)	(in.)	(in.)	(kips)	(kips)	(ksi)	(kip-ft.)	(Kips)
7.00	4.90		5.375	74.125	3.890	N/A		66.7	72.5	340.4	0.0	0.652	2,241	370.92
11.00	5.26		5.279	74.221	4.17	N/A		66.8	72.4	315.1	0.0	0.604	3,393	562.38
16.50	5.81		5.158	74.342	4.60	N/A		66.9	72.4	280.7	0.0	0.539	4,755	788.12
22.00	5.81		5.158	74.342	4.60	N/A		66.9	72.4	246.7	0.0	0.473	5,897	97.74
27.50	6.73		5.000	74.500	5.31	N/A		67.1	72.2	213.4	0.0	0.41	6,821	1133.68
33.0	6.73		5.000	74.500	5.31	N/A		67.1	72.2	180.6	0.0	0.347	7,535	1252.35
38.50	6.73		5.000	74.500	5.31	N/A		67.1	72.2	148.3	0.0	0.285	8,063	1340.11
44.00	6.73		5.000	74.500	5.31	N/A		67.1	72.2	116.7	0.0	0.224	8,381	1392.96
49.50	6.73		5.000	74.500	5.31	N/A		67.1	72.2	85.7	0.0	0.165	8,494	1411.75
54.50	6.73		5.000	74.500	5.31	N/A		67.1	72.2	118.4	0.0	0.228	8,456	1405.43
55.00	6.73		5.000	74.500	5.31	N/A		67.1		121.3	0.0	0.233	8,440	1402.77
60.50	6.73		5.000	74.500	5.31	N/A		67.1		153.5	0.0	0.295	8,163	1356.73
66.00	6.73		5.000	74.500	5.31	N/A		67.1		185.7	0.0	0.357	7,690	1278.12
71.50	6.73		5.000	74.500	5.31	N/A		67.1		217.9	0.0	0.419	7,027	1167.92
77.00	6.73		5.000	74.500	5.31	N/A		67.1		250.0	0.0	0.481	6,180	1027.15
82.50	6.73		5.000	74.500	5.31	N/A		67.1		282	0.0	0.542	5,158	857.29
88.00	5.81		5.000	74.500	4.60	N/A		66.9		313.8	0.0	0.602	3,966	657.35
93.5	5.81	15.31	5.000	74.500	8.58	N/A		68.0		345.4	0.0	0.663	-393	-65.23
99	4.90	15.31	5.000	74.500	8.58	N/A		68.0		376.8	0.0	0.724	-1,535	-254.77
120.5	4.90	15.31	5.000	74.500	8.58	N/A		68.0		3.6.6	0.0	0.762	-2,489	-413.11

Table 3.12. (Ref. AASHTO-LRFD Table 5.7-2) - Shear Analysis at Different Sections

Dist,[1]	$A_{ps}f_{po}$[2,13]	εs (14)	θ[15]	β[15]	V_c	S_{Reqd}[17]	S_{Provd}[17]	Vs(comp.)	φVn	φVn/Vu	T[16]	Nom. Flex. Res.	Nom. Flex.>T?
(ft.)	(kips)	(strain)	(deg)		(kips)	(in.)	(in.)	(kips)	(kips)		(kips)	(kips)	
7.00	926.1	0.00000	29.0	4.8	215.4	19.3	6.0	196.2	370.5	1.09	N/A	1147.1	N/A
11.00	994.1	0.00000	29.0	4.8	215.3	23.3	16.0	195.9	370.0	1.17	1017.3	1357.3	OK
16.50	1098.10	0.00000	29.00	4.80	215.16	32.40	18.00	174.20	350.40	1.25	1193.70	1484.70	OK
22.00	1098.1	0.00076	31.67	3.05	136.91	20.5	20.0	140.9	250.1	1.01	1307.7	1532.3	OK
27.50	1272.00	0.00039	30.36	3.72	166.31	41.8	20.0	147.8	282.4	1.32	1412.2	1718.0	OK
33.0	1272.00	0.00084	31.92	2.95	132.01	40.5	24.0	115.8	222.8	1.23	1481.4	1764.1	OK
38.50	1272.00	0.00112	32.94	2.60	166.49	55.4	24.0	111.4	204.9	1.38	1508.4	1781.3	OK
44.00	1272.00	0.00123	33.32	2.49	111.47	145	24.0	109.8	198.9	1.70	1506.7	1781.3	OK
49.50	1272.00	0.00117	33.10	2.56	114.31	139.3	24.0	110.7	202.3	2.36	1472.9	1781.3	OK
54.50	1272.00	0.00131	33.58	2.42	108.31	112.6	24.0	108.7	195.1	1.65	1521.7	1781.3	OK
55.00	1272.00	0.00131	33.959	2.42	108.31	98.6	24.0	108.7	195.1	1.61	1523.8	1781.3	OK
60.50	1272.00	0.00124	33.33	2.49	111.34	44.5	24.0	109.7	198.7	1.29	1532.5	1781.3	OK
66.00	1272.00	0.00100	32.49	2.75	122.90	32.6	24.0	113.3	212.4	1.14	1513	1781.3	OK
71.50	1272.00	0.00059	31.07	3.33	148.86	30.9	24.0	119.8	241.5	1.11	1470.3	1781.3	OK
77.00	1272.00	0.00002	29.08	4.72	210.95	46.6	24.0	129.7	306.1	1.22	1409.7	1755.7	OK
82.50	1272.00	0.00000	29.00	4.80	214.73	31.7	24	130.3	310.4	1.1	1305.1	1709.7	OK
88.00	1272.00	0.00000	29.00	4.80	215.16	23.6	11	285	450.2	1.43	1029.3	1523.6	OK
93.5	1098.1[2]	0.00092	32.34	2.83	126.89	10.7	11.0[11]	250.1	339.3	0.98 [11]	N/A	1476.0	N/A

Dist,(1)	Apsfpo(2,13)	εs (14)	θ(15)	β(15)	Vc	SReqd(17)	SProvd(17)	Vs(comp.)	φVn	φVn/Vu	T(16)	Nom. Flex. Res.	Nom. Flex.>T?
99	926.1[(2)]	0.00142	33.98	2.32	103.98	8.2	7.0	267.8	242.6	1.13	N/A	1295.5	N/A
120.5	926.1[(2)]	0.00182	35.38	2.03	90.76	7.0	6.0	407.2	448.2	1.13	N/A	1106.1	N/A

Notes:

(1) Distance measured from the centerline of the end support. Calculations for Span 1 are shown.

(2) Pre-stressing steel is on the compression side of the section in the negative moment region of the girder (intermediate pier region). This pre-stressing steel is ignored where the area of steel in an equation is defined asthe area of steel on the tension side of the section.

(3) Area of continuity reinforcement, i.e., the longitudinal reinforcement of the deck slab within the effective flange width of the girder in the girder negative moment region.

(4) Distance from the centroid of the tension steel reinforcement to the extreme tension fiber of the section. In the positive moment region, this is the distance from the centroid of pre-stressing strands to the bottom of the pre-stressed beam. In the negative moment region, this is the distance from the centroid of the longitudinal deck slab reinforcement to the top of the structural deck slab (ignore the thickness of the integral wearing surface).

(5) Effective depth of the section equals the distance from the centroid of the tension steel reinforcement to the extreme compression fiber of the section. In the positive moment region, this is the distance from the centroid of the pre-stressing strands to the top of the structural deck slab (ignore the thickness of the integral wearing surface). In the negative moment region, this is the distance from the centroid of the longitudinal deck slab reinforcement the bottom of the pre-stressed beam. The effective depth is calculated as the total depth of the section (which equals the depth of precast section, 72 in. + structural deck thickness, 7.5 in. = 79.5 in.) minus the quantity defined in note (4) above.

(6) Distance from the extreme compression fiber to the neutral axis calculated assuming rectangular behavior using Eq. S5.7.3.1.1-4. Pre-stressing steel, effective width of slab and slab compressive strength are considered in the positive moment region. The slab longitudinal reinforcement, width of the girder bottom (compression) flange and girder concrete strength are considered in the negative moment region.

(7) Distance from the extreme compression fiber to the neutral axis calculated assuming T-section behavior using Eq. S5.7.3.1.1-3. Only applicable if the rectangular section behavior proves untrue.

(8) Effective depth for shear calculated using S5.8.2.9.

(9) Maximum applied factored load effects obtained from the beam load analysis.

(10) Vertical component of pre-stressing which is 0.0 for straight strands.

(11) The applied shear stress, v_u, calculated as the applied factored shear force divided by the product of the web width, b, and effective shear depth, d_v.

(12) Only the controlling case (positive moment or negative moment) is shown.

(13) In the positive moment region, the parameter f_{po} is taken equal to $0.7f_{pu}$ of the pre-stressing steel as allowed by S5.8.3.4.2. This value is reduced within the transfer length of the strands to account for the lack of full development.

(14) Value of the parameter ε_s calculated using Eq. S5.8.3.4.2-4. The quantity εs is always taken as positive. Where ε_s is negative, it shall be taken as 0.

(15) The values of β and θ as calculated according to the general procedure in S5.8.3.4.2.

(16) Force in longitudinal reinforcement including the effect of the applied shear (S5.8.3.5).

(17) $S_{req'd}$ is the theoretical required spacing of the shear stirrup reinforcement at each analysis location, in inches. $S_{prov'd}$ is the chosen stirrup spacing at each analysis location, in inches. In cases where the theoretical spacing exceeds 24", the maximum spacing allowed of24"controls.

(18) Since a crack must cross through the height of the girder, it must cross multiple stirrups in the process. At this analysis point the chosen stirrup spacing changes from 7 in to 11 in. If an average spacing of 9 in is used to investigate shear strength at this location, the capacity is computed as 389 kips, resulting in an adequate design.

Design Step 5.7.2 Shear Analysis for a Section in the Positive Moment Region. Sample Calculations

AASHTO-LRFD Design Step 5.7.2.1

Determine the effective depth for shear, dv

dv = effective shear depth taken as the distance, measured perpendicular to the neutral axis, between the resultants of the tensile and compressive forces due to flexure; it need not be taken to be less than the greater of 0.9de or 0.72h (S5.8.2.9)

Depth of the precast beam = 72 in

Structural slab thickness = 7.5 in

h total depth of beam (in.) depth of the precast beam + structural slab thickness = 72 +7.5 = 79.5 in. (notice that the depth of the haunch was ignored in this calculation)

de = distance from the extreme compression fiber to the center of gravity of the pre-stressing steel at the section (in.). From AASHTO-LRFD Figure 2-6,
= 79.5 – 5.375 = 74.125 in.

Assuming rectangular section behavior with no compression steel or mild tension reinforcement, the distance from the extreme compression fiber to the neutral axis, c, may be calculated as:

c = Apsfpu /[α1 f'cβ1b + kAps(fpu/dp)] (S5.7.3.1)

α1 = 0.85 for 4 ksi slab concrete (not exceeding 10 ksi) (S5.7.2.2)

β1 = 0.85 for 4 ksi slab concrete (not exceeding 4 ksi) (S5.7.2.2)

b = effective flange width = 116 in. (calculated in Section 2.2)

Area of pre-stressing steel that is bonded at the section under consideration (see AASHTO-LRFD Figure 2-5) = 32X0.153 = 4.896 in^2

c = 4.896X270/[(0.85X4X0.85X116) +0.28X4.896(15.31/74.125)]
= 3.89 in. < structural slab thickness = 7.5 in.

The assumption of the section behaving as a rectangular section is correct.
Depth of compression block, a = β1c = 0.85X3.89 = 3.3 in.

Distance between the resultants of the tensile and compressive forces due to flexure:

de – a/2 = 74.125– 3.3/2
= 72.5 in. (1)
0.9de = 0.9 x74.125
= 66.71 in. (2)
0.72h = 0.72X79.5
= 57.24 in. (3)
dv = largest of (1), (2) and (3) = 72.5 in.

Notice that 0.72h is always less than the other two values for all sections of this beam. This value is not shown in AASHTO-LRFD Table 3.10 (Ref. AASHTO-LRFD Table 5.7-1) for clarity.

AASHTO-LRFD Design Step 5.7.2.2

Minimum Required Transverse Reinforcement

Limits on maximum factored shear stresses for sections without transverse reinforcement are presented in S5.8.2.4. Traditionally, transverse reinforcement satisfying the minimum transverse reinforcement requirements of S5.8.2.5 is provided along the full length of the beam.

Minimum transverse reinforcement, Av:

f'c = compressive strength of the web concrete = 6 ksi
fy = yield strength of the transverse reinforcement = 60 ksi
Assume that #4 bars are used for the stirrups. Av = area of 2 legs of a #4 bar = 0.4 in^2

Substitute 0.4 in^2 to determine "s", the maximum allowable spacing of #4 bars (2-leg stirrups).

$0.4 \geq 0.0316 \text{ x } (2.449) \text{ x } (8/60)s = 0.010318$
$s \leq 38.7650$

AASHTO-LRFD Design Step 5.7.2.3

Maximum Spacing for Transverse Reinforcement

The maximum spacing of transverse reinforcement is determined in accordance with S5.8.2.7.

Depending on the level of applied factored shear stress, vu, the maximum permitted spacing, smax, is determined as:

- If vu < 0.125f'c, then:

smax = 0.8dv < 24 in. (S5.8.2.7-1)

- If vu ≥ 0.125f'c, then:

smax = 0.4dv < 12 in. (S5.8.2.7-2)

For the section under consideration, vu is determined using Equation S5.8.2.9-1. vu = 0.653 ksi= 0.109f'c. Therefore, the maximum permitted spacing.

AASHTO-LRFD Design Step 5.7.2.4

Shear Strength

The shear strength provided by the concrete, Vc, is calculated using the following equation:

Vc = 0.0316β√f'c bv dv (S5.8.3.3-3)

The values of β and the shear cracking inclination angle, θ, are determined using the procedure outlined in S5.8.3.4.2.

The parameter εs is the net longitudinal tensile strain in the section at the centroid of the tension reinforcement. For sections containing at least the minimum transverse reinforcement calculated above, εs may be calculated using the following equations:

$$\varepsilon_s = \frac{\left(\frac{|M_u|}{d_v} + 0.5N_u + |V_u - V_p| - A_{ps}f_{po}\right)}{E_sA_s + E_pA_{ps}} \qquad \text{(S5.8.3.4.2-4)}$$

If the value of εs from Eq. S5.8.3.4.2-4 is negative, the strain may be recomputed or conservatively taken as 0. For this example, the simplification is used and negative strains will be taken as 0.

For this example, the value of both the applied factored axial load, Nu, and the vertical component of pre-stressing, Vp, are taken equal to 0.

For the section under consideration:

Vu = maximum applied factored shear = 340.4 kips

Mu = maximum factored moment at the section = 2,241 k-ft, which must be ≥ Vudv = (340.4X72.5)/12 = 2057 k-ft.

Use Mu = 2,241 k-ft

Note that the maximum live load moment and the maximum live load shear at any section are likely to result from two different locations of the load along the length of the bridge. Conducting the shear analysis using the maximum factored shear and the concurrent factored moment is permitted. However, most computer programs list the maximum values of the moment and the maximum value of the shear without listing the concurrent forces. Therefore, hand calculations and most design computer programs typically conduct shear analysis using the maximum moment value instead of the moment concurrent with the maximum shear. This results in a conservative answer.

According to S5.8.3.4.2, fpo is defined as follows:

fpo = a parameter taken as the modulus of elasticity of the pre-stressing tendons multiplied by the locked-in difference in strain between the pre-stressing tendons and the surrounding concrete (ksi). For the usual levels of pre-stressing, a value of 0.7fpu will be appropriate for both pre-tensioned and post-tensioned members.

For pre-tensioned members, multiplying the modulus of elasticity of the pre-stressing tendons by the locked in difference in strain between the pre-stressing tendons and the surrounding concrete yields the stress in the strands when the concrete is poured around them, i.e., the stress in the strands immediately prior to transfer. For pre-tensioned members, SC5.8.3.4.2 allows fpo to be taken equal to the jacking stress. This value is typically larger than 0.7fpu. Therefore, using 0.7fpu is more conservative since it results in a larger value of εs.

For this example, fpo is taken as 0.7fpu.

fpo = 0.7(270) = 189 ksi (SC5.8.3.4.2)

Nu = 0 k

Notice that, as required by Article S5.8.3.4.2, within the transfer length, fpo shall be increased linearly from zero at the location where the bond between the strands and concrete commences to its full value at the end of the transfer length.

As, Es, Aps and Eps are the area of mild tension reinforcement (0.0), modulus of elasticity of mild reinforcement (29,000 ksi), area of pre-stressing steel (4.896 in^2) and modulus of elasticity of the pre-stressing strands (28,500 ksi), respectively.

Modulus of elasticity of mild reinforcement = 29,000 ksi

Area of pre-stressing steel = 4.896 in^2

Modulus of elasticity = 28,500 ksi

εs = [(2241/6.04) + (0.5 X 0) +340.4 - (4.89 X 189)] / [28500 X4.896]
= -0.00152494 < 0

Therefore, use $\varepsilon s = 0$
The value of θ is found using S5.8.3.4.2-3:

$\theta = 29 + 3500\varepsilon s = 29 + 0 = 29°$

Aps = area of pre-stressed steel at the section on the flexural side of the member.
= 32x 0.153
= 4.896 in^2
dv = 72.5 in. 6.04 ft.

The value of β is found using S5.8.3.4.2-1:

$\beta = 4.8/(1+750\varepsilon s) = 4.8$ (S5.8.3.4.2-1)

Calculate the shear resistance provided by the concrete, Vc.

Vc = $0.0316\beta\sqrt{f'c}$ bv dv (S5.8.3.3-3)
β= 4.8
f'c= 2.449
bv= 8
dv= 72.5
Vc = 0.0316 x 4.8 x 2.449 x 8 x 72.5 = 215.4 k

Calculate the shear resistance provided by the transverse reinforcement (stirrups), Vs.
Vs = [Av fy dv(cot θ + cot α)sin α]/s (S5.8.3.3-4)

Assuming the stirrups are placed perpendicular to the beam longitudinal axis at 16 in. spacing and are comprised of #4 bars, each having two legs:

Av = area of shear reinforcement within a distance "s" (in^2)
= 2 (area of #4 bar)
= 2 (0.2)
= 0.4 in^2
s = 16 in.

α = angle between the stirrups and the longitudinal axis of the beam
= 90 degrees
Vs = [0.4(60) x (72.5) x (cot 29.0)]/16 = 196.2 k

The nominal shear resistance, Vn, is determined as the lesser of:

Vn = Vc + Vs + Vp (S5.8.3.3-1)
Vn = 0.25 f'c bv dv + Vp (S5.8.3.3-2)

Notice that the purpose of the limit imposed by Eq. S5.8.3.3-2 is to ensure that the concrete in the web will not crush prior to yield of the transverse reinforcement.

Vp = 0 for straight strands
Vn = lesser of:
Vc + Vs + Vp = 215.4 +196.2 +0 = 411.6 k

and

0.25f'cbvdv + Vp = (0.25X6X8X72.5) +0 = 870 k

Therefore, Vn = 411.6 k
The resistance factor, ϕ, for shear in normal weight concrete is 0.9. (S5.5.4.2.1)
Shear factored resistance, Vr:

Vr = ϕVn (S5.8.2.1-2)
= 0.9x 411.6
= 370.44 k > Vu = 340.4 k OK

Check if transverse reinforcement is required (S5.8.2.4):
Transverse reinforcement shall be provided in pre-stressed concrete beams where:

Vu > 0.5φ(Vc + Vp)

All of the terms have been defined above, therefore:
0.5 x (0.9) x (215.4 + 0) = 97 k < 340.4 k transverse reinforcement is required.

3.3.7.3. AASHTO-LRFD Design Step 5.7.3 Shear Analysis for Sections in the Negative Moment Region

The critical section for shear near the intermediate pier may be determined using the same procedure as shown in Design Steps 3.3.7.1 and 3.3.7.2 (AASHTO-LRFD Design

Step 5.7.1 and 5.7.2) for a section near the end support. Calculations for a section in the negative moment region are illustrated below for the section at 99 ft. from the centerline of the end bearing. This section is not the critical section for shear and is used only for illustrating the design process.

Sample Calculations: Section 99 ft. from the centerline of end bearings.

AASHTO-LRFD Design Step 5.7.3.1

Difference in shear analysis in the positive and negative moment regions.

1) For the pier (negative moment) regions of precast simple span beams made continuous for live load, the pre-stressing steel near the piers is often in the compression side of the beam. The term Aps in the equations for εs is defined as the area of pre-stressing steel on the tension side of the member. Since the pre-stressing steel is on the compression side of the member, this steel is ignored in the analysis. This results in an increase in εx and, therefore, a decrease in the shear resistance of the section. This approach gives conservative results and is appropriate for hand calculations.

A less conservative approach is to calculate εsas the average longitudinal strain in the web. This requires the calculation of the strain at the top and bottom of the member at the section under consideration at the strength limit state. This approach is more appropriate for computer programs.

The difference between the two approaches is insignificant in terms of the cost of the beam. The first approach requires more shear reinforcement near the ends of the beam. The spacing of the stirrups in the middle portion of the beam is often controlled by the maximum spacing requirements and, hence, the same stirrup spacing is often required by both approaches.

It is beneficial to use the second approach in the following situations:

- Heavily loaded girders where the first approach results in congested shear reinforcement
- Analysis of existing structures where the first approach indicates a deficiency in shear resistance.

2) In calculating the distance from the neutral axis to the extreme compression fiber "c", the following factors need to be considered:

- The compression side is at the bottom of the beam. The concrete strength used to determine "c" is that of the precast girder.
- The width of the bottom flange of the beam is substituted for "b", the width of the member.

- The area of the slab longitudinal reinforcement over the intermediate pier represents the reinforcement on the tension side of the member. The area and yield strength of this reinforcement should be determined in advance.

The first approach is used in this example.

AASHTO-LRFD Design Step 5.7.3.2

Determine the Effective Depth for Shear, dv

h = 72+ 7.5 = 79.5 in. (notice that the depth of the haunch was ignored in this calculation)

The center of gravity of the deck slab longitudinal reinforcement from the top of the structural thickness of the deck= 3.98 in. (see design step 3.3.6.5.1 AASHTO-LRFD design step 5.6.5.1).

de = 79.5 –3.98 = 75.52 in.

The area of longitudinal slab reinforcement within the effective flange width of the beam is 15.31 in^2 (see design step 5.6.5.1).

Yield strength of the slab reinforcement 60 ksi

Assuming rectangular section behavior with no compression or pre-stressing steel, the distance:

C = Asfy / (α1 f'cβ1b) (S5.7.3.1.1-4)

where:

α1 = 0.85 for 6 ksi beam concrete (S5.7.2.2)

β1 = 0.75 for 6 ksi beam concrete (S5.7.2.2)

b = precast beam bottom (compression) flange width (in.)

= 28 in.

f'c = 6 ksi

c = 15.31 x60/[0.85 x6x0.75 x28]

= 8.58 in. ≈ thickness of the beam bottom flange (8 in.)

Therefore, the assumption of the section behaving as a rectangular section is considered correct.

Notice that if the value of "c" is significantly larger than the beam bottom flange thickness, a rectangular behavior may be used after adjusting the beam bottom flange width to account for the actual beam area in compression. However, if "c" is not significantly larger than the beam bottom flange thickness, the effect on the results will be minor and the analysis may be continued without adjusting the beam bottom flange width. This reasoning is used in this example.

Depth of compression block, $a = \beta 1c = 0.75X8.58 = 6.44$ in.

Distance between the resultants of the tensile and compressive forces due to flexure:

$= de - a/2$
$= 75.52 - 6.44/2$
$= 72.3$ in (1)
$0.9de = 0.9 \times 75.52$
$= 67.97$ in. (2)
$0.72h = 0.72 \times 79.5$
$= 57.24$ in. (3)
dv = largest of (1), (2) and (3) = 72.3 in.

Notice that 0.72h is always less than the other two values for all sections of this beam. This value is not shown in Table 3.10 (Ref. AASHTO-LRFD Table 5.7-1) for clarity.

AASHTO LRFD Design Step 5.7.3.3

Shear Stress on Concrete

From AASHTO-LRFD Table 3.3-4 (Ref. AASHTO-LRFD Table 5.3-4), the factored shear stress at this section,

$Vu = 376.8$ kips

$\phi = 0.9$ (shear) (S5.5.4.2.1)

bv = width of web = 8 in.

From Article S5.8.2.9, the shear stress on the concrete is:

$vu = (Vu - \phi Vp)/(\phi bvdv)$
$vu = (376.8 - 0)/[0.9X8X72.3)] = 0.724$ ksi
$vu / f'c = 0.724 / 6 = 0.1207$

AASHTO-LRFD Design Step 5.7.3.4

Minimum Required Transverse Reinforcement

Maximum allowable spacing for #4 stirrups with two legs per stirrup was calculated in para 1.2.7.2.2 (design step 3.3.7.2.2 AASHTO-LRFD design step 5.7.2.2).

$s \leq 38.7700$ in.

AASHTO-LRFD Design Step 5.7.3.5

Maximum Spacing for Transverse Reinforcement

The maximum spacing of transverse reinforcement is determined in accordance with S5.8.2.7. Depending on the level of applied factored shear stress, vu, the maximum permitted spacing, smax, is determined as:

- If vu < 0.125f'c, then:

  smax = 0.8dv < 24.0 in. (S5.8.2.7-1)

- If vu ≥ 0.125f'c, then:

  smax = 0.4dv < 12.0 in. (S5.8.2.7-2)

For the section under consideration, vu = 0.1207f'c.
Therefore, the maximum permitted spacing,
smax = 0.8dv
= 0.8x 72.3
= 57.84 in. > 24 in. NG
Assume maximum permitted stirrup spacing = 24 in.

AASHTO-LRFD Design Step 5.7.3.6

Shear Resistance

For sections in the negative moment region of the beam, calculate εx, using Eq. S5.8.3.4.2-4 and assume there is no pre-stressing steel.

$$\varepsilon_s = \frac{\left(\frac{|M_u|}{d_v} + 0.5N_u + |V_u - V_p| - A_{ps}f_{po}\right)}{(E_sA_s + E_pA_{ps})} \qquad \text{(S5.7.3.4.2-4)}$$

For this example, the value of both the applied factored axial load, Nu, and the vertical component of pre-stressing, Vp, are taken equal to 0.

Vu = maximum applied factored shear from Table 3.3-4 (Ref. AASHTO-LRFD Table 5.3-4)
= 376.8 kips
Mu = maximum applied factored moment from Table 3.3-2 (Ref. AASHTO-LRFD Table 5.3-2)
= -1,535 k-ft

Notice that the term Mu/dv represents the force in the tension reinforcement due to the applied factored moment. Therefore, Mu/dv is taken as a positive value regardless of the sign of the moment.

fpo = 0 ksi at this location (pre-stressing force ignored)
As = area of longitudinal reinforcement in the deck at this section
= 15.31 in^2

Note that the area of deck longitudinal reinforcement used in this calculation is the area of the bars that extend at least one development length beyond the section under consideration. If the section lies within the development length of some bars, these bars may be conservatively ignored or the force in these bars be prorated based on the ratio between the full and available development length. Consideration should also be given to adjusting the location of the center of gravity of the reinforcement to account for the smaller force in the bars that are not fully developed.

dv = 72.3 in. (6.03 ft.)
Es = 29,000 ksi
Eps = 28,500 ksi
Aps = area of pre-stressing steel on the tension side of the member
= 0.0 in^2

Substitute these variables in Eq. S5.8.3.4.2-4 to determine εs:

εs = [1535 x 12/72.3 + (376.8 – 0) – 0]/[(29000) x (15.31) + 0]
= 0.001422

Determine the values of θ and β using Equations S5.8.3.4.2-1 and S5.8.3.4.2-3.

β = 4.8/[1+750X(0.001422)] = 2.32 (S5.8.3.4.2–1)

$\theta = 29 + 3500X(0.001422) = 33.98°$ (S5.8.3.4.2–3)

Calculate the shear resistance provided by concrete, Vc:

$Vc = 0.0316\ \beta\ \sqrt{f'c}\ bv\ dv$ (S5.8.3.3-3)

$Vc = 0.0316 \times (2.32) \times (2.449) \times (8) \times (72.3) = 103.9$ k

Calculate the shear resistance provided by the transverse reinforcement (stirrups), Vs:

$Vs = [Av\ fy\ dv\ (\cot\theta + \cot\alpha)\sin\alpha]/s$ (S5.8.3.3-4)

Assuming the stirrups are placed perpendicular to the beam longitudinal axis at 7 in. spacing and are comprised of #4 bars, each having two legs:

$Av = 2(\text{area of \#4 bar}) = 2 \times (0.2)$
$= 0.4\ in^2$
$s = 7$ in.
$\alpha = 90$ degrees $Vs = 0.4X(60) \times (72.3) \times (\cot 33.98)/7 = 367.8$ k

The nominal shear resistance, Vn, is determined as the lesser of:

$Vn = Vc + Vs + Vp$ (S5.8.3.3-1) $Vn = 0.25f'cbvdv + Vp$ (S5.8.3.3-2)

Notice that the purpose of the limit imposed by Eq. S5.8.3.3-2 is intended to eliminate excessive shear cracking.

$Vp = 0$ for straight strands

Vn is taken as the lesser of:

$Vc + Vs + Vp = 103.9 + 367.8 + 0 = 471.7$ k

and

$0.25f'cbvdv + Vp = 0.25x\ (6) \times (8) \times (72.3) + 0 = 867.6$ k

Therefore, $Vn = 471.7$ k

The resistance factor, ϕ, for shear in normal weight concrete = 0.9 (S5.5.4.2.1)

Factored shear resistance:

$Vr = \phi Vn$
$= 0.9x\ (471.7)$
$= 424.53\ k > Vu = 367.8\ k$ OK

3.3.7.4. AASHTO-LRFD Design Step 5.7.4 Factored Splitting Resistance (S5.10.10.1)

The bursting resistance of the pre-tensioned anchorage zones is calculated according to S5.10.10.1 at the service limit state.

$Pr = fsAs$ (S5.10.10.1-1)

where:

fs = stress in the steel not exceeding 20 ksi
As = total area of vertical reinforcement located within the distance h/4 from the end of the beam (in^2)
H = overall depth of the precast member (in.)

The resistance shall not be less than 4% of the pre-stressing force at transfer.

From para 1.2.4.4 (design step 3.3.4.4 design step 5.4.4):

Pre-stressing force at transfer at end of beam

$= 32X(0.153)X(188.8)$
$= 924.4$ kips

Determine the required area of steel to meet the minimum resistance using fs = 20 ksi (max).

Therefore, $0.04 \times 924.4 = 20 \times (As)$
$As = 1.85\ in^2$

Since one stirrup is 0.4 in^2 (includes 2 legs), determine the number of stirrups required = 1.85/0.4 = 4.63 Say 5 stirrups required

These stirrups must fit within h/4 distance from the end of the beam.

h/4 = 72/4 = 18 in

Use 5 stirrups at 3 in. spacing as shown in Figure 3.36 (Ref. AASHTO-LRFD Figure 5.7-1).

3.3.7.5. AASHTO-LRFD Design Step 5.7.5 Confinement Reinforcement (S5.10.10.2)

For the distance of 1.5d [1.5(72/12) = 9 ft.] from the end of the beam, reinforcement shall be placed to confine the pre-stressing steel in the bottom flange.

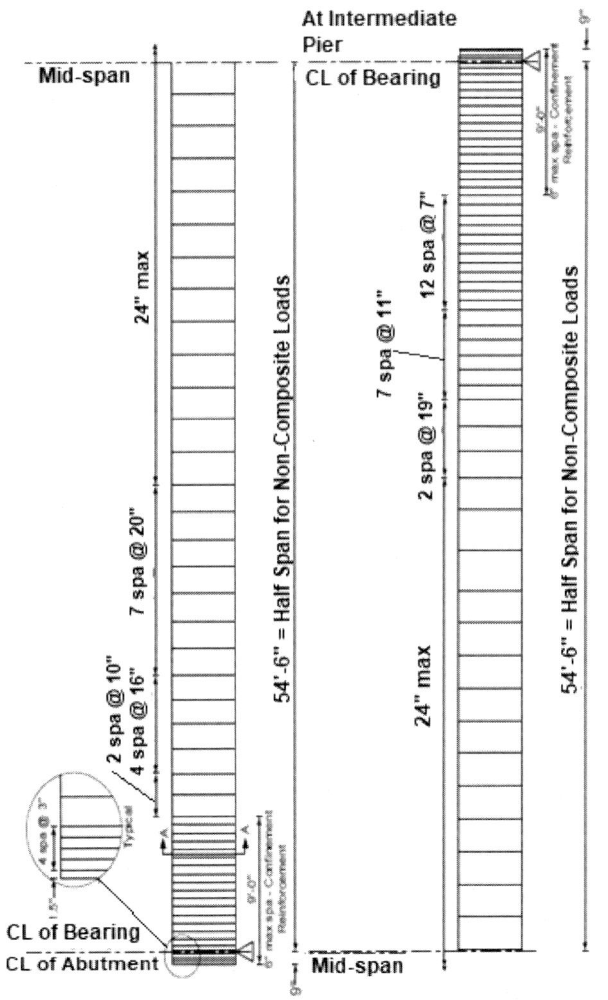

Figure 3.36. Beam transverse reinforcements. Source: Figure 5.7-1 from AASHTO LRFD Bridge Design Specifications, FHWA Office of Bridges and Structures, 1200 New Jersey Ave., SE, Washington, DC 20590.

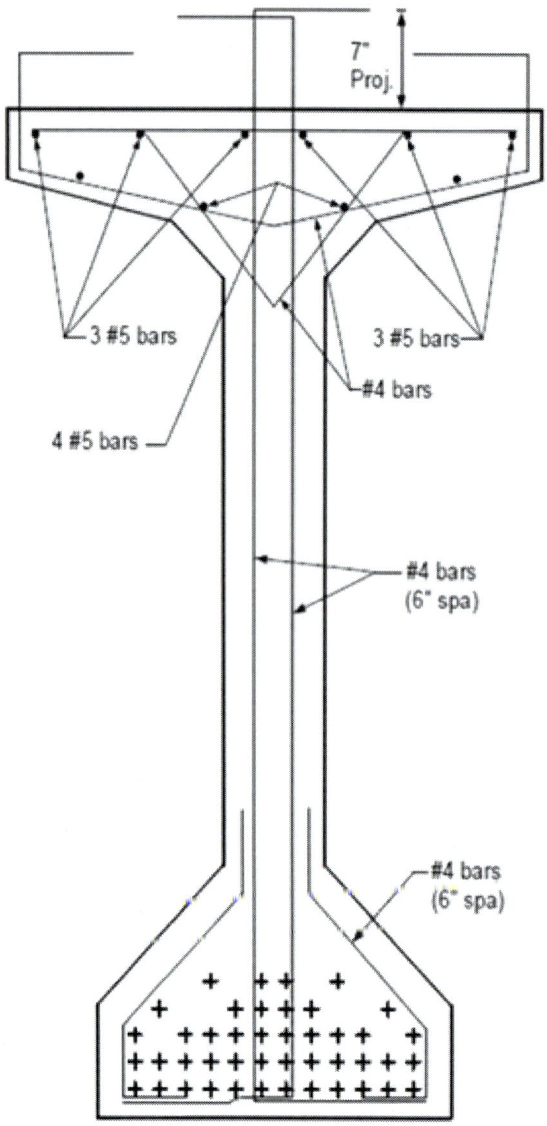

Figure 3.37. Section A-A from Figure 5.7-1, beam cross section near the girder ends. Source: Figure 5.7-2 from AASHTO LRFD Bridge Design Specifications, FHWA Office of Bridges and Structures, 1200 New Jersey Ave., SE, Washington, DC 20590.

The reinforcement is required to be not less than No. 3 deformed bars, with spacing not exceeding 6.0 in. and shaped to enclose the strands. The stirrups required to resist the applied shear and to satisfy the maximum stirrup requirements are listed in Table 3.10 (Ref. AASHTO-LRFD Table 5.7-1) for different sections. The maximum required spacing shown in Table 3.10 (Ref. AASHTO-LRFD Table 5.7-1) in the end zones of the beam are greater than 6 in. For a beam where all strands are located in the bottom flange, two different approaches may be utilized to provide the required confinement reinforcement:

1) Reduce the stirrup spacing in the end zone (1.5d) to not greater than 6 in.
2) Place the main vertical bars of the stirrups at the spacing required by vertical shear analysis. Detail the vertical bars in the bottom of the beam to enclose the pre-stressing and place these bars at a spacing not greater than 6 in. within the end zones. The stirrups and the confinement bars in this approach will not be at the same spacing and pouring of the concrete may be difficult.

For a beam where some strands are located in the web approach (1) should be used.

For this example, approach (1) was used. This is the basis for the stirrup distribution shown in Figure 3.36 (Ref. AASHTO-LRFD Figure 5.7-1).

3.3.7.6. Design Step 5.7.6 Force in the Longitudinal Reinforcement Including the Effect of the Applied Shear (S5.8.3.5)

In addition to the applied moment, Mu, the following force effects contribute to the force in the longitudinal reinforcement:

- Applied shear forces, Vu
- Vertical component of the pre-stressing force
- Applied axial force, Nu
- The shear force resisted by the transverse reinforcement, Vs

To account for the effect of these force effects on the force in the longitudinal reinforcement, S5.8.3.5 requires that the longitudinal reinforcement be proportioned so that at each section, the tensile capacity of the reinforcement on the flexural tension side of the member, taking into account any lack of full development of that reinforcement, is greater than or equal to the force (Apsfps + Asfy) calculated as:

$$A_{ps}f_{ps} + A_s f_y = \frac{|M_u|}{d_v \varphi_v} + 0.5\frac{N_u}{\varphi_c} + \left(\left|\frac{V_u}{\varphi_v} - V_p\right| - 0.5V_s\right)\cot\theta$$

(S5.8.3.5-1)

where:

Vs = shear resistance provided by the transverse reinforcement at the section under investigation as given by Eq. S5.8.3.3-4, except Vs shall not be taken as greater than Vu/ϕ (kips)

θ = angle of inclination of diagonal compressive stresses used in determining the nominal shear resistance of the section under investigation as determined by S5.8.3.4 (degrees)

ϕ = resistance factors are taken from S5.5.4.2 as appropriate for moment, shear, and axial resistance.

This check is required for sections located no less than a distance equal to dvcot θ from the support. The values for the critical section for shear near the end support are substituted for dv and θ.

0.5x(72.5) cot 29 = 65.4 in. ≈ 5.45 ft.

The check for tension in the longitudinal reinforcement must be performed for all sections no closer than 5.45 ft. from the support.

Sample calculation: Section at 11.0 ft. from the centerline of bearing at the end support.

Using information from Table 3.11 (Ref. AASHTO-LRFD Table 5.7-1):

Force in the longitudinal reinforcement at nominal flexural resistance, (Apsfps + Asfy)

Mu = 3.393
Dv = 72.5
ϕv = 0.9
Nu = 0
ϕc = 1
Vu = 315.1
Vp = 0
Vs = 195.9
θ = 29
(Aps fps + As fy) = ((3.393x12)/(72.5x0.9))+0.5x0/
1(+((|(315.1/0.9)-0|)-(0.5x195.9))) x (cot(29)) = 455.47

From Table 3.4 and 3.5 (Ref. AASHTO-LRFD Table 5.5-1), the maximum strand resistance at this section at the nominal moment resistance is 1,295.50 kips > (Aps fps + As fy) = 455.47 kips OK

3.3.7.7. AASHTO-LRFD Design Step 5.7.7 Horizontal Shear between the Beam and Slab

Sample calculations at 11 ft. from the centerline of bearing on the abutment (11 ft. – 9 in. from girder end)

Horizontal shear forces develop along the interface between the concrete girders and the deck. As an alternative to the classical elastic strength of materials approach, the value of these forces per unit length of the girders at the strength limit state can be taken as:

Vhi = V1/dv (SC5.8.4.2-7)

where:

Vhi = factored horizontal shear per unit length of the girder (kips/length)

V1 = the factored vertical shear (kips) = 315.1 k

From Table 3.11 (Ref. AASHTO-LRFD Table 5.7-2)

dv = distance between the centroid of the tension steel and the mid-thickness of the slab to compute a factored interface shear (in.).

= 72 +7.5/2 – 5.279 = 70.47 in.

See Table 3.11 (Ref. AASHTO-LRFD Table 5.7-2)

Vhi = 315.1/70.47 = 4.47 k/in.

Stirrup spacing at this location = 18 in.

Assume that the stirrups extend into the deck. In addition, assume that there is another #4 bar with two legs extending into the deck as shown in Figure 3.37 (Ref. AASHTO-LRFD Figure 5.7-2).

Area of reinforcement passing through the interface between the deck and the girder, Avf.

Avf = 4 #4 bars

= 4 x (0.2)

= 0.8 in^2

Avf per unit length of beam = 0.8/18 = 0.044 in^2/in. of beam length.

Check if the minimum interface shear reinforcement may be waived (S5.8.4.4)

For cast-in-place concrete decks on clean concrete girder surfaces free of laitance, the minimum interface shear reinforcement requirements may be waived provided that the beam top flange surface is roughened to an amplitude of 0.25 inches where the interface stress vui (Eq. S5.8.4.2-1) is less than 0.210 ksi and all transverse shear reinforcement required by S5.8.2.5 is extended across the beam/slab interface and adequately anchored in the slab.

Area of interface per unit length= 42

Shear stress on the interface = Vhi/Area of interface per unit length

= 4.47/42

= 0.106 ksi < 0.21 ksi

Therefore, the minimum reinforcement requirement may be waived.

Table 3.13. (Ref. AASHTO-LRFD Table 5.7-2) - Interface Shear Calculations

Dist.	**d_v**	**$V_1 = V_u$**	**Max. Stirrup Spacing**	**Interface Reinf., A_{vf}**	**Horiz. Shear, V_h**	**Nominal Resistance**	**Factored Resistance**	**Resist./ Applied Load**
(ft.)	(in.)	(kips)	(in.)	(in²/in.)	(k/in.)	(k/in.)	(k/in.)	>1.0 OK
7.00	70.4	340.4	46.0	0.050	4.84	14.8	13.3	OK
11.00	70.5	315.1	18.0	0.044	4.47	14.4	13.0	OK
16.50	70.6	280.7	21.0	0.038	3.98	14.0	12.6	OK
22.00	70.6	146.7	21.0	0.040	3.49	14.2	12.7	OK
27.50	70.8	213.4	24.0	0.033	3.02	13.7	12.4	OK
33.0	70.8	180.6	24.0	0.033	2.55	13.7	12.4	OK
38.50	70.8	148.3	24.0	0.033	2.10	13.7	12.4	OK
44.00	70.8	116.7	24.0	0.033	1.65	13.7	12.4	OK
49.50	70.8	85.70	24.0	0.033	1.21	13.7	12.4	OK
54.50	70.8	118.4	24.0	0.033	1.67	13.7	12.4	OK
55.00	70.8	121.3	24.0	0.033	1.71	13.70	12.4	OK
60.50	70.8	153.5	24.0	0.033	2.17	13.70	12.4	OK
66.00	70.8	185.7	24.0	0.033	2.62	13.70	12.4	OK
71.50	70.8	217.9	24.0	0.033	3.08	13.70	12.4	OK
77.00	70.8	250.0	24.0	0.033	3.53	13.70	12.4	OK
82.50	70.8	282.0	21.0	0.033	3.99	13.70	12.4	OK
88.00	70.6	313.8	19.0	0.042	4.45	14.30	12.9	OK
93.5	70.5	345.4	11.0	0.073	4.57	16.10	14.5	OK
99	70.5	376.8	8.0	0.0100	4.99	17.80	16.0	OK
120.5	70.5	396.6	7.0	0.0114	5.25	18.60	16.7	OK

Note that the interface shear stress is only approximately half of the limiting value of 0.210 ksi. In typical design, the minimum shear interface reinforcement requirement would be waived. For this example, in order to provide a complete reference, the minimum reinforcement requirement will not be waived, and calculation will continue.

bv = 42

fy = 60

Check the minimum interface shear reinforcement

Avf ≥ 0.05Acv/fy = 0.05bv/fy (S5.8.4.4-1) = 0.05 x 42/60 = 0.035 in²/in. of beam length

0.035 in²/in. < As provided OK

Note that the minimum interface shear reinforcement, Avf, also need not exceed the lesser of the amount determined using Eq. S5.8.4.4-1 (0.035 in^2/in) and the amount needed to resist 1.33Vui/φ as determined using Eq. S5.8.4.1-3.

Factored interface shear resistance (S5.8.4.1)

The interface shear resistance of the interface has two components. The first component is due to the adhesion between the two surfaces. The second component is due to the friction. In calculating friction, the force acting on the interface is taken equal to the compression force on the interface plus the yield strength of the reinforcement passing through the interface. The nominal shear resistance of the interface plane, Vni, is calculated using S5.8.4.1.

Vni = cAcv + μ(Avffy + Pc) (S5.8.4.1-1)

where:

Vni = nominal shear friction resistance (kips)

Acv = area of concrete engaged in shear transfer (in^2) Avf = area of shear reinforcement crossing the shear plane (in^2)

fy = yield strength of reinforcement (ksi) (The strength of reinforcing steel crossing the interface shall not be taken as greater than 60.0 ksi)

c = cohesion factor specified in S5.8.4.3 (ksi)

μ = friction factor specified in S5.8.4.3

Pc = permanent net compressive force normal to the shear plane (kips)

(If force is tensile, Pc = 0.0 k)

Calculate the nominal shear resistance per unit length of beam.

For cast-in-place concrete slab on clean concrete girder surfaces, free of laitance with surface roughened to an amplitude of 0.25 in.:

c = 0.28 ksi (S5.8.4.3)

μ = 1

K1 = 0.3

K2 = 1.8 ksi

Ignore compression on the interface from loads on the deck: Pc = 0

Acv = 42 in^2/in. of beam length

Avf = 0.044 in^2/in. of beam length

fy = 60 ksi

Therefore,

Vni = 0.28 x42 + 1x [0.044x 60 + 0]
= 14.4 k/in. of beam length

3.4. Computer Applications for the Design of PSC I-Girder Bridge Superstructure in AASHTO LRFD

The Computer Applications for the design of PSC I-Girder Bridge Superstructure in AASHTO LRFD, BS Eurocode II and IRC 112/IRC 6 are described in detail in the softcopy volume 'User's Guide for Hands on Practice with Computer Applications' for the book 'Computer Aided Bridge Engineering' is available for download from web site 'www.roadbridgedesign.com'.

References

ASTRA Pro software, User's Guide, Tutorial Videos, *User's Manual and Design Manual*, Techsoft Engineering Services, Web site: www.techsoftglobal.com and www.roadbridgedesign.com.

Bridge Engineering Handbook, Second Edition, Superstructure Design, Edited by Wai Fah Chen and Lian Duan, Published by CRC Press, Taylor & Francis Group, 6000 Broken Sound Parkway, NW, Suite 300, Boca Raton, FL 33487, Web site: www.crcpress.com

Design Specifications, Customary U.S. Units, Sixth Edition 2012, American Association of State Highway and Transportation Officials. 444 North Capitol Street, NW, Suite 249, Washington, DC 20001, Phone 202-624-5800 / Fax 202-624-5806, Web site: www.transportation.org

Chapter 4

DESIGN OF BRIDGE ABUTMENTS

ABSTRACT

This chapter describes the step wise design procedure for integral abutment with structural reinforcement details by describing the general conditions and common practices, design criteria, bridge length limits, soil conditions, skew angle, alignment and geometry, grade, girder details, arrangement of piles under pile cap, dynamic loads allowance in pile design, construction sequence, negative moment connection between the integral abutment and the superstructure, wing walls, approach slab, expansion joints and bearing pads.

Note: The various formulas used in the design may be seen from the design Excel Worksheet of software ASTRA Pro by downloading from web site www.techsoftglobal.com under 'downloads'.

In the Excel Worksheets the values in 'Red Color', are Design Input Data which may be changed by the User, but in Tab Pages the 'Red and Blue colored' values are either taken from other pages or calculated values and should not be changed by the user. For any query write to techsoftinfra@gmail.com.

4.1. GENERAL

4.1.1. Integral Abutment Design in AASHTO-LRFD Design Step 7.1

4.1.1.1. General Considerations and Common Practices

Integral abutments are used to eliminate expansion joints at the end of a bridge. They often result in "jointless bridges" and serve to accomplish the following desirable objectives:

- Long-term serviceability of the structure
- Minimal maintenance requirements
- Economical construction
- Improved aesthetics and safety considerations

A jointless bridge concept is defined as any design procedure that attempts to achieve the goals listed above by eliminating as many expansion joints as possible. The ideal jointless bridge, for example, contains no expansion joints in the superstructure, substructure or deck.

Integral abutments are generally founded on one row of piles made of steel or concrete. The use of one row of piles reduces the stiffness of the abutment and allows the abutment to translate parallel to the longitudinal axis of the bridge. This permits the elimination of expansion joints and movable bearings. Because the earth pressure on the two end abutments is resisted by compression in the superstructure, the piles supporting the integral abutments, unlike the piles supporting conventional abutments, do not need to be designed to resist the earth loads on the abutments.

When expansion joints are completely eliminated from a bridge, thermal stresses must be relieved or accounted for in some manner. The integral abutment bridge concept is based on the assumption that due to the flexibility of the piles, thermal stresses are transferred to the substructure by way of a rigid connection, i.e., the uniform temperature change causes the abutment to translate without rotation. The concrete abutment contains sufficient bulk to be considered as a rigid mass. A positive connection to the girders is generally provided by encasing girder ends in the reinforced concrete back wall. This provides for full transfer of forces due to thermal movements and live load rotational displacement experienced by the abutment piles.

4.1.1.2. Design Criteria

Neither the AASHTO-LRFD specifications nor the AASHTO-standard specifications contain detailed design criteria for integral abutments. In the absence of universally accepted design criteria, many states have developed their own design guidelines. These guidelines have evolved over time and rely heavily on past experience with integral abutments at a specific area. There are currently two distinctive approaches used to design integral abutments:

One group of states design the piles of an integral abutment to resist only gravity loads applied to the abutment. No consideration is given to the effect of the horizontal displacement of the abutment on the pile loads and/or pile resistance. This approach is simple and has been used successfully. When the bridge is outside a certain range set by the state, e.g., long bridges, other considerations are taken into account in the design.

The second approach accounts for effects of different loads, in additional to gravity loads, when calculating pile loads. It also takes into account the effect of the horizontal movements on the pile load resistance. One state that has detailed design procedures following this approach is Pennsylvania.

The following discussion does not follow the practices of a specific state; it provides a general overview of the current state-of-practice.

4.1.1.3. Bridge Length Limits

Most states set a limit on the bridge length of jointless bridges beyond which the bridge is not considered a "typical bridge" and more detailed analysis is taken into account. Typically, the bridge length is based on assuming that the total increase of the bridge length under uniform temperature change from the extreme low to the extreme high temperature is 4 inches. This means that the movement at the top of the pile at each end is 2 inches or, when the bridge is constructed at the median temperature, a 1 inch displacement in either direction. This results in a maximum bridge length of 600 ft for concrete bridges and 400 ft. for steel bridges at locations where the climate is defined as "Moderate" in accordance to S3.12.2.1. The maximum length is shorter for regions defined as having a "cold" climate.

4.1.1.4. Soil Conditions

The above length limits assume that the soil conditions at the bridge location and behind the abutment are such that the abutment may translate with relatively low soil resistance. Therefore, most jurisdictions specify select granular fill for use behind integral abutments. In addition, the fill within a few feet behind the integral abutment is typically lightly compacted using a vibratory plate compactor (jumping jack). When bedrock, stiff soil and/or boulders exist in the top layer of the soil (approximately the top 12 to 15 ft.), it is typically required that oversized holes be drilled to a depth of approximately 15 ft.; the piles are then installed in the oversized holes. Subsequently, the holes are filled with sand. This procedure is intended to allow the piles to translate with minimal resistance.

4.1.1.5. Skew Angle

Earth pressure acts in a direction perpendicular to the abutments. For skewed bridges, the earth pressure forces on the two abutments produce a torque that causes the bridge to twist in plan. Limiting the skew angle reduces this effect. For skewed, continuous bridges, the twisting torque also results in additional forces acting on intermediate bents. In addition, sharp skews are suspected to have caused cracking in some abutment back-walls due to rotation and thermal movements. This cracking may be reduced or eliminated by limiting the skew. Limiting the skew will also reduce or eliminate design uncertainties, backfill compaction difficulty and the additional design and details that would need to be worked out for the abutment U-wing walls and approach slab.

Currently, there are no universally accepted limits on the degree of skew for integral abutment bridges.

4.1.1.6. Horizontal Alignment and Bridge Plan Geometry

With relatively few exceptions, integral abutments are typically used for straight bridges. For curved superstructures, the effect of the compression force resulting from the earth pressure on the abutment is a cause for concern. For bridges with variable width, the difference in the length of the abutments results in unbalanced earth pressure forces if the two abutments are to move the same distance. To maintain force equilibrium, it is expected that the shorter abutment will deflect more than the longer one. This difference should be considered when determining the actual expected movement of the two abutments as well as in the design of the piles and the expansion joints at the end of the approach slabs (if used).

4.1.1.7. Vertical Grade

Some jurisdictions impose a limit on the maximum vertical grade between abutments. These limits are intended to reduce the effect of the abutment earth pressure forces on the abutment vertical reactions.

4.1.1.8. Girder Types, Maximum Depth and Placement

Integral abutments have been used for bridges with steel I-beams, concrete I-beams, concrete bulb tees and concrete spread box beams.

Deeper abutments are subjected to larger earth pressure forces and, therefore, less flexible. Girder depth limits have been imposed by some jurisdictions based on past successful practices and are meant to ensure a reasonable level of abutment flexibility. Soil conditions and the length of the bridge should be considered when determining maximum depth limits. A maximum girder depth of 6 ft. has been used in the past. Deeper girders may be allowed when the soil conditions are favorable and the total length of the bridge is relatively short.

4.1.1.9. Type and Orientation of Piles

Integral abutments have been constructed using steel H-piles, concrete-filled steel pipe piles and reinforced and pre-stressed concrete piles. For H-piles, there is no commonly used orientation of the piles. In the past, H-piles have been placed both with their strong axis parallel to the girder's longitudinal axis and in the perpendicular direction. Both orientations provide satisfactory results.

4.1.1.10. Consideration of Dynamic Load Allowance in Pile Design

Traditionally, dynamic load allowance is not considered in foundation design. However, for integral abutment piles, it may be argued that the dynamic load allowance should be considered in the design of the top portion of the pile. The rationale for this requirement is that the piles are almost attached to the superstructure, therefore, the top portions of the piles do no benefit from the damping effect of the soil.

4.1.1.11. Construction Sequence

Typically, the connection between the girders and the integral abutment is made after the deck is poured. The end portion of the deck and the back wall of the abutment are usually poured at the same time. This sequence is intended to allow the dead load rotation of the girder ends to take place without transferring these rotations to the piles.

Two integral abutment construction sequences have been used in the past:

- One-stage construction: In this construction sequence, two piles are placed adjacent to each girder, one pile on each side of the girder. A steel angle is connected to the two piles and the girder is seated on the steel angle. The abutment pier cap (the portion below the bottom of the beam) and the end diaphragm or back wall (the portion encasing the ends of the beams) are poured at the same time. The abutment is typically poured at the time the deck in the end span is poured.
- Two-stage construction: Stage 1: A pile cap supported on one row of vertical piles is constructed. The piles do not have to line up with the girders. The top of the pile cap reaches the bottom of the bearing pads under the girders. The top of the pile cap is required to be smooth in the area directly under the girders and a strip approximately 4 in. wide around this area. Other areas are typically roughened (i.e., rake finished).

  Stage 2: After pouring the entire deck slab, except for the portions of the deck immediately adjacent to the integral abutment (approximately the end 4 ft. of the deck from the front face of the abutment) the end diaphragm (back wall) encasing the ends of the bridge girders is poured. The end portion of the deck is poured simultaneously with the end diaphragm.

4.1.1.12. Negative Moment Connection between the Integral Abutment and the Superstructure

The rigid connection between the superstructure and the integral abutment results in the development of negative moments at this location. Some early integral abutments showed signs of deck cracking parallel to the integral abutments in the end section of the deck due to the lack of proper reinforcement to resist this moment. This cracking was prevented by specifying additional reinforcement connecting the deck to the back (fill) face of the abutment. This reinforcement may be designed to resist the maximum moment that may be transferred from the integral abutment to the superstructure. This moment is taken equal to the sum of the plastic moments of the integral abutment piles. The section depth used to design these bars may be taken equal to the girder depth plus the deck thickness. The length of the bars extending into the deck is typically specified by the bridge owner. This length is based on the length required for the superstructure dead load positive moment to overcome the connection negative moment.

4.1.1.13. Wing Walls

Typically, U-wing walls (wing walls parallel to the longitudinal axis of the bridge) are used in conjunction with integral abutments. A chamfer (typically 1 ft.) is used between the abutment and the wing walls to minimize concrete shrinkage cracking caused by the abrupt change in thickness at the connection.

4.1.1.14. Approach Slab

Bridges with integral abutments were constructed in the past with and without approach slabs. Typically, bridges without approach slabs are located on secondary roads that have asphalt pavements. Traffic and seasonal movements of the integral abutments cause the fill behind the abutment to shift and to self compact. This often caused settlement of the pavement directly adjacent to the abutment.

Providing a reinforced concrete approach slab tied to the bridge deck moves the expansion joint away from the end of the bridge. In addition, the approach slab bridges cover the area where the fill behind the abutment settles due to traffic compaction and movements of the abutment. It also prevents undermining of the abutments due to drainage at the bridge ends. Typically, approach slabs are cast on polyethylene sheets to minimize the friction under the approach slab when the abutment moves.

The approach slab typically rests on the abutment at one end and on a sleeper slab at the other. The approach slab differs from typical roadway pavement since the soil under the approach slab is more likely to settle unevenly resulting in the approach slab bridging a longer length than expected for roadway pavement. Typically, the soil support under the approach slab is ignored in the design and the approach slab is designed as a one-way slab bridging the length between the integral abutment and the sleeper slab. The required length of the approach slab depends on the total depth of the integral abutment.

The sleeper slab should be placed outside the area where the soil is expected to be affected by the movement of the integral abutment. This distance is a function of the type of fill and the degree of compaction.

Due to the difference in stiffness between the superstructure and the approach slab, the interface between the integral abutment and the approach slab should preferably allow the approach slab to rotate freely at the end connected to the abutment. The reinforcement bars connecting the abutment to the approach slab should be placed such that the rotational restraint provided by these bars is minimized.

A contraction joint is placed at the interface between the approach slab and the integral abutment. The contraction joint at this location provides a controlled crack location rather than allowing a random crack pattern to develop.

4.1.1.15. Expansion Joints

Typically, no expansion joints are provided at the interface between the approach slab and the roadway pavement when the bridge total length is relatively small and the roadway uses flexible pavement. For other cases, an expansion joint is typically used.

4.1.1.16. Bearing Pads

Plain elastomeric bearing pads are placed under all girders when the integral abutment is constructed using the two-stage sequence described above. The bearing pads are intended to act as leveling pads and typically vary from ½ to ¾ in. thick. The pad length parallel to the girder's longitudinal axis varies depending on the bridge owner's specifications and the pad length in the perpendicular direction varies depending on the width of the girder bottom flange and the owner's specifications. It is recommended to block the area under the girders that is not in contact with the bearing pads using backer rods. Blocking this area is intended to prevent honeycombing of the surrounding concrete. Honeycombing will take place when the cement paste enters the gap between the bottom of girder and the top of the pile cap in the area under the girders not in contact with the bearing pads.

4.1.1.17. AASHTO-LRFD Design Step 7.1.1 - Gravity Loads

Interior girder: un-factored loads (See AASHTO LRFD Table 4.3-3 for girder end shears)

Non-composite:

Girder = 61.6 k
Slab and haunch = 62.2 k
Exterior diaphragm = 2.5 k
Total NC = 126.3 k

Composite:

Parapets= 8.9 k
Future wearing surface = 12 k

Live load:

Maximum truck per lane (without impact or distribution factors) = 64.42 k
Minimum truck per lane (without impact or distribution factors) = -6.68 k
Maximum lane per lane = 30.81 k
Minimum lane per lane = -4.39 k

Exterior girder: un-factored loads
(See AASHTO LRFD Table 4.3-7 for girder end shears)
Non-composite:

Girder = 61.6 k
Slab and haunch = 55.1 k
Exterior diaphragm = 1.3 k
Total NC = 118 k

Composite:

Parapets = 8.9 k
Future wearing surface = 8.1 k

Live load:

Maximum truck per lane (without impact or distribution factors) = 64.42 k
Minimum truck per lane (without impact or distribution factors) = -6.68 k
Maximum lane per lane = 30.81 k
Minimum lane per lane = -4.39 k

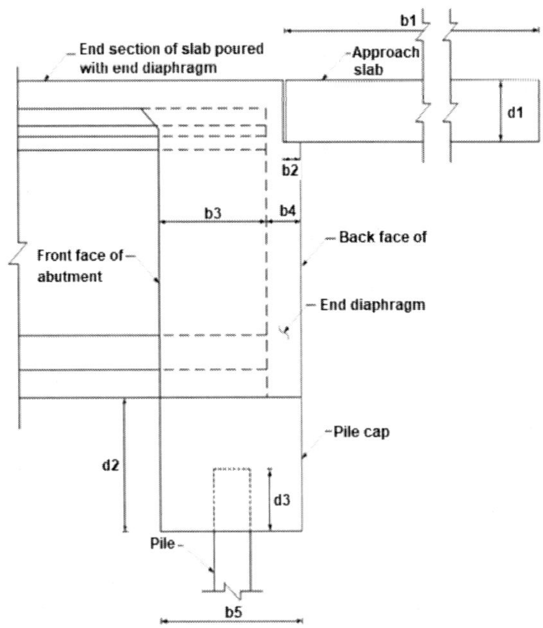

Figure 4.1. General view of an integral abutment showing dimensions used for the example. Source: Figure 7.1-1 from AASHTO LRFD Bridge Design Specifications, FHWA Office of Bridges and Structures, 1200 New Jersey Ave., SE, Washington, DC 20590.

b1 = 25ft 300 in

b2 = 0.5ft 6in

b3 = 2ft 24in

b4 = 1ft 12in

b5 = 3ft 36in

d1 = 1.5ft 18 in

d2 = 3.25 ft 39 in

d3 = 1.5ft 18 in

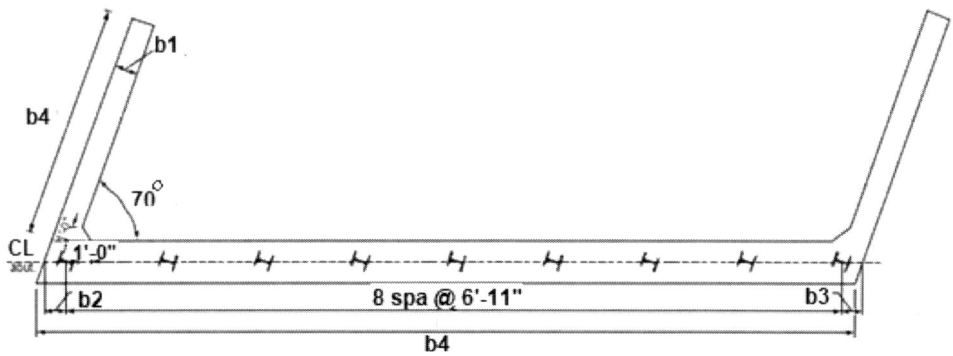

Figure 4.2. Plan view of the integral abutment. Source: Figure 7.1-2 from AASHTO LRFD Bridge Design Specifications, FHWA Office of Bridges and Structures, 1200 New Jersey Ave., SE, Washington, DC 20590.

b1 = 1.6875 ft 20.25 in

b2 = 1.792 ft 21.504 in

b3 = 1.792 ft 21.504 in

b4 = 58.93 ft 707.16 in

d1 = 15 ft 180 in

b1 = 14 ft 168 in

b2 = 1 ft 12 in

b3 = 3 ft 36 in

d1 = 2 ft 24 in

d2 = 8.3125 ft 99.75 in

d3 = 7.0625 ft 84.75 in

d4 = 3.25 ft 39 in

d5 = 10.3125 ft 123.75 in

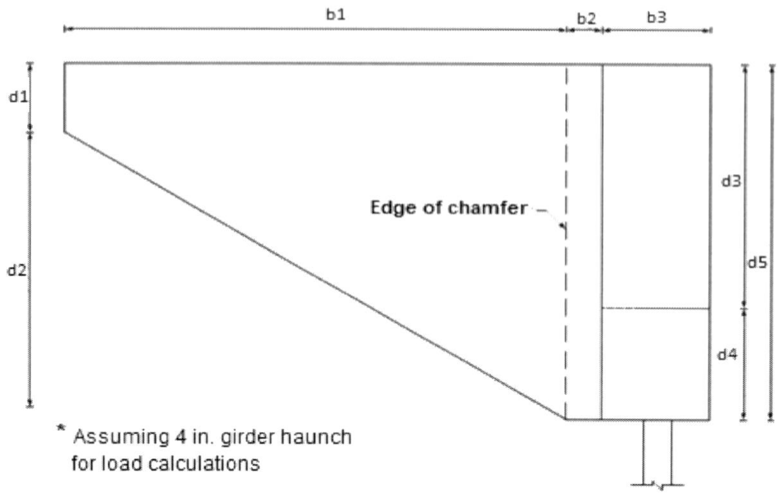

Figure 4.3. Elevation view of integral abutment and tapered wing wall. Source: Figure 7.1-3 from AASHTO LRFD Bridge Design Specifications, FHWA Office of Bridges and Structures, 1200 New Jersey Ave., SE, Washington, DC 20590.

In the next section, "w" and "P" denote the load per unit length and the total load, respectively. The subscripts denote the substructure component. Dimensions for each component are given in Figures 4.1 (AASHTO 7.1-1) through 4.3 (AASHTO 7.1-3).

Pile cap: un-factored loading

Pile cap length along the skew = 55.354/cos 20

= 58.91

$W_{cap} - 3.25X(3)X(0.15)$

= 1.46 k/ft

OR'

$P_{cap} = 1.46X(58.91)$

= 86.01 k

Concrete weight from the end diaphragm (approximate, girder volume not removed):

Un-factored loading

Assuming bearing pad thickness of ¾ in., girder height of 72 in., haunch thickness of 4 in., and deck thickness of 8 in.:

Pad thickness = 0.75 in

Girder height = 72 in

Haunch thickness = 4 in

Deck thickness = 8 in

$W_{end\ dia}= 3((0.75 +72 +4+8)/12)(0.15)$

= 3.18 k/ft

OR'

$P_{end\ dia}$= 3.18X(58.91)

= 187.33 k

Wing wall: un-factored load

A_{wing} = (123.75/12)X(15) -½ (14)X(99.75/12) = 96.5 ft2

Wing wall thickness= parapet thickness at the base = 20.25 in. (given in Section 4)

Wing wall weight= 96.5X(20.25/12)X(0.15) = 24.43 k

Chamfer weight= (123.75/12)X(1)X(1)X(0.15)/2 = 0.77 k

Notice that the chamfer weight is insignificant and is not equal for the two sides of the bridge due to the skew. For simplicity, it was calculated based on a right angle triangle and the same weight is used for both sides.

Weight of two wingwalls plus chamfer= 2X(24.43 + 0.77) = 50.4 k

Parapet weight = 0.65 k/ft (given in Section 4.2) Parapet length on wingwall and abutment

= 15 + 3/ sin 70 = 8.19 ft.

$P_{parapet}$= 2X(0.65)X(18.19)

= 23.65 k total weight

Approach slab load acting on the integral abutment: un-factored loading.

Approach slab length = 25 ft

Approach slab width between parapets = 58.91 – 2((20.25/12)/sin 70)

= 55.32 ft

Self weight of the approach slab:

$W_{approach\ slab}$ = ½ X(25)X(1.5)X(0.15)

= 2.81 k/ft OR'

$P_{approach\ slab}$ = 2.81X(55.32)

= 155.4 k

Future wearing surface acting on the approach slab (assuming 25 psf):

W_{FWS} = ½ X(0.025)X(25)

= 0.31 k/ft

OR'

P_{FWS} = 0.31X55.32

= 17.15 k

Live load on the approach slab, reaction on integral abutment:

$P_{lane\ load}$ = ½ X(0.64)X(25) S3.6124
= 8 k (one lane)

Notice that one truck is allowed in each traffic lane and that the truck load is included in the girder reactions. Therefore, no trucks were assumed to exist on the approach slab and only the uniform load was considered.

4.1.1.18. AASHTO-LRFD Design Step 7.1.2 - Pile Cap Design

The girder reactions, interior and exterior, are required for the design of the abutment pile cap. Notice that neither the piles nor the abutment beam are infinitely rigid. Therefore, loads on the piles due to live loads are affected by the location of the live load across the width of the integral abutment. Moving the live load reaction across the integral abutment and trying to maximize the load on a specific pile by changing the number of loaded traffic lanes is not typically done when designing integral abutments. As a simplification, the live load is assumed to exist on all traffic lanes and is distributed equally to all girders in the bridge cross section. The sum of all dead and live loads on the abutment is then distributed equally to all piles supporting the abutment.

The maximum number of traffic lanes allowed on the bridge based on the available width (52 ft. between gutter lines) is:

N_{lanes} = 52 ft./12 ft.
Per lane = 4.33 say 4 lanes
Number of Girders = 6

Factored dead load plus live load reactions for one interior girder, Strength I limit state controls (assume the abutment is poured in two stages as discussed earlier):

Maximum reaction Stage I:
$P_{SI(I)}$= 1.25 X(girder + slab + haunch)
= 1.25 x 126.3
= 158 k

Notice that construction loads should be added to the above reaction if construction equipment is allowed on the bridge before pouring the back wall (Stage II).

Maximum reaction for Final Stage:

Including the dynamic load allowance (for design of the pile cap top portion of the piles):

$P_{FNL(I)}$= 1.25 x (DC) + 1.50 x (DW) + 1.75 x (LL + IM) x (Nlanes)/Ngirders = 1.25X(126.3 + 8.9) +1.5X(12) +1.75X(1.33X(64.42) +30.81)X4/6 = 323 k

Without the dynamic load allowance (for design of the lower portion of the piles):

$P_{FNL(I)}$ = 298.3 k

Factored dead load plus live load reactions for one exterior girder, Strength I limit state controls:

Maximum reaction Stage I:

$P_{SI(E)}$ = 1.25X118
= 147.5 k

Notice that construction loads should be added to the above reaction.

Maximum reaction for final stage:

Including the dynamic load allowance:

$P_{FNL(E)}$ = 1.25 x (DC) + 1.50 x (DW) + 1.75 x (LL + IM) x (Nlanes)/Ngirders = 1.25 x (118 +8.9) + 1.5 x (8.1) +1.75 x (1.33 x (64.42) +30.81) x4/6 = 306.7 k

Without the dynamic load allowance:

$P_{FNL(E)}$= 281.8 k

4.1.1.19. AASHTO-LRFD Design Step 7.1.3 Design of Piles

Typically, integral abutments may be supported on end bearing piles or friction piles. Reinforced and pre-stressed concrete piles, concrete-filled steel pipe piles or steel H-piles may be used. Steel H-piles will be used in this example.

Typically, the minimum distance between the piles and the end of the abutment, measured along the skew, is taken as 1'-6" and the maximum distance is usually 2'-6". These distances may vary from one jurisdiction to another. The piles are assumed to be embedded 1'-6" into the abutment. Maximum pile spacing is assumed to be 10 ft. The minimum pile spacing requirements of S10.7.1.5 shall apply.

- From S10.7.1.5, the center-to-center pile spacing shall not be less than the greater of 30.0 in. or 2.5 pile diameters (or widths). The edge distance from the side of any pile to the nearest edge of the footing shall be greater than 9.0 in.
- According to S10.7.1.5, where a reinforced concrete beam is cast-in-place and used as a bent cap supported by piles, the concrete cover at the sides of the piles

shall be greater than 6.0 in., plus an allowance for permissible pile misalignment, and the piles shall project at least 6.0 in. into the cap. This provision is specifically for bent caps, therefore, keep 1'-6" pile projection for integral abutment to allow the development of moments in the piles due to movements of the abutment without distressing the surrounding concrete.

From Figure 4.1 (AASHTO 7.1-2), steel H-piles are shown to be driven with their weak axis perpendicular to the centerline of the beams. As discussed earlier, piles were also successfully driven with their strong axis perpendicular to the centerline of the beams in the past.

According to S10.7.4.1, the structural design of driven concrete, steel, and timber piles must be in accordance with the provisions of Sections S5, S6, and S8 respectively. Articles S5.7.4, S5.13.4, S6.15, S8.4.13, and S8.5.2.2 contain specific provisions for concrete, steel, and wood piles. Design of piles supporting axial load only requires an allowance for unintended eccentricity. For the steel H-piles used in this example, this has been accounted for by the resistance factors in S6.5.4.2 for steel piles.

General Pile Design

As indicated earlier, piles in this example are designed for gravity loads only.

Generally, the design of the piles is controlled by the minimum capacity as determined for the following cases:

- Case A - Capacity of the pile as a structural member according to the procedures outlined in S6.15. The design for combined moment and axial force will be based on an analysis that takes the effect of the soil into account.
- Case B - Capacity of the pile to transfer load to the ground.
- Case C - Capacity of the ground to support the load.

For piles on competent rock, only Case A needs to be investigated.

4.1.1.20. AASHTO-LRFD Design Step 7.1.3.1 Pile Compressive Resistance

Pile Compressive Resistance (S6.15 and S6.9.2)

The factored resistance of components in compression, Pr, is taken as:

$Pr = \phi Pn$ (S6.9.2.1-1)

where:

Pn = nominal compressive resistance specified in S6.9.4 and S6.9.5 (kip)

ϕc = resistance factor for axial compression, steel only as specified in S6.5.4.2

$\phi c = 0.5$ for H-piles assuming severe driving conditions

Check the width/thickness requirements per S6.9.4.2. Assume HP12x53 piles.
Slenderness of plates must satisfy:

$$\frac{b}{t} \leq k\sqrt{\frac{E}{F_y}} \qquad \text{(S6.9.4.2-1)}$$

where:

k =plate buckling coefficient as specified in AASHTO LRFD Table S6.9.4.2-1

k= 0.56 for flanges and projecting legs or plates

b =width of plate equals one-half of flange width as specified in AASHTO LRFD Table S6.9.4.2-1 (in.)

Flange width = 12.045 in

b = 12.045/2 = 6.02 in.

t = flange thickness (in.)

t = 0.435 in.

b/t = 6.02/0.435 = 13.8

E = 29000

fy = 36

$$k\sqrt{\frac{E}{F_y}} = 0.56\sqrt{\frac{29{,}000}{36}}$$

$$k\sqrt{\frac{E}{F_y}} = 15.9$$

15.9 > 13.8 OK

Therefore, use S6.9.4.1 to calculate the compressive resistance.

(Notice that the b/t ratio for the webs of HP sections is always within the limits of AASHTO LRFD Table S6.9.4.2-1 for webs and, therefore, need not be checked.)

For piles fully embedded in soil, the section is considered continuously braced and Eq. S6.9.4.1-1 is reduced to:

Pn = FyAs.

As = 15.5 in^2

Pn = 36X15.5

= 558 k

Therefore, the factored resistance of components in compression, Pr, is taken as:

$P_r = \varphi P_n \quad = 0.5x558 = 279$ k

The above capacity applies to the pile at its lower end where damage from driving may have taken place. At the top of the pile, higher resistance factors that do not account for damage may be used. For piles designed for gravity loads only, as in this example, the resistance at the lower end will always control due to the lower resistance factor regardless if the dynamic load allowance is considered in determining the load at the top of the pile or not (notice that the dynamic load allowance is not considered in determining the load at the bottom of the pile).

4.1.1.21. AASHTO-LRFD Design Step 7.1.3.2 Determine of Number of Piles

Determine the Number of Piles Required

Maximum total girder reactions for Stage I (detailed calculations of girder reactions shown earlier):

$P_{SI\ (Total)} = (2X147.5) + (4X158) = 927$ k

Maximum total girder reaction for final stage not including the dynamic load allowance (detailed calculations of girder reactions shown earlier):

$P_{FNL(Total)} = (2X281.8) + (4X298.3) = 1756.8$ k

Maximum factored DL + LL on the abutment, Strength I limit state controls:

PStr. I = PFNL (Total) + 1.25(DC) + 1.50(DW) + 1.75 (LLmax) (Nlanes)
= 1756.8 +1.25X(86.01+187.33+50.4+23.65+155.4)+1.5X(17.15)+1.75X(8)X(4)
= 2467 k

where:

"PFNL (Total)" is the total factored DL + LL reaction of the bridge girders on the abutment.

"DC" includes the weight of the pile cap, diaphragm, wing walls, approach slab and parapet on the wing walls.

"DW" includes the weight of the future wearing surface on the approach slab.

"LLmax" is the live load reaction from the approach slab transferred to the abutment (per lane)

"Nlanes" is the maximum number of traffic lanes that fit on the approach slab, 4 lanes.

Therefore, the number of piles required to resist the applied dead and live loads is:

Npiles = PStr. I/Pr
= 2467/279
= 8.84 piles, say 9 piles

4.1.1.22. AASHTO-LRFD Design Step 7.1.3.3 Pile Spacing

Pile Spacing

Total length of the pile cap = 58.91 ft.
Assume pile spacing is 6'-11" = 6.917 ft.
which provides more than the recommended edge distance of 1'-6" for the piles.

Pile end distance = (58.91 – (8X6.917))/2
= 1.8 ft (1'-9 ½ ")

4.1.1.23. AASHTO-LRFD Design Step 7.1.4 Back-Wall Design

The thickness of the abutment back-wall is taken to be 3 ft.

Design of the pier cap for gravity loads.

For an integral abutment constructed in two stages, the abutment is designed to resist gravity loads as follows:

- Case A - The first stage of the abutment, i.e., the part of the abutment below the bearing pads, is designed to resist the self weight of the abutment, including the diaphragm, plus the reaction of the girders due to the self weight of the girder plus the deck slab and haunch.
- Case B - The entire abutment beam, including the diaphragm, is designed under the effect of the full loads on the abutment.
  Instead of analyzing the abutment beam as a continuous beam supported on rigid supports at pile locations, the following simplification is common in conducting these calculations and is used in this example:
- Calculate moments assuming the abutment beam acting as a simple span between piles and then taking 80% of the simple span moment to account for the continuity. The location of the girder reaction is often assumed at the mid-span for moment calculations and near the end for shear calculations. This assumed position of the girders is meant to produce maximum possible load effects. Due

to the relatively large dimensions of the pile cap, the required reinforcement is typically light even with this conservative simplification.

Required Information:

Concrete compressive strength, $f'c = 3$ ksi
Reinforcing steel yield strength, $Fy = 60$ ksi
Pile spacing = 6.917 ft

Case A

The maximum factored load due to the girders and slab (from the interior girder):

$Pu = 1.5 \times 126.3$
$= 189.45$ k

Factored load due to the self weight of the pile cap and diaphragm:

$wu = 1.5X(1.46 + 3.18)$
$= 6.96$ k/ft

Notice that only dead loads exist at this stage. The 1.5 load factor in the above equations is for Strength III limit state, which does not include live loads.

Flexural Design for CASE A

The maximum positive moment, Mu, assuming a simple span girder, is at mid-span between piles. The simple span moments are reduced by 20% to account for continuity:

$Mu = Pul/4 + wul2/8$
$= 0.8 \times (189.45 \times 6.917/4 + (6.96 \times 6.917)\text{^}2/8) = 295.4$ k-ft

Determine the required reinforcing at the bottom of the pile cap.

$Mr = \phi Mn$ (S5.7.3.2.1-1)

The nominal flexural resistance, Mn, is calculated using Eq. (S5.7.3.2.2-1).

$Mn = Asfy(ds - a/2)$ (S5.7.3.2.2-1)

where:

As =area of non-pre-stressed tension reinforcement (in^2), notice that available room will only allow four bars, two on either side of the piles. Use 4 #8 bars.

= 4 X 0.79

= 3.16 in^2

fy = specified yield strength of reinforcing bars (ksi) fy = 60 ksi

ds = distance from the extreme compression fiber to the centroid of the non-pre-stressed tensile reinforcement (in.)

Depth of pile cap = 3.25 ft

Bottom cover = 3 in

Bar diameter = 1 in

= depth of pile cap – bottom cover – ½ diameter bar

ds = (3.25X12)-3 - ½X1

= 35.5 in.

A = cβ1, depth of the equivalent stress block (in.) = Asfy/0.85f'cb

(S5.7.3.1.1-4)

= (3.16 x60)/(0.85 x3 x3 x12)

= 2.07 in.

Mn = (3.16X60)X(35.5 – 2.07/2)/12 = 544.5 k-ft

Therefore,

Mr = 0.9 x 544.5

= 490 k-ft > 295.4 OK

Negative moment over the piles is taken equal to the positive moment. Use the same reinforcement at the top of the pile cap as determined for the bottom (4 #8 bars).

By inspection:

- Mr > 4/3(Mu). This means the minimum reinforcement requirements of S5.7.3.3.2 are satisfied.
- The depth of the compression block is small relative to the section effective depth. This means that the maximum reinforcement requirements of S5.7.3.3.1 are satisfied.

Shear Design for Case A

The maximum factored shear due to the construction loads assuming the simple span condition and girder reaction at the end of the span:

$Vu = Pu + wul/2$
$= 189.45 + (6.96 x6.917)/2$
$= 213.5$ k

The factored shear resistance, Vr, is calculated as:

$Vr = \phi Vn$ (S5.8.2.1-2)

The nominal shear resistance, Vn, is calculated according to S5.8.3.3 and is the lesser of:

$Vn = Vc + Vs$ (S5.8.3.3-1)

Or,

$Vn = 0.25f'cbvdv$ (S5.8.3.3-2)

where:

$Vc = 0.0316\beta\sqrt{f'c}bvdv$ (S5.8.3.3-3)

β = factor indicating ability of diagonally cracked concrete to transmit tension as specified in S5.8.3.4
β − 2
f'c = specified compressive strength of the concrete (ksi) f'c = 3.0 ksi
bv = effective shear width taken as the minimum web width within the depth dv as determined in S5.8.2.9 (in.)
bv = 36 in.
dv = effective shear depth as determined in S5.8.2.9 (in.)

S5.8.2.9 states that dv is not to be taken less than the greater of 0.9de or 0.72h

dv = ds– a/2
= 35.5- (2.07/2)
= 34.47 in.
0.9ds = 0.9X35.5
= 31.95 in
0.72h = 0.72X(3.25X12)
= 28.08 in

Therefore, dv should be taken as 34.47 in

Vc= (0.0316X2)X√3X(36X34.47)
= 135.8 k

Assuming shear reinforcement is #5 @ 10 in. spacing perpendicular to the pier cap longitudinal axis.

Vs = Avfydv/s (S5.8.3.3-4)

where:

Bar area = 0.31 in^2
Av = area of shear reinforcement within a distance "s" (in^2) = 2 legs(0.31)
= 0.62 in^2
s = spacing of stirrups (in.) = 10 in
Vs = (0.62x60x34.47)/10 = 128.2 k

The nominal shear resistance, Vn, is taken as the smaller of:

Vn = 135.8 +128.2
= 264 k
Or,
Vn = 0.25x3x36x34.47 = 930.7 k

Therefore, use the shear resistance due to the concrete and transverse steel reinforcement.

Vr= ϕVn
= 0.9 x264
= 237.6 k > 213.5 OK

Case B

The maximum factored load due to all applied dead and live loads which include the approach slab, live load on approach slab, etc. The load due to the wing walls is not included since its load minimally affects the responses at the locations where girder reactions are applied.

Point load:

P_{Str-I} = maximum factored girder reaction calculated earlier

$P_{Str\text{-}I} = 323$ K

Notice that the 323 k assumes that the live load is distributed equally to all girders. This approximation is acceptable since this load is assumed to be applied at the critical location for moment and shear. Alternately, the maximum reaction from the tables in Section 5.3 may be used.

Distributed load:

W_{Str}-I = 1.25 x (cap self wt. + end diaph. + approach slab) + 1.5 x (approach FWS) + 1.75 x (approach slab lane load) x (Nlanes)/Labutment
= 1.25x(1.46+ 3.18 + 2.81) +(1.5x0.31) + (1.75x8x4)/58.93 = 10.73 k/ft

Flexural Design for CASE B

The maximum positive moment is calculated assuming the girder reaction is applied at the midspan between piles and taking 80% of the simple span moment.

Mu = 0.8x[((323x6.917)/4) +((10.73x(6.917)^2)/8)] = 498.2 k-ft

Determine the required reinforcing at the bottom of the pile cap.

Mr = ϕMn (S5.7.3.2.1-1)

Mn = As x fy x (ds – a/2) (S5.7.3.2.2-1)

where:

As = use 4 #8 bars
= 4x0.79
= 3.16 in^2
fy = 60 ksi

Total depth of int. abut. (no haunch) = 119.75 in

ds = total depth of int. abut. (no haunch) – bottom cover – ½ bar diameter
= 119.75 –3– ½x1
= 116.25 in

a= Asfy/0.85f'cb (S5.7.3.1.1-4)

= (3.16x60)/(0.85x3x(3x12)) = 2.07 in.

$Mn = (3.16x60)x(116.25 – 2.07/2)/12 = 1820$ k-ft

Therefore,
$Mr = 0.9x1820$
$= 1638$ k-ft > 498.2 OK

Negative moment over the piles is taken equal to the positive moment. Use the same reinforcement at the top of the abutment beam as determined for the bottom (4 #8 bars).

By inspection:

- Mr > 4/3(Mu).
- The depth of the compression block is small relative to the section effective depth.

Shear Design for CASE B

Assume the girder reaction is adjacent to the pile.
The maximum factored shear due to all applied loading:

$Vu= Pu + wul/2$
$= 323+ 10.73x6.917/2$
$= 360.1$ K

The factored shear resistance, Vr, is calculated as:

$$Vr = \phi Vn \qquad \text{(S5.8.2.1-2)}$$

The nominal shear resistance, Vn, is calculated according to S5.8.3.3 and is the lesser of:

$$Vn = Vc + Vs \qquad \text{(S5.8.3.3-1)}$$

Or,

$$Vn = 0.25 \times f'c \times bv \times dv \qquad \text{(S5.8.3.3-2)}$$

where:

$$Vc = 0.0316\, \beta \times \sqrt{f'c} \times bv \times dv \qquad \text{(S5.8.3.3-3)}$$

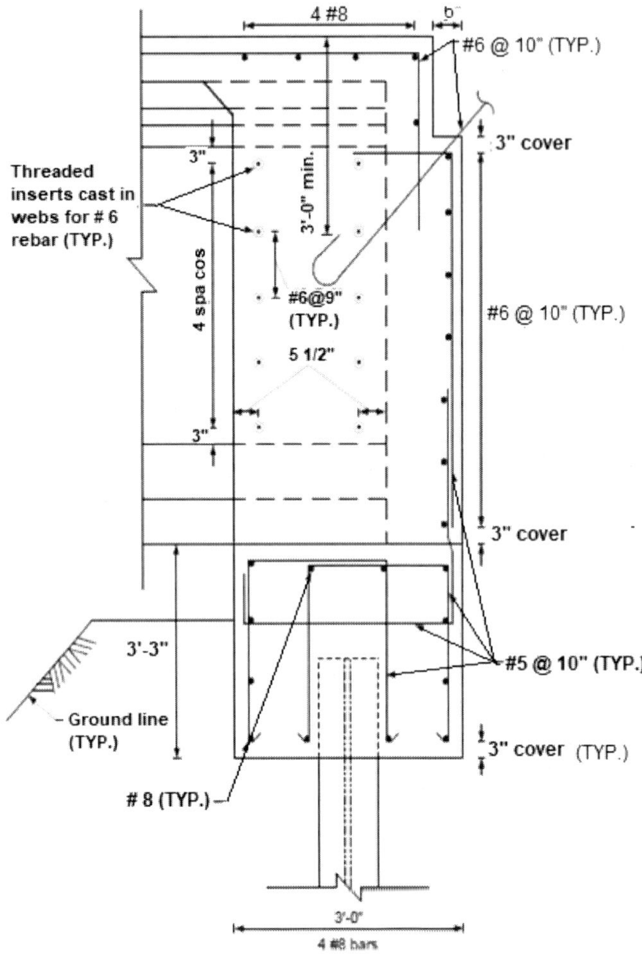

Figure 4.4. Integral abutment reinforcement, girder and pile exist at the same section. Source: Figure 7.1-4 from AASHTO LRFD Bridge Design Specifications, FHWA Office of Bridges and Structures, 1200 New Jersey Ave., SE, Washington, DC 20590.

$\beta = 2$

f'c = 3 ksi

bv= 36 in.

de = 116.25 in. (calculated earlier)

dv = de – a/2

dv = 116.25 – (2.07/2) = 115.2 in.

0.9 x de = 0.9x116.25 = 104.6 in.

0.72 x h = 0.72x119.75 = 86.22 in.

Therefore, dv should be taken as 115.2 in.

The nominal shear resistance, Vn, is taken as the lesser of:

$Vc = 0.0316x2x(\sqrt{3})x36x115.2$
$= 454$ K

Notice that Vc is large enough, relative to the applied load, that the contribution of the transverse shear reinforcement, Vs, is not needed.

Or,

$Vn = 0.25x3x36x115.2$
$= 3110.4$ K

Therefore, use the shear resistance due to the concrete, Vc
$Vr = \phi Vn$
$= 0.9x454$
$= 408.6 \text{ K} > 360.1$ OK

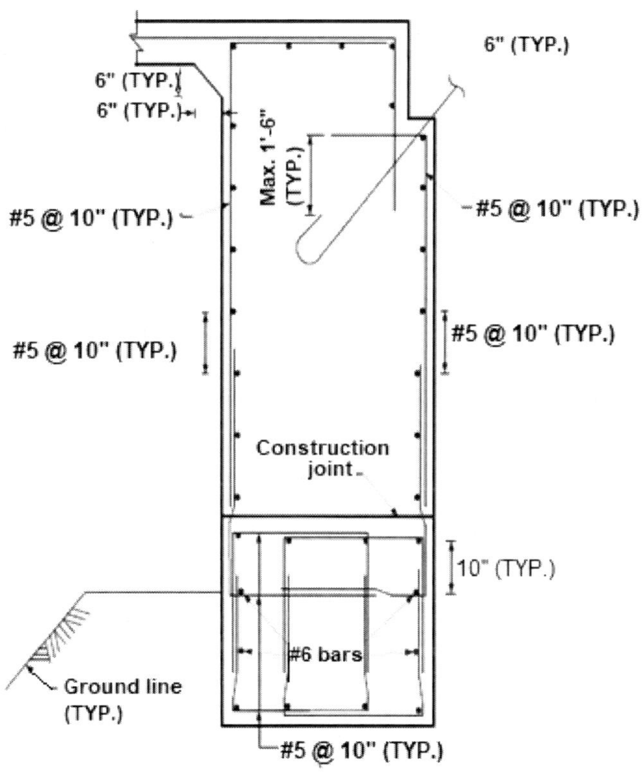

Figure 4.5. Integral abutment reinforcement, no girder and no pile at the section. Source: Figure 7.1-5 from AASHTO LRFD Bridge Design Specifications, FHWA Office of Bridges and Structures, 1200 New Jersey Ave., SE, Washington, DC 20590.

Typical reinforcement details of the abutment beam are shown in Figures Figure 4.4 (AASHTO-LRFD Figure 7.1-4) through Figure 4.7 (AASHTO-LRFD Figure 7.1-7).

Notice that bar shapes vary depending on the presence of girders and/or piles at the section.

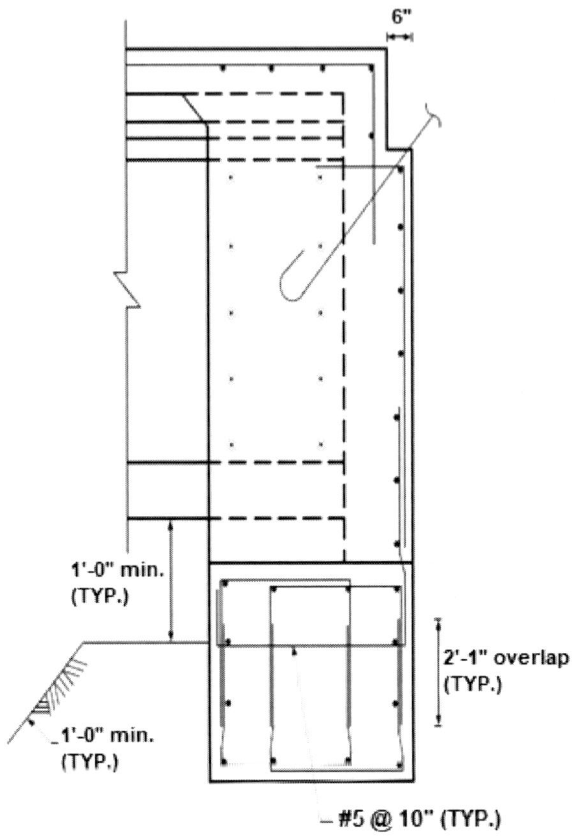

Figure 4.6. Integral abutment reinforcement, girder, no pile at the section. Source: Figure 7.1-6 from AASHTO LRFD Bridge Design Specifications, FHWA Office of Bridges and Structures, 1200 New Jersey Ave., SE, Washington, DC 20590.

Calculate the adequacy of the back wall to resist passive pressure due to the abutment backfill material.

Passive earth pressure coefficient, kp = (1 + sin φ)/(1 – sin φ)

(Notice that kp may also be obtained from AASHTO LRFD Figure S3.11.5.4-1)

wp = ½ γz2kp (S3.11.5.1-1)

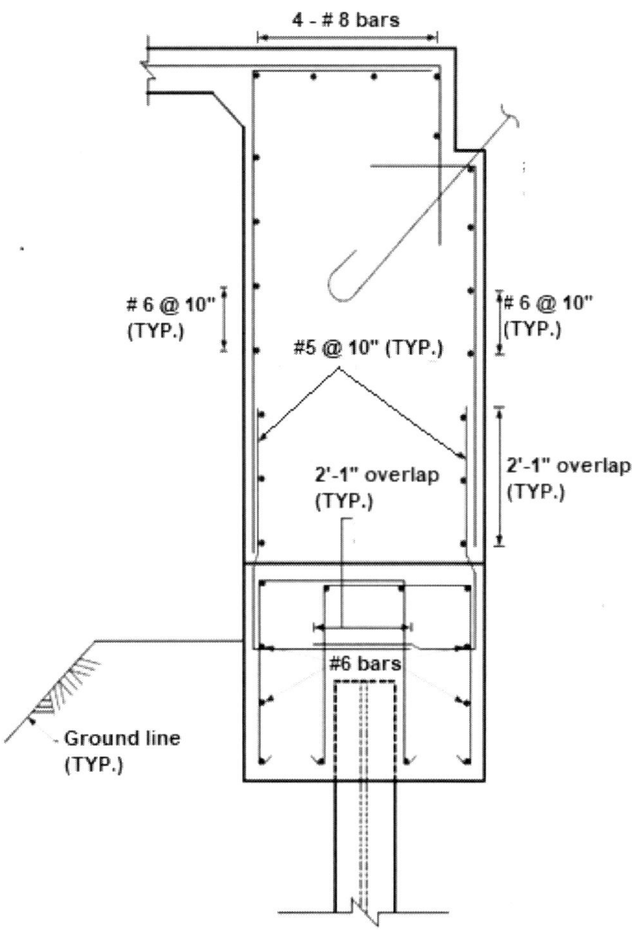

Figure 4.7. Integral abutment reinforcement, pile without girder. Source: Figure 7.1-7 from AASHTO LRFD Bridge Design Specifications, FHWA Office of Bridges and Structures, 1200 New Jersey Ave., SE, Washington, DC 20590.

where:

wp = passive earth pressure per unit length of back wall (k/ft)

γ = unit weight of soil bearing on the back wall (kcf) = 0.13 kcf

Z = height of the back wall from the bottom of the approach slab to the bottom of the pile cap (ft.)

= slab + haunch + girder depth + bearing pad thickness + pile cap depth – approach slab thickness

= (8/12) + (4/12) +6 + (0.75/12) +3.25 – 1.5 = 8.81 ft.

φ = internal friction of backfill soil assumed to be 30°

wp = ½ X(0.13)X(8.81)^2[(1 + sin 30)/(1 – sin 30)] = 15.1 k/ft of wall

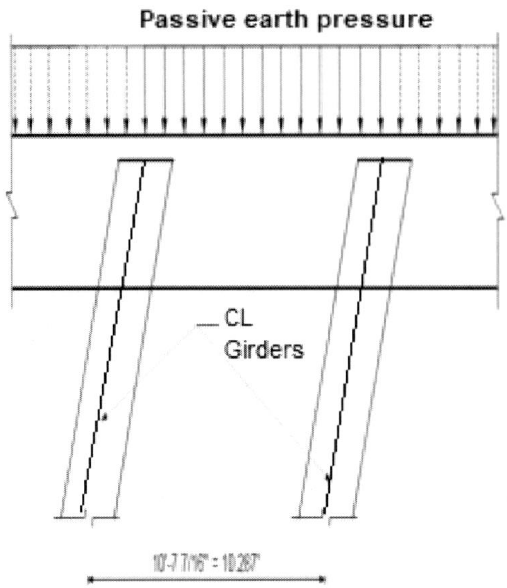

Figure 4.8. Passive earth pressure applied to back wall. Source: Figure 7.1-8 from AASHTO LRFD Bridge Design Specifications, FHWA Office of Bridges and Structures, 1200 New Jersey Ave., SE, Washington, DC 20590.

Notice that developing full passive earth pressure requires relatively large displacement of the structure (0.01 to 0.04 of the height of the structure for cohesion-less fill). The expected displacement of the abutment is typically less than that required to develop full passive pressure.

However, these calculations are typically not critical since using full passive pressure is not expected to place high demand on the structure or cause congestion of reinforcement.

No load factor for passive earth pressure is specified in the LRFD specifications. Assume the load factor is equal to that of the active earth pressure ($\phi = 1.5$).

$\phi = 1.5$
wu = ϕEHwp = 1.5X15.1
= 22.65 k/ft of wall

The back wall acts as a continuous horizontal beam supported on the girders, i.e., with spans equal to the girder spacing along the skew.

Mu ≅ wul2/8
= 22.65X(9.667/cos 20)X2/8 = 300 k-ft/ft

Calculate the nominal flexural resistance, Mr, of the back wall.

$Mr = \phi Mn$ (S5.7.3.2.1-1)

and

$Mn = Asfy(ds - a/2)$ (S5.7.3.2.2-1)

where:

As = area of the longitudinal reinforcement bars at front face (tension side) of the abutment (9 #6 bars)
= 9X0.44 = 3.96 in^2
Fy = 60 ksi
Width of Back wall = 3 ft
Concrete Cover= 3 ft
Vertical Bar Dia = 0.625 ft
Bar Dia. = 0.75 ft
ds = width of back wall – concrete cover – vertical bar dia. – ½ bar dia.
= (3x12) –3 – 0.625 – ½ (0.75) = 32 in

A = As x fy/0.85 x f'c x b (S5.7.3.1.1-4)

where "b" is the height of the component
= (3.96x60)/(0.85x3x119.75) = 0.78 in
Mn = (3.96x60)x(32 – 0.78/2)/12 = 626 k-ft/ft

Therefore, the factored flexural resistance, where ϕ = 0.9 for flexure (S5.5.4.2.1), is taken as:

Mr = 0.9X626
= 563 k-ft/ft > 300 k-ft/ft OK

By inspection:

- Mr > 4/3(Mu).
- The depth of the compression block is small relative to the depth.

Check shear for the section of back wall between girders:

Vu = Pul/2S = 9.667
= 22.65x(9.667/sin 20)/2
= 320.0945241

116.5 k/ft

The factored shear resistance, Vr, is calculated as:

$Vr = \phi Vn$ (S5.8.2.1-2)

The nominal shear resistance, Vn, is calculated according to S5.8.3.3 and is the lesser of:

$Vn = Vc + Vs$ (S5.8.3.3-1)

Or,

$Vn = 0.25 \times f'c \times bv \times dv$ (S5.8.3.3-2)

where:

$Vc = 0.0316 \times \beta \times \sqrt{f'c} \times bv \times dv$ (S5.8.3.3-3)

$\beta = 2$
$f'c = 3$ ksi
bv = effective horizontal beam width taken as the abutment depth (in.)
= 119.75 in
dv = de – a/2
= 32 – (0.78/2)
= 31.61 in
0.9 x de = 0.9X32
= 28.8 in
0.72h = 0.72X36
= 25.92 in

Therefore, dv should be taken as 31.61 in.

Ignore the contribution of the transverse reinforcement to the shear resistance (i.e., Vs = 0), Vn is taken as the smaller of:

$Vc = 0.0316 \times 2 \times (\sqrt{3}) \times 119.75 \times 31.61 = 414.4$ k/ft

Or,

$Vn = 0.25 \times 3 \times 119.75 \times 31.61 = 2839$ k/ft

Therefore, use the shear resistance due to the concrete, Vc.

$Vr = \phi Vn$
$= 0.9 \times 414.4$
$= 373$ k/ft > 116.5 k/ft OK

4.1.1.24. AASHTO-LRFD Design Step 7.1.5 Wing Wall Design

There is no widely accepted method of determining design loads for the wing walls of integral abutments. The following design procedure will result in a conservative design as it takes into account maximum possible loads.

Two load cases are considered:

Load Case 1:

The wing wall is subjected to passive earth pressure. This case accounts for the possibility of the bridge moving laterally and pushing the wing wall against the fill. It is not likely that the displacement will be sufficient to develop full passive pressure. However, there is no available method to determine the expected pressure with certainty. This load case is considered under strength limit state.

Load Case 2:

The wing wall is subjected to active pressure and collision load on the parapet. Active pressure was considered instead of passive to account for the low probability that a collision load and passive pressure will exist simultaneously. This load case is considered at the extreme event limit state, i.e., $\phi = 1.0$ (AASHTO LRFD Table S3.4.1-1)

Required information:

Angle of internal friction of fill, $\varphi = 30$ degrees

Coefficient of active earth pressure,
$ka = (1 - \sin \varphi)/(1 + \sin \varphi) = 0.33$

Coefficient of passive earth pressure,
$kp = [(1 + \sin \varphi)/(1 - \sin \varphi)] / 3$
$ka/kp = 0.33/3 = 0.11$

Load Case 1

From Figure 4.9 (AASHTO-LRFD Figure 7.1-9) and utilizing properties of a right angle pyramid volume = (1/3) x (base area)(height) and the center of gravity (applied at a distance measured from the vertical leg of the right angle pyramid) = ¼ base length].

Moment at the critical section for moment under passive pressure:

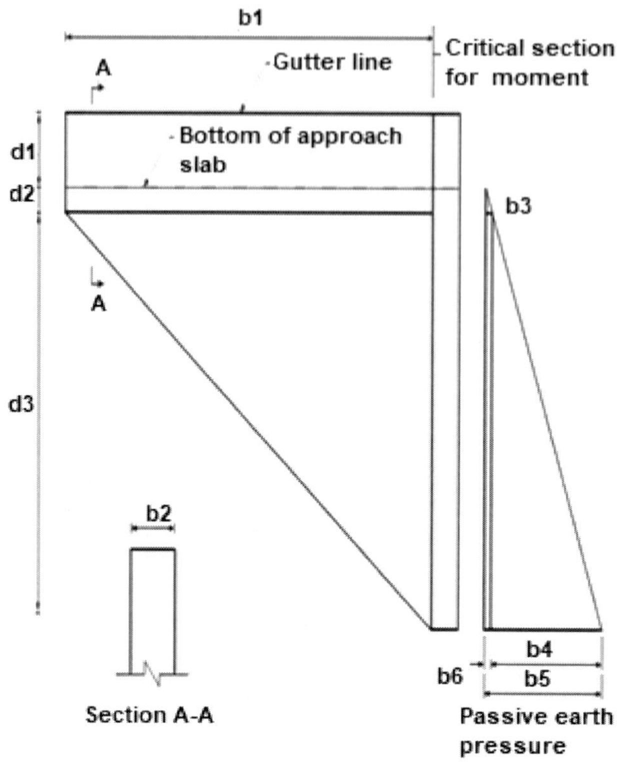

Figure 4.9 - Critical section for moment under passive pressure. Source: Figure 7.1-9 from AASHTO LRFD Bridge Design Specifications, FHWA Office of Bridges and Structures, 1200 New Jersey Ave., SE, Washington, DC 20590.

b1 = 14 ft
b2 = 1.6875 ft
b3 = 2 k/ft2
b4 = 3.24 ft
b5 = 3.44 k/ft2
b6 = 0.2 ft
d1 = 1.5 ft
d2 = 0.5 ft
d3 = 8.31 ft

Mp = [0.2x14x0.5x(14/2)] +[0.2x14x(8.31/2)x(14/3)]
+[(1/3)x3.24x8.31x(14/2)x(14/4)] = 284 k-ft

Minimum required factored flexural resistance,

Mr = 284 k-ft.
Mr = ϕMn (S7.3.2.1-1)

where:

Mn = nominal resistance (k-ft)

= Mp

$\Phi = 0.9$ for flexure at the strength limit state (S5.5.4.2)

Min. required Mn = 284/0.9 = 316 k-ft

Load Case 2

Moment on the critical section for moment under active pressure:

Ma = 0.11x284 = 31.2 k-ft

Moment from collision load on the parapet:

From SA13.2 for Test Level 5, the crash load on the parapet is equal to 124 kips and is applied over a length of 8 ft.

Load on the parapet = 124 kips

Applied length = 8 ft

Maximum collision moment on the critical section:

M = 124x(14 – 8/2) = 1240 k-ft

Total moment for Load Case 2,

$M_{total} = 1240 + 31.2 = 1271.2$ k-ft

The minimum required factored flexural resistance,

Mr = 1271.2 k-ft

$Mr = \phi Mn$ (S5.7.3.2.1-1)

where: $\phi = 1$ for flexure at the extreme event limit state

Min. required

Mn = 1271.2/1 = 1271.2 k-ft

From the two cases of loading:

Mn required = 1,271.20 k-ft

Develop a section that provides the minimum nominal flexural resistance.

Required information:

Assuming reinforcement of #8 @ 6 in. Number of bars within the 10.3125 ft. height of the wing wall = 22 bars

Parapet thickness at base = 20.25 in

Section thickness = Parapet thickness at base = 20.25 in.

Concrete cover =3 in.

The nominal flexural resistance, Mn, is taken as:

Mn = As x fy x (ds – a/2) (S5.7.3.2.2-1)

where:

ds = section thickness – cover – ½ bar diameter = 20.25 – 3 – ½ x1 = 16.75 in.
bar dia. = 1 in
bar area = 0.79 in^2
As = 22x0.79 = 17.38 in^2
fy = 60
f'c = 3

a = Asfy/0.85f'cb (S5.7.3.1.1-4)

= 17.38 x60/(0.85 x3 x123.75) = 3.3 in.
Mn = Asfy(ds – a/2)
= 17.38 x 60 x(16.75 – 3.3/2)/12
= 1312 k-ft > 1,271.20 OK

Secondary reinforcement of the wing wall is not by design, it is only meant for shrinkage. Use #6 @ 12 in. spacing as shown in Figure 4.10 (AASHTO-LRFD Figure 7.1-10).

b1 = 14 ft
b2 = 1.6875 ft
b3 = 2 k/ft2
b4 = 3.24 ft
b5 = 3.44 k/ft2
b6 = 0.2 ft
d1 = 1.5 ft
d2 = 0.5 ft
d3 = 8.3125 ft

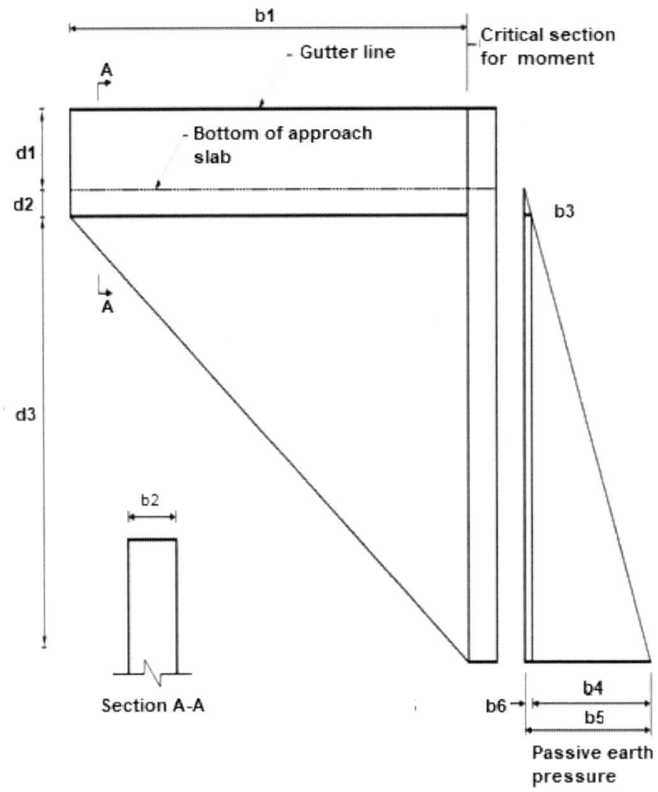

Figure 4.10. Wing wall dimensions. Source: Figure 7.1-10 from AASHTO LRFD Bridge Design Specifications, FHWA Office of Bridges and Structures, 1200 New Jersey Ave., SE, Washington, DC 20590.

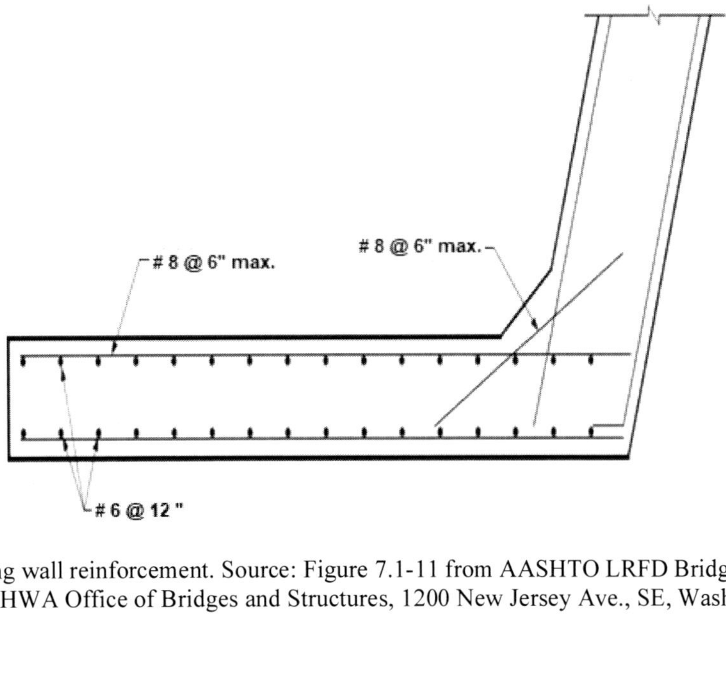

Figure 4.11. Wing wall reinforcement. Source: Figure 7.1-11 from AASHTO LRFD Bridge Design Specifications, FHWA Office of Bridges and Structures, 1200 New Jersey Ave., SE, Washington, DC 20590.

4.2. Computer Applications for the Design of Bridge Abutment in AASHTO LRFD

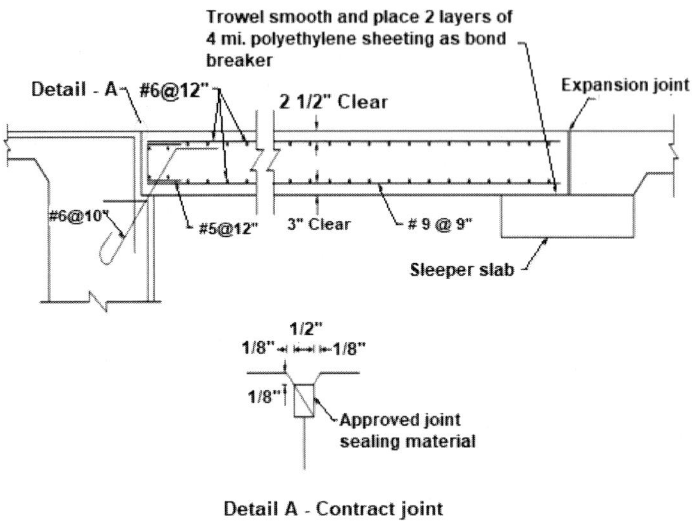

Figure 4.12.

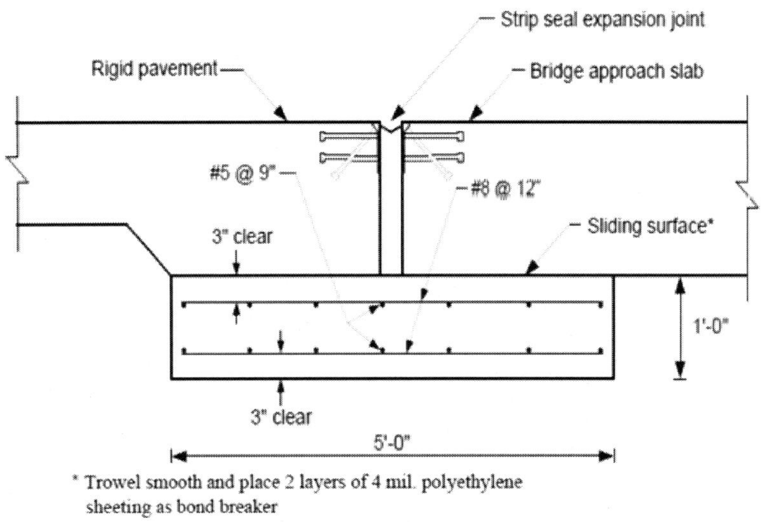

Figure 4.13.

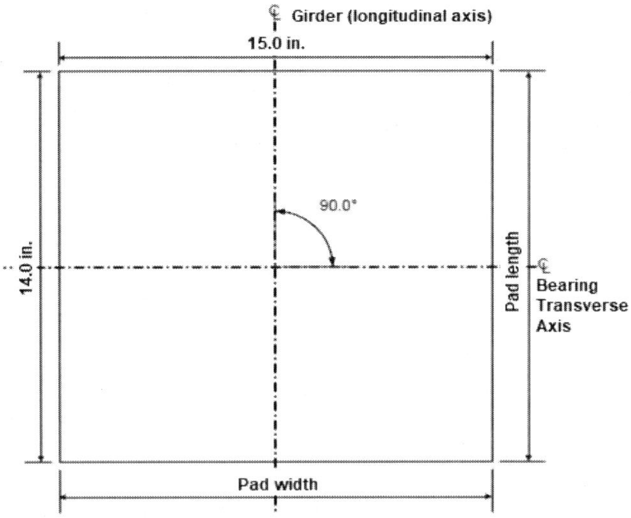

Figure 4.14.

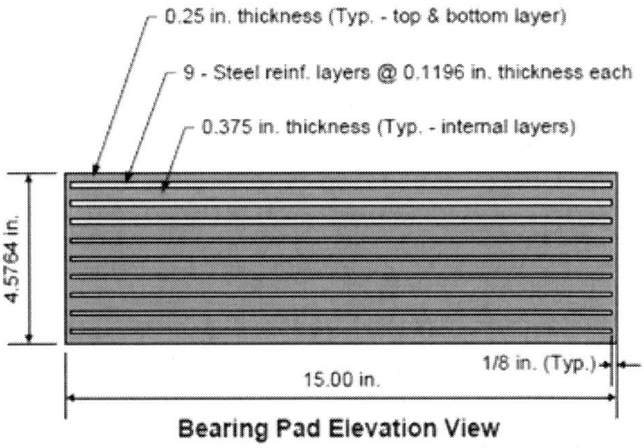

Figure 4.15.

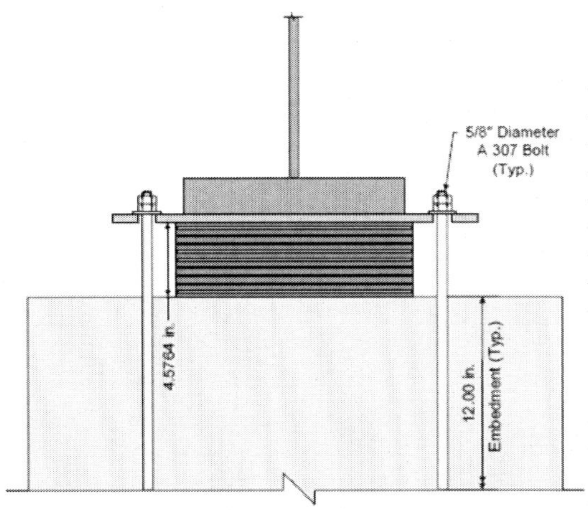

Figure 4.16.

The Computer Applications for the design of bridge abutments in AASHTO LRFD, BS Eurocode II and IRC 112/IRC 6are described in detail in the softcopy volume 'User's Guide for Hands on Practice with Computer Applications' for the book 'Computer Aided Bridge Engineering' is available for download from web site 'www. roadbridgedesign.com'.

REFERENCES

ASTRA Pro software, User's Guide, Tutorial Videos, *User's Manual and Design Manual,* Techsoft Engineering Services, Web site: www.techsoftglobal.com and www.roadbridgedesign.com

Bridge Engineering Handbook, Second Edition, Superstructure Design, Edited by Wai Fah Chen and Lian Duan, Published by CRC Press, Taylor & Francis Group, 6000 Broken Sound Parkway, NW, Suite 300, Boca Raton, FL 33487, Web site: www.crcpress.com

British Standard, BS 5400-3: 2000, *Steel, Concrete and Composite Bridges Đ,* Part 3: Code of practice for design of steel bridges

British Standard, BS 5400-4: 1990, *Steel, Concrete and Composite Bridges* — Part 4: Code of practice for design of concrete bridges

British Standard, *Design Manual for Roads and Bridges,* Volume 1, Section 3, Part 14, BD 37/01, Loads For Highway Bridges, Published by: British Standards Institution.

British Standard, *Eurocode 2: Design of Concrete Structures,* BS EN 1992-1-1:2004, Published by: British Standards Institution.

Design Specifications, Customary U.S. Units, Sixth Edition 2012, American Association of State Highway and Transportation Officials. 444 North Capitol Street, NW, Suite 249, Washington, DC 20001, Phone 202-624-5800 / Fax 202-624-5806, Web site: www.transportation.org

Chapter 5

DESIGN OF BRIDGE PIERS

ABSTRACT

This chapter describes the step wise design procedure for Pier-Pier cap-Pile Cap-Piles-Footings with structural reinforcement details by describing about the general conditions and common practices, design criteria, bridge length limits, soil conditions, skew angle, alignment and geometry, girder layouts, arrangement of piles under pile cap, construction sequence, maximum positive &negative moments and reinforcement, transverse reinforcement for compression members, limits for reinforcement, control of cracking by distribution of reinforcement, shear analysis and foundation soil bearing resistance at the Strength Limit State. This chapter also dealt with design of bearing pads. Based on the design criteria selecting optimum bearing type from its properties, the design is described by considering compute shape factor, compressive stress, compressive deflection, shear deformation, rotation or combined compression and rotation, stability, reinforcement, anchorage for fixed bearings and drawing schematic of final bearing design.

5.1. GENERAL

Piers are the medium which act as an integral part of the load path between the superstructure and the foundation. Piers are designed to resist the vertical loads, as well as the horizontal loads from the bridge superstructure. These horizontal loads are not resisted by the abutments. The configuration of the fixed and expansion bearings, the bearing types and the relative stiffness of all of the piers are determined by the magnitude of the superstructure loads applied to each pier. To estimate the horizontal loads applied at each pier, one must consider the entire system of piers and abutments and not just an individual pier.

The piers shall also resist wind loads, ice loads, water pressures and vehicle impact, the loads applied directly to them. Design of bridges by considering staged construction, whether new or rehabilitation, is to satisfy the requirements of LRFD for each construction stage and by utilizing the same load factors, resistance factors, load combinations, etc. as required for the final configuration.

Note: The various formulas used in the design may be seen from the design Excel Worksheet of software ASTRA Pro by downloading from web site www.techsoftglobal.com under 'downloads'. The values in 'Red and Blue Color' in Excel Worksheet, are Design Input Data which may be changed by the User, but in Tab Pages the colored values are taken from other pages and should not be changed by the user. For any query write to techsoftinfra@gmail.com.

5.2. Pier Design in AASHTO-LRFD

5.2.1. AASHTO-LRFD Design Step 7.2 Intermediate Pier Design

(This section was updated in 2015)

In the following sections, the word "pier" is used to refer to the intermediate pier or intermediate bent. The values in 'Red Color' are Design Input Data by the User.

5.2.1.1. Dead Load

Notice that the LRFD specifications include a maximum and minimum load factor for dead load. The intent is to apply the maximum or the minimum load factors to all dead loads on the structure. It is not required to apply maximum load factors to some dead loads and minimum load factors simultaneously to other dead loads to obtain the absolute maximum load effects.

5.2.1.2. Live Load Transmitted from the Superstructure to the Substructure

Accurately determining live load effects on intermediate piers has always represented an interesting problem. The live load case of loading to produce the maximum girder reactions on the substructure varies from one girder to another and, therefore, the case of loading that maximizes live load effects at any section of the substructure also vary from one section to another. The equations used to determine the girder live load distribution produce the maximum possible live load distributed to a girder without consideration to the live load distributed concurrently to the surrounding girders. This is adequate for girder design but is not sufficient for substructure design. Determining the concurrent girder reactions requires a three-dimensional modeling of the structure. For typical structures, this will be cumbersome and the return, in terms of more accurate results, is not justifiable. In the past, different jurisdictions opted to incorporate some simplifications in the application of live loads to the substructure and these procedures, which are independent of the design specifications, are still applicable under the

AASHTO-LRFD specifications. The goal of these simplifications is to allow the substructure to be analyzed as design a two-dimensional frame. One common procedure is as follows:

- Live load reaction on the intermediate pier from one traffic lane is determined. This reaction from the live load uniform load is distributed over a 10 ft. width and the reaction from the truck is applied as two concentrated loads 6 ft. apart. This means that the live load reaction at the pier location from each traffic lane is a line load 10 ft. wide and two concentrated loads 6 ft. apart. The loads are assumed to fit within a 12 ft. wide traffic lane. The reactions from the uniform load and the truck may be moved within the width of the traffic lane, however, neither of the two truck axle loads may be placed closer than 2 ft. from the edge of the traffic lane.
- The live load reaction is applied to the deck at the pier location. The load is distributed to the girders assuming the deck acts as a series of simple spans supported on the girders. The girder reactions are then applied to the pier. In all cases, the appropriate multiple presence factor is applied.
- First, one lane is loaded. The reaction from that lane is moved across the width of the bridge. To maximize the loads, the location of the 12 ft. wide traffic lane is assumed to move across the full width of the bridge between gutter lines. Moving the traffic lane location in this manner provides for the possibility of widening the bridge in the future and/or eliminating or narrowing the shoulders to add additional traffic lanes. For each load location, the girder reactions transmitted to the pier are calculated and the pier itself is analyzed.
- Second, two traffic lanes are loaded. Each of the two lanes is moved across the width of the bridge to maximize the load effects on the pier. All possible combinations of the traffic lane locations should be included.
- The calculations are repeated for three lanes loaded, four lanes loaded and so forth depending on the width of the bridge.
- The maximum and minimum load effects, i.e., moment, shear, torsion and axial force, at each section from all load cases are determined as well as the other concurrent load effects, e.g., maximum moment and concurrent shear and axial loads. When a design provision involves the combined effect of more than one load effect, e.g., moment and axial load, the maximum and minimum values of each load effect and the concurrent values of the other load effects are considered as separate load cases. This results in a large number of load cases to be checked. Alternatively, a more conservative procedure that results in a smaller number of load cases may be used. In this procedure, the envelopes of the load effects are determined. For all members except for the columns and footings, the maximum values of all load effects are applied simultaneously. For columns and footings,

two cases are checked, the case of maximum axial load and minimum moment and the case of maximum moment and minimum axial load.

This procedure is best suited for computer programs. For hand calculations, this procedure would be cumbersome. In lieu of this lengthy process, a simplified procedure used satisfactorily in the past may be utilized.

5.2.1.3. Load Combinations

The live load effects are combined with other loads to determine the maximum factored loads for all applicable limit states. For loads other than live, when maximum and minimum load factors are specified, each of these two factored loads should be considered as separate cases of loading. Each section is subsequently designed for the controlling limit state.

5.2.1.4. Temperature and Shrinkage Forces

The effects of the change in superstructure length due to temperature changes and, in some cases, due to concrete shrinkage are typically considered in the design of the substructure.

In addition to the change in superstructure length, the substructure member lengths also change due to temperature change and concrete shrinkage. The policy of including the effects of the substructure length change on the substructure forces varies from one jurisdiction to another. These effects on the pier cap are typically small and may be ignored without measurable effect on the design of the cap. However, the effect of the change in the pier cap length may produce a significant force in the columns of multiple column bents. This force is dependent on:

- The length and stiffness of the columns: higher forces are developed in short, stiff columns
- The distance from the column to the point of equilibrium of the pier (the point that does not move laterally when the pier is subjected to a uniform temperature change): Higher column forces develop as the point of interest moves farther away from the point of equilibrium. The point of equilibrium for a particular pier varies depending on the relative stiffness of the columns. For a symmetric pier, the point of equilibrium lies on the axis of symmetry. The column forces, due to the pier cap length changes, are higher for the outer columns of multi-column bents. These forces increase with the increase in the width of the bridge.

5.2.1.5. Torsion

Another force effect that some computer design programs use in pier design is the torsion in the pier cap. This torsion is applied to the pier cap as a concentrated torque at the girder locations. The magnitude of the torque at each girder location is calculated differently depending on the source of the torque.

- Torque due to horizontal loads acting on the superstructure parallel to the bridge longitudinal axis: The magnitude is often taken equal to the horizontal load on the bearing under the limit state being considered multiplied by the distance from the point of load (ft.) above the deck surface.
- Torque due to non-composite dead load on simple spans made continuous for live load: Torque at each girder location is taken equal to the difference between the product of the non-composite dead load reaction and the distance to the mid-width of the cap for the two bearings under the girder line being considered.

According to SC5.8.2.1, if the factored torsion moment is less than one-quarter of the factored pure torsion cracking moment, it will cause only a very small reduction in shear capacity or flexural capacity and, hence, can be neglected. For pier caps, the magnitude of the torsion moments is typically small relative to the torsion cracking moments and, therefore, is typically ignored in hand calculations.

For the purpose of this example, a computer program that calculates the maximum and minimum of each load effect and the other concurrent load effects was used. Load effects due to substructure temperature expansion/contraction and concrete shrinkage were not included in the design. The results are listed in Appendix C. Selected values representing the controlling case offloading are used in the sample calculations.

5.2.1.6. Superstructure Dead Load

These loads can be obtained from Section 5.2 of the superstructure portion of this design example. Summary of the un-factored loading applied vertically at each bearing (12 bearings total, 2 per girder line):

Girders (E/I) 61.6 k
Deck slab and haunch (E) 55.1 k
Deck slab and haunch (I) 62.2 k
Intermediate diaphragm (E) 1.3 k
Intermediate diaphragm (I) 2.5 k
Parapets (E/I) 14.8 k
Future wearing surface (E) 13.4 k
Future wearing surface (I) 19.9 k

(E) – exterior girder
(I) – interior girder

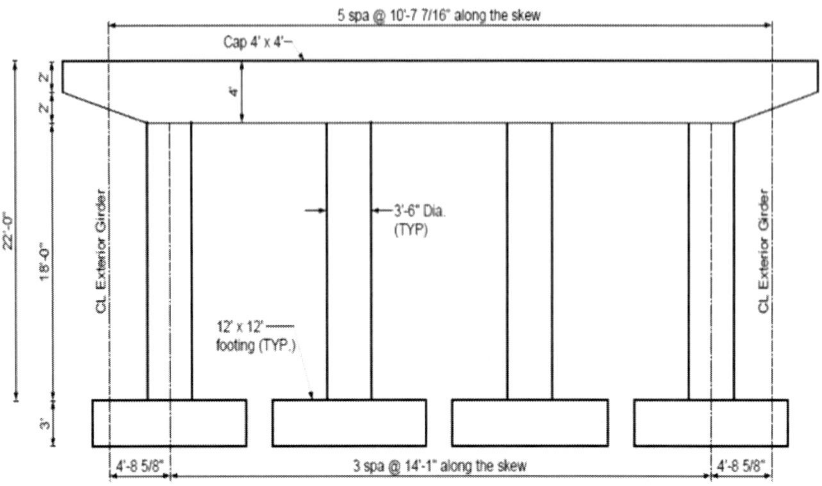

Figure 5.1. General Pier Dimensions. Source: Figure 7.2-1 from AASHTO LRFD Bridge Design Specifications, FHWA Office of Bridges and Structures, 1200 New Jersey Ave., SE, Washington, DC 20590.

Pier Cap Un-Factored Dead Load

Wcap = (cap cross-sectional area)(unit weight of concrete)

Varying cross-section at the pier cap ends:

cap1 = varies linearly from 2x(2)x(0.15) = 0.6 k/ft
to 4x(4)x(0.15) = 2.4 k/ft constant cross-scction:

Wcap2 = 4x(4)x(0.15)= 2.4 k/ft

Or

Pcap = 2.4(45.75) + [(2 + 4)/2](0.15)(13.167) = 115.7 k

Single Column Un-Factored Dead Load

Column cross sectional area = Wcolumn = (column cross sectional area) x (unit weight of concrete)

$= \pi \times (1.75)^2 x(0.150)= 1.44$ k/ft

Or

Pcolumn = 1.44x(18)= 25.92 k

Single Footing Un-Factored Dead Load

Wfooting = (footing cross sectional area)(unit weight of concrete)

= 12 x 12 x 0.150= 21.6 k/ft

Or

Pfooting = 21.6x(3)= 64.8 k

Live Load from the Superstructure

Use the output from the girder live load analysis to obtain the maximum un-factored live load reactions for the interior and exterior girder lines.

Summary of HL-93 live load reactions, without distribution factors or impact, applied vertically to each bearing (truck pair + lane load case governs for the reaction at the pier, therefore, the 90% reduction factor from S3.6.1.3.1 is applied):

Maximum truck = 59.50 k
Minimum truck = 0.00 k
Maximum lane = 43.98 k
Minimum lane = 0.00 k

Braking Force (BR) (S3.6.4)

According to the specifications, the braking force shall be taken as the greater of:

25 percent of the axle weight of the design truck or design tandem

Or

5 percent of the design truck plus lane load or 5 percent of the design tandem plus lane load.

The braking force is placed in all design lanes which are considered to be loaded in accordance with S3.6.1.1.1 and which are carrying traffic headed in the same direction. These forces are assumed to act horizontally at a distance of 6 ft. above the roadway surface in either longitudinal direction to cause extreme force effects. Assume the example bridge can be a one-way traffic bridge in the future. The multiple presence factors in S3.6.1.1.2 apply.

BR_1 = 0.25(32 + 32 + 8)(4 lanes)(0.65)/1 fixed support = 46.8 k

Or

$BR_{2A} = 0.05[72 + (110 + 110)(0.64)] = 10.64$ k

$BR_{2B} = 0.05$ x $[(25 + 25) + 220$ x $(0.64)] = 9.54$ k

where the subscripts are defined as:

1 – use the design truck to maximize the braking force

2A – check the design truck + lane

2B – check the design tandem + lane

Therefore, the braking force will be taken as 46.8 k (3.9 k per bearing or 7.8 k per girder) applied 6 ft. above the top of the roadway surface.

Deck thickness = 0.667 ft Haunch thickness = 0.333 ft Girder depth = 6 ft

Moment arm = 6 ft. + deck thickness + haunch + girder depth = 6 + 0.667 + 0.333 + 6 = 13 ft. above the top of the bent cap

Apply the moment 2(3.9)(13) = 101.4 k-ftat each girder location.

Wind Load on Superstructure (S3.8.1.2)

The pressures specified in the specifications are assumed to be caused by a base wind velocity, VB., of 100 mph.

Wind load is assumed to be uniformly distributed on the area exposed to the wind. The exposed area is the sum of all component surface areas, as seen in elevation, taken perpendicular to the assumed wind direction. This direction is varied to determine the extreme force effects in the structure or in its components. Areas that do not contribute to the extreme force effect under consideration may be neglected in the analysis.

Base design wind velocity varies significantly due to local conditions. For small or low structures, such as this example, wind usually does not govern.

Pressures on windward and leeward sides are to be taken simultaneously in the assumed direction of wind.

The direction of the wind is assumed to be horizontal, unless otherwise specified in S3.8.3. The design wind pressure, in KSF, may be determined as:

PD = PB(VDZ/VB)2 (S3.8.1.2.1-1)

= PB (VDZ2/10,0)

where:

PB = base wind pressure specified in AASHO LRFD Table S3.8.1.2.1-1 (ksf)

Since the bridge component heights are less than 30 ft. above the ground line, VB is taken to be 100 mph.

Wind Load Transverse to the Superstructure

$F_{TSuper} = p_{wT}$ x (H_{wind}) x $[(L_{back} + L_{ahead})/2]$
Girder height = 6 ft
Haunch height = 0.333 ft
Deck height = 0.667 ft
Parapet height = 3.5 ft

where:

H_{wind} = The exposed superstructure height (ft.)
= Girder + Haunch + Deck + Parapet
= 6 + 0.333 +0.667 + 3.5= 10.5 ft.

p_{wT} = Transverse wind pressure values (ksf) = PB (use AASHO LRFD Table S3.8.1.2.2-1)

L_{back} = Span length to the deck joint, or end of bridge, back station from pier (ft.) = 110 ft.

L_{ahead} = Span length to the deck joint, or end of bridge, ahead station from pier (ft.) = 110 ft.

$F_{T\ Super}$ = 0.05x(10.5)x[(110 + 110)/2] = 57.8 k (0 degrees)
0.044x(1155) = 50.8 k (15 degrees)
0.041x(1155) = 47.4 k (30 degrees)
0.033x(1155) = 38.1 k (45 degrees)
0.017x(1155) = 19.6 k (60 degrees)

Wind Load along Axes of Superstructure (Longitudinal Direction)

The longitudinal wind pressure loading induces forces acting parallel to the longitudinal axis of the bridge.

$F_{L\ Super} = p_{wL\ x} (H_{wind})$ x $(L_{back} + L_{ahead})/n_{fixed\ piers}$

where:

H_{wind} = 10.5 ft.

p_{wL} = Longitudinal wind pressure values (ksf) = PB (use AASHO LRFD Table S3.8.1.2.2-1)

L_{back} = 110 ft.

L_{ahead} = 110 ft.

$F_{L\ Super}$ = 0 x (10.5) x [(110 + 110)]/1 = 0 k (0 degrees)
= 0.006 x (2310)/1 = 13.9 k (15 degrees)
= 0.012 x (2310)/1 = 27.7 k (30 degrees)

= 0.016 x (2310)/1 = 37 k (45 degrees)
= 0.019 x (2310)/1 = 43.9 k (60 degrees)

The transverse and longitudinal pressures should be applied simultaneously.

Resultant Wind Load along Axes of Pier

The transverse and longitudinal superstructure wind forces, which are aligned relative to the superstructure axis, are resolved into components that are aligned relative to the pier axes.

Load perpendicular to the plane of the pier:

skew = 20 deg

$F_{L\ Pier} = FL_{Super} \cos(\theta_{skew}) + F_{T\ Super} \sin(\theta_{skew})$

At 0 degrees:

$F_{L\ Pier} = 0 \cos 20 + 57.8 \sin 20 = 19.8$ k

At 60 degrees:

$F_{T\ Pier} = 43.9 \cos 20 + 19.6 \sin 20 = 48$ k

Load in the plane of the pier (parallel to the line connecting the columns):

$F_{L\ Pier} = F_{L\ Super} \cos(\theta_{skew}) + F_{T\ Super} \cos(\theta_{skew})$

At 0 degrees: $F_{L\ Pier} = 0 \sin 20 + 57.8 \cos 20 = 54.3$ k

At 60 degrees:

$F_{T\ Pier} = 43.9 \sin 20 + 19.6 \cos 20 = 33.4$ k

The superstructure wind load acts at 10.5/2 = 5.25 ft. from the top of the pier cap.

The longitudinal and transverse forces applied to each bearing are found by dividing the forces above by the number of girders. If the support bearing line has expansion bearings, the FL Super component in the above equations is zero.

Wind Load on Substructure (S3.8.1.2.3)

The transverse and longitudinal forces to be applied directly to the substructure are calculated from an assumed base wind pressure of 0.040 ksf (S3.8.1.2.3). For wind directions taken skewed to the substructure, this force is resolved into components perpendicular to the end and front elevations of the substructures. The component perpendicular to the end elevation acts on the exposed substructure area as seen in end elevation, and the component perpendicular to the front elevation acts on the exposed areas and is applied simultaneously with the wind loads from the superstructure.

$W_{wind\ on\ sub} = W_{cap} + W_{column}$

Transverse wind on the pier cap (wind applied perpendicular to the longitudinal axis of the superstructure):

cap width = 4 ft
W_{cap} = 0.04 x (cap width)= 0.04 x (4)
= 0.16 k/ft of cap height

Longitudinal wind on the pier cap (wind applied parallel to the longitudinal axis of the superstructure):

Cap length along the skew = 58.93 ft
Wcap = 0.04 x (cap length along the skew) = 0.04 x 58.93
= 2.36 k/ft of cap height

Transverse wind on the end column, this force is resisted equally by all columns:

Column diameter = 3.5 ft $n_{columns}$ = 4
WT, column = 0.04 x (column diameter)/$n_{columns}$= 0.04 x 3.5/4
= 0.035 k/ft of column height above ground

Longitudinal wind on the columns, this force is resisted by each of the columns individually:

$W_{L,\ column}$ = 0.04(column diameter) = 0.04 x 3.5
= 0.14 k/ft of column height above ground

There is no wind on the footings since they are assumed to be below ground level.
Total wind load on substructure:

$W_{T\ wind\ on\ sub}$ =0.16 + 0.035
= 0.2 k/ft $W_{L\ wind\ on\ sub}$
= 2.36 + 0.14 = 2.5 k/ft

Wind on Live Load (S3.8.1.3)

When vehicles are present, the design wind pressure is applied to both the structure and vehicles. Wind pressure on vehicles is represented by an interruptible, moving force of 0.10 klf acting normal to, and 6.0 ft. above, the roadway and is transmitted to the structure.

When wind on vehicles is not taken as normal to the structure, the components of normal and parallel force applied to the live load may be taken as follows with the skew angle taken as referenced normal to the surface.

Use AASHO LRFD Table S3.8.1.3-1 to obtain FW values,

FT Super = FWT(Lback + Lahead)/2
FT Super = 0.1x(110 + 110)/2 = 11 k (0 degrees)
= 0.088x(110) = 9.68 k (15 degrees)
= 0.082x(110) = 9.02 k (30 degrees)
= 0.066x(110) = 7.26 k (45 degrees)
= 0.034x(110) = 3.74 k (60 degrees)
$F_{L\ Super} = F_{WL}(L_{back} + L_{ahead})/n_{fixed\ piers}$
$F_{L\ Super}$ = 0(110 + 110)/1 = 0 k (0 degrees)
= 0.012x (220) = 2.64 k (15 degrees)
= 0.024x (220) = 5.28 k (30 degrees)
= 0.032x (220) = 7.04 k (45 degrees)
= 0.038x (220) = 8.36 k (60 degrees)
$F_{W\ LL}$ = 11 k (transverse direction, i.e., perpendicular to longitudinal axis of the superstructure)

Temperature Force (S3.12.2)

Due to the symmetry of the bridge superstructure, no force is developed at the intermediate bent due to temperature expansion/shrinkage of the superstructure.

Shrinkage (S3.12.4)

Due to the symmetry of the bridge superstructure, no force is developed at the intermediate bent due to shrinkage of the superstructure.

Load Combinations

Figures 5.2 (AASHTO-LRFD Figure 7.2-2) and 5.3 (AASHTO-LRFD Figure 7.2-3) show the un-factored loads applied to the bent from the superstructure and wind.

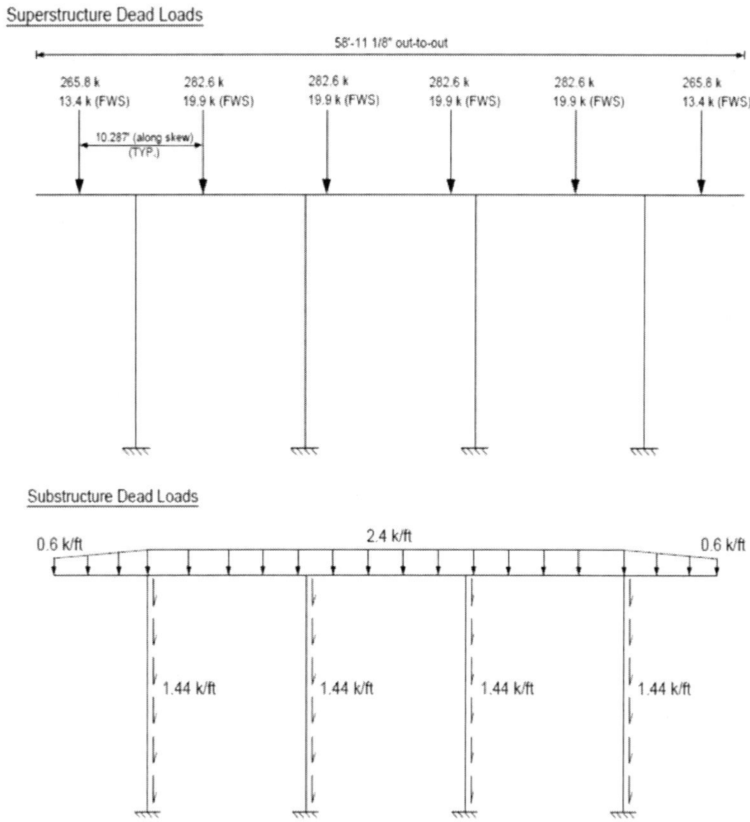

Figure 5.2. Super- and Substructure Applied Dead Loads. Source: Figure 7.2-2 from AASHTO LRFD Bridge Design Specifications, FHWA Office of Bridges and Structures, 1200 New Jersey Ave., SE, Washington, DC 20590.

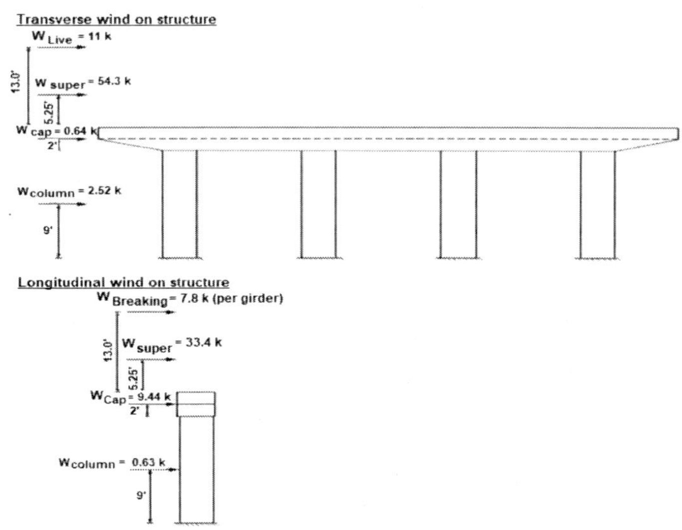

Figure 5.3. Wind and Braking Loads on Super- and Substructure. Source: Figure 7.2-3 from AASHTO LRFD Bridge Design Specifications, FHWA Office of Bridges and Structures, 1200 New Jersey Ave., SE, Washington, DC 20590.

5.2.2. AASHTO-LRFD Design Step 7.2.2 Pier Cap Design

Required information:
General (these values are valid for the entire pier cap):

f'c = 3.0 ksi
β1 = 0.85
fy = 60 ksi
Cap width = 4 ft.
Cap depth = 4 ft. (varies at ends)
No. stirrup legs = 6
Stirrup diameter = 0.625 in. (#5 bars)
Stirrup area = 0.31 in^2 (per leg)
Stirrup spacing = varies along cap length
Side cover = 2 in. (AASHO LRFD Table S5.12.3-1)

Cap bottom flexural bars:

No. of bars in bottom row, positive region = 9 (#8 bars)
Positive region bar diameter = 1.0 in.
Positive region bar area, As = 0.79 in^2
Bottom cover = 2 in. (AASHO LRFD Table S5.12.3-1)

Table 5.1. (AASHTO-LRFD Table 7.2-1) – Strength I Limit State for Critical Locations in the Pier Cap(Maximum Positive Moment, Negative Moment and Shear)

	Location*	**Un-factored Responses**				**Str-I**
		DC	DW	LL + IM	BR	
Max Pos M (k-ft)	44.65 ft.	147.5	37.1	437.9	5.2	1,015.50
Max Neg M (k-ft)	6.79 ft.	-878.5	-84.9	-589	-1.9	-2,259.40
Max Shear M (k)	34.96 ft.	292.9	39.5	210.4	2.8	798.3

*measured from the end of the cap.

Notes:

DC: Superstructure dead load (girders, slab and haunch, diaphragms, and parapets) plus the substructure dead load (all components).

DW: Dead load due to the future wearing surface LL + IM: Live load + impact transferred from the superstructure.

BR: Braking load transferred from the superstructure Str-I: Load responses factored using Strength I limit state load factors.

Cap top flexural bars:

No. of bars in top row, negative region = 14 (7 sets of 2 #9 bars bundled horizontally)
Negative region bar diameter = 1.128 in.
Negative region bar area, As = 1.0 in^2
Top cover = 2 in. (AASHO LRFD Table S5.12.3-1)

From the analysis of the different applicable limit states, the maximum load effects on the cap were obtained. These load effects are listed in Table 5.1 (AASHTO-LRFD Table 7.2-1). The maximum factored positive moment occurs at 44.65 ft. from the cap end under Strength I limit state.

5.2.3. AASHTO-LRFD Design Step 7.2.2.1 - Pier Cap Flexural Resistance (S5.7.3.2)

5.2.3.1. Pier Cap Flexural Resistance (S5.7.3.2)

The factored flexural resistance, Mr, is taken as:

$$Mr = \phi Mn \qquad (S5.7.3.2.1\text{-}1)$$

where:

ϕ= Flexural resistance factor as specified in S5.5.4.2
= 0.9
Mn= Nominal resistance (k-in)

For calculation of Mn, use the provisions of S5.7.3.2.3 which state, for rectangular sections subjected to flexure about one axis, where approximate stress distribution specified in S5.7.2.2 is used and where the compression flange depth is not less than "c" as determined in accordance with Eq. S5.7.3.1.1-3, the flexural resistance Mn may be determined by using Eq. S5.7.3.1.1-1 through S5.7.3.2.2-1, in which case "bw" is taken as "b".

Rectangular section behavior is used to design the pier cap. The compression reinforcement is neglected in the calculation of the flexural resistance.

5.2.4. AASHTO-LRFD Design Step 7.2.2.2 - Maximum Positive Moment and Reinforcement

5.2.4.1. Maximum Positive Moment

Applied Strength I moment, Mu = 1,015.50 k-ft
Applied Service I moment, Ms = 653.3 k-ft (from computer software)

Axial load on the pier cap is small, therefore, the effects of axial load is neglected in this example.

5.2.4.2. Check Positive Moment Resistance (Bottom Steel)

Calculate the nominal flexural resistance according to S5.7.3.2.3.

Mn = As x fy x (ds – a/2) (S5.7.3.2.2-1)

Determine ds, the corresponding effective depth from the extreme fiber to the centroid of the tensile force in the tensile reinforcement.

ds = cap depth – CSG_b

where:

Cover = 2 in
Stirrup diameter = 0.625 in
Bar diameter = 1 in
Bar area = 0.79 in
CGSb = distance from the centroid of the bottom bars to the bottom of the cap (in.)
= Cover + Stirrup diameter + (½) xBar diameter
= 2 + 0.625 + (½) x (1)
= 3.125 in.
ds = 4(12) – 3.125
= 44.875 in.
$(n_{bars\ Tension})$ = 9
As = $(n_{bars\ Tension})$ x (As_{bar})
= 9 x (0.79)
= 7.1 in^2

Determine "a" using Eq. S5.7.3.1.1-4

a = Asfy/0.85f'cb (S5.7.3.1.1-4)

= 7.1 x 60/[0.85 x 3 x 4 x 12] Capdepth = 4 ft
= 3.48 in.

Calculate the nominal flexural resistance, M_n

M_n = As x fy x (ds – a/2) (S5.7.3.2.2-1)

= 7.1 x 60 x [44.875 – (3.48/2)]/12
= 1531 k-ft

Therefore, the factored flexural resistance, Mr, can be calculated as follows:

Mr = 0.9x(1531)
= 1378 k-ft > Mu = 1,015.50 k-ft OK

Limits for Reinforcement (S5.7.3.3)

Check if the section is over-reinforced.

The maximum amount of non-pre-stressed reinforcement shall be such that:

$c/de \leq 0.42$ (S5.7.3.3.1-1)

where:

c = a/β1
= 3.48/0.85
= 4.1 in.
de = ds
= 44.875 in.
c/de = 4.1/44.875
= 0.091 <0.42 OK

Check the Minimum Reinforcement Requirements (S5.7.3.3.2)

Unless otherwise specified, at any section of a flexural component, the amount of non-pre-stressed tensile reinforcement must be adequate to develop a factored flexural resistance, Mr, at least equal to the lesser of:

1.2 x Mcr = 1.2 x f_r x S

where:

$f_r = 0.24\sqrt{f'c}$ (S5.4.2.6)

= 0.24 √3
= 0.42 ksi
S = bh2/6
= 4 x (12) x $[4 \text{ x } (12)]^{2/6}$
= 18,432 in3
1.2 x Mcr = 1.2(0.42)(18432)/12
= 774.1 k-ft Or
1.33 x Mu = 1.33x (1015.5)
= 1,351 k-ft

Minimum required section resistance = 774 k-ft
Provided section resistance = 1,378 k-ft>774.1 k-ft OK

Check the flexural reinforcement distribution (S5.7.3.4)

Check allowable stress, f_s

$f_{s,\ allow}$ = Z/[(dcA)1/3] ≤ 0.6fy (S5.7.3.4-1)

where:
Z = crack width parameter (k/in)
= 170 k/in (moderate exposure conditions are assumed)
dc = Distance from the extreme tension fiber to the center of the closest bar (in.)
= Clear cover + Stirrup diameter + (½) xBar diameter

The cover on the bar under investigation cannot exceed 2.0 in., therefore, the stirrup diameter is not taken into account for dc is:

Bar diameter = 1 in
dc = 2 + (½) x(1)
= 2.5 in
A = Area having the same centroid as the principal tensile reinforcement and bounded by the surfaces of the cross-section and a straight line parallel to the neutral axis, divided by the number of bars (in^2)
n_{bars} = 2 x dc x (cap width)/n_{bars}
= 2 x (2.5) x (48)/9
= 26.7 in^2
fs, allow = Z/[$(dc \text{ x } A)^{1/3}$]
= 170/[(2.5) x (26.7)]^(1/3)

= 41.91

0.6 x (60) = 36 ksi, $f_{s, allow}$ = 36

Therefore, 41.91 > 36

Check service load applied steel stress, $f_{s, actual}$

Modular ratio, n = 9

f'c = 3 ksi

For 3 ksi concrete, the modular ratio, n = 9 (see S6.10.3.1.1b or calculate by dividing the steel modulus of elasticity by the concrete and rounding up as required by S5.7.1)

Assume the stresses and strains vary linearly.

From the load analysis of the bent:

Dead load + live load positive service

Load moment = 653.3 k-ft

The transformed moment of inertia is calculated assuming elastic behavior, i.e., linear stress and strain distribution. In this case, the first moment of area of the transformed steel on the tension side about the neutral axis is assumed equal to that of the concrete in compression.

Assume the neutral axis at a distance "y" from the compression face of the section.

The section width = 48 in.

Transformed steel area = (total steel bar area)(modular ratio)

= 7.1x9 = 63.9 in^2

By equating the first moment of area of the transformed steel about that of the concrete, both about the neutral axis:

63.9 x (44.875 – y) = 48y x (y/2)

Solving the equation results in y = 9.68 in.

Itransformed = Ats x $(ds – y)^2 + by^3/3$

= 63.9 x $(44.875 – 9.68)^2$ + 48 x $(9.68)^3/3$

= 93,665 in4

Stress in the steel, fs, actual = (Msc/I)n, where M is the moment action on the section.

fs,actual = [653.3 x 12 x 35.195/93664.79] x 9

= 26.51 ksi < fs, allow = 36 ksi OK

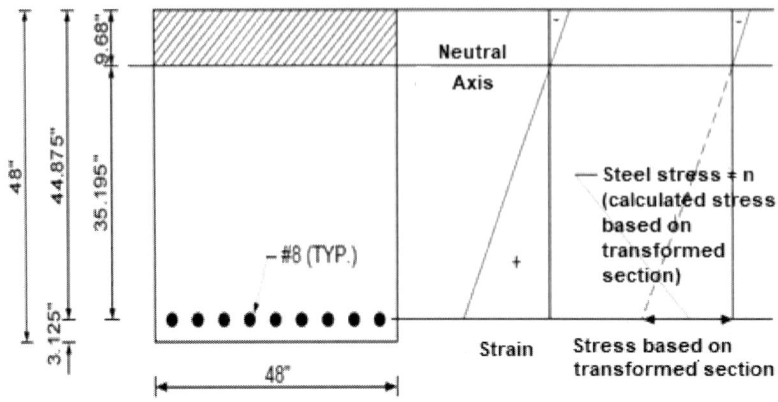

Figure 5.4. Crack Control for Positive Reinforcement under Service Load. Source: Figure 7.2-4 from AASHTO LRFD Bridge Design Specifications, FHWA Office of Bridges and Structures, 1200 New Jersey Ave., SE, Washington, DC 20590.

5.2.5. AASHTO-LRFD Design Step 7.2.2.3 - Maximum Negative Moment and Reinforcement

5.2.5.1. Maximum Negative Moment

From the bent analysis, the maximum factored negative moment occurs at 6.79 ft. from the cap edge under Strength I limit state:

Applied Strength I moment, Mu = -2,259.40 k-ft

Applied Service I moment, Ms = -1,572.40 k-ft (from computer analysis)

5.2.5.2. Check Negative Moment Resistance (Top Steel)

Calculate M_n using Eq. S5.7.3.2.2-1.

Determine ds, the corresponding effective depth from the extreme fiber to the centroid of the tensile force in the tensile reinforcement. The compressive reinforcement is neglected in the calculation of the nominal flexural resistance.

ds = Cap depth – CGSt

where:

CGSt = distance from the centroid of the top bars to the top of the cap (in.)

= cover + stirrup diameter + ½ bar diameter

= 2 + 0.625 + ½ x (1.128)

= 3.19 in.

ds = 4 x (12) – 3.19

= 44.81 in.

$n_{bars\ Tension} = 14$
$As = (n_{bars\ Tension}) \times (A_{s\ bar}) = 14 \times 1 = 14\ in^2$

Determine "a" using Eq. S5.7.3.1.1-4

$a = As \times fy/0.85f'cb$ (S5.7.3.1.1-4)

$= 14 \times (60)/[(0.85 \times 3 \times 4 \times 12)]$
$= 6.86$ in.

Calculate the nominal flexural resistance, Mn
$Mn = 14 \times (60) \times [44.81 - (6.86/2)]/12 = 2{,}897$ k-ft

Therefore, the factored flexural resistance, Mr:
$Mr = 0.9(2896.6)$
$= 2{,}607$ k-ft $> |Mu| = 2{,}259.40$ k-ft OK

5.2.5.3. Limits for Reinforcement (S5.7.3.3)

Check if the section is over-reinforced.
The maximum amount of non-pre-stressed reinforcement shall be such that:

$c/de \leq 0.42$ (S5.7.3.3.1-1)

where:

$c = a/\beta 1$
$= 6.86/0.85$
$= 8.07$ in.
$de = ds$
$= 44.81$ in.
$c/de = 8.07/44.81$
$= 0.18 < 0.42$ OK

5.2.5.4. Check Minimum Reinforcement (S5.7.3.3.2)

Unless otherwise specified, at any section of a flexural component, the amount of non-pre-stressed tensile reinforcement shall be adequate to develop a factored flexural resistance, Mr, at least equal to the lesser of:

$1.2Mcr = 1.2frS$

where:

$fr = 0.24\sqrt{f'c}$ (S5.4.2.6)

$= 0.24\sqrt{3} = 0.42$ ksi

$S = bh^2/6$

$= 4 \times (12) \times [4 \times (12)]^2/6 = 18432$ in 3

$1.2 \times Mcr = 1.2 \times (0.42) \times (18432)/12$

$= 774.14$ k-ft

Or

$1.33 \times Mu = 1.33 \times (-2259.4)$

= Absolute(-3005.002) k-ft

= 3005.002 k-ft

Minimum required section resistance = 774.14 k-ft

Provided section resistance = 2,607 k-ft > 774.1 k-ft OK

Check the flexural reinforcement distribution (S5.7.3.4)

Check the allowable stress, fs

$fs, allow = Z/[(dc \times A)^{1/3}] \leq 0.6 \times fy$ (S5.7.3.4-1)

where:

Z = 170 k/in. (moderate exposure conditions are assumed)

$dc = 2 + ½(1.128)$

= 2.56 in.

A = area having the same centroid as the principal tensile reinforcement and bounded by the surfaces of the cross-section and a straight line parallel to the neutral axis, divided by the number of bars (in^2).

$n_{bars} = 14$

$A = 2 \times dc \times (\text{cap width})/n_{bars}$

$= 2 \times 2.56 \times 48/14$

$= 17.6\ in^2$

fs, allow = Z/[(dcA)1/3]

= 170/[2.56 x (17.6)]^(1/3)

= 47.8 ksi > 0.6(60) = 36 ksi OK

therefore, use fs, allow = 36 ksi

5.2.5.5. Check the Service Load Applied Steel Stress, fs, Actual

For 3 ksi concrete, the modular ratio, n = 9 Assume the stresses and strains vary linearly.

From the load analysis of the bent:

Dead load + live load negative service load moment = -1,572.40 k-ft

The transformed moment of inertia is calculated assuming elastic behavior, i.e., linear stress and strain distribution. In this case, the first moment of area of the transformed steel on the tension side about the neutral axis is assumed equal to that of the concrete in compression.

Assume the neutral axis at a distance "y" from the compression face of the section.

Section width = 48 in.

Total steel bar area = 14 in^2

Transformed steel area = (total steel bar area) x (modular ratio) = 14x9 = 126 in^2

By equating the first moment of area of the transformed steel about that of the concrete, both about the neutral axis:

$$126 \times (44.81 - y) = 48 \times (y) \times (y/2)$$

Solving the equation results in y = 12.9 in.

$$I_{transformed} = Ats(ds - y)^2 + by^3/3$$
$$= 126x(44.81 - 12.9)^2 + (48x(12.9^3))/3 = 1,62,646 \text{ in}^4$$

Stress in the steel, fs, actual = (Msc/I)n, where M is the moment action on the section.

$$f_{s,actual} = [\text{Absoulte}(-1572.4) \times (12) \times (31.91)/162646] \times 9$$
$$= 33.3 \text{ ksi} < \text{fs, allow} = 36 \text{ ksi OK}$$

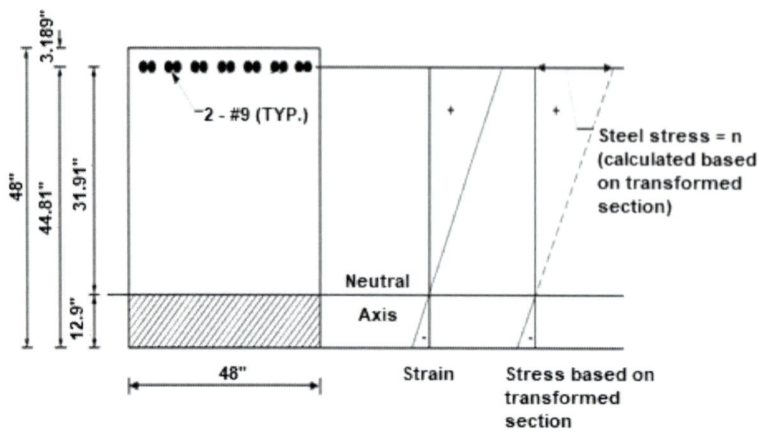

Figure 5.5. Crack Control for Negative Reinforcement under Service Load. Source: Figure 7.2-5 from AASHTO LRFD Bridge Design Specifications, FHWA Office of Bridges and Structures, 1200 New Jersey Ave., SE, Washington, DC 20590.

5.2.6. AASHTO-LRFD Design Step 7.2.2.4 Check Minimum Temperature and Shrinkage Steel

Check minimum temperature and shrinkage steel (S5.10.8).

Reinforcement for shrinkage and temperature stresses is provided near the surfaces of the concrete exposed to daily temperature changes and in structural mass concrete. Temperature and shrinkage reinforcement is added to ensure that the total reinforcement on exposed surfaces is not less than that specified below.

Using the provisions of S5.10.8.2,

As, min1 = 0.11Ag/fy (S5.10.8.2-1)

where:

Ag = gross area of section (in^2)
= [4 x (12)]^2= 2304 in^2
As, min1 = 0.11 x (2304)/60
= 4.22 in^2

This area is to be divided between the two faces, i.e., 2.1 in^2 per face. Shrinkage and temperature reinforcement must not be spaced farther apart than 3.0 times the component thickness or 18.0 in.

Use 4 #7 bars per face.

As provided = 4 x (0.6)
= 2.4 in^2> 2.11 in^2 OK

5.2.7. AASHTO-LRFD Design Step 7.2.2.5 - Skin Reinforcement

5.2.7.1. Skin Reinforcement (S5.7.3.4)

If the effective depth, de, of the reinforced concrete member exceeds 3 ft., longitudinal skin reinforcement is uniformly distributed along both side faces of the component for a distance of d/2 nearest the flexural tension reinforcement. The area of skin reinforcement (in^2/ft of height) on each side of the face is required to satisfy:

$$Ask \geq 0.012 \times (de - 30) \leq (As + Aps)/4 \qquad (S5.7.3.4\text{-}4)$$

where:

Aps = area of pre-stressing (in^2)
de = flexural depth taken as the distance from the compression face of the centroid of the steel, positive moment region (in.) = 44.875
As = 14
Ask = 0.012 x (44.875 – 30)
= 0.179 in^2/ft ≤ 14/4 3.5 in^2/ft
Required Ask per face = 0.179 x (4) = 0.72 in^2< 2.4 in^2 provided OK

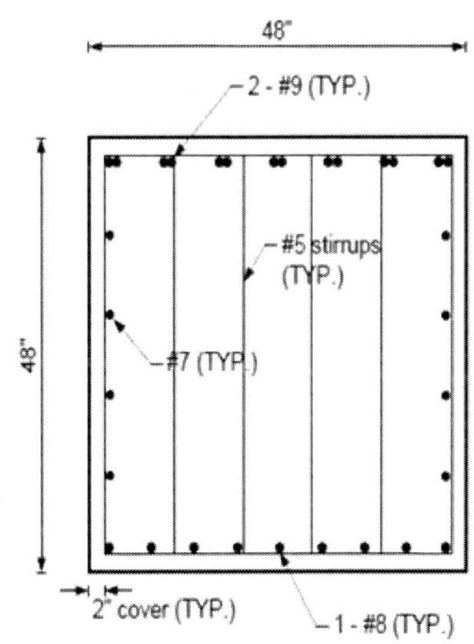

Figure 5.6. Cap Cross-Section. Source: Figure 7.2-6 from AASHTO LRFD Bridge Design Specifications, FHWA Office of Bridges and Structures, 1200 New Jersey Ave., SE, Washington, DC 20590.

5.2.8. AASHTO-LRFD Design Step 7.2.2.6 - Maximum Shear

5.2.8.1. Maximum Shear

From analysis of the bent, the maximum factored shear occurs at 34.96 ft. from the cap end under Strength I limit state:

Shear, Vu = 798.3 k

Calculate the nominal shear resistance using S5.8.3.3.
The factored shear resistance, Vr

$Vr = \phi Vn$ (S5.8.2.1-2)

where:
ϕ = 0.9, shear resistance factor as specified in S5.5.4.2
Vn = nominal shear resistance (k)
The nominal shear resistance, Vn, shall be determined as the lesser of:

$Vn = Vc + Vs + Vp$ (S5.8.3.3-1)

Or

$Vn = 0.25f'cbvdv + Vp$ (S5.8.3.3-2)

where:
Vc = Shear resistance due to concrete (k)

$= 0.0316\beta\sqrt{f'c}bvdv$ (S5.8.3.3-3)

where:
bv = Effective web width taken as the minimum web width within the depth dv as determined in S5.8.2.9 (in.)
= 48 in.
dv = Effective shear depth as determined in S5.8.2.9 (in.). It is thedistance, measured perpendicular to the neutral axis between the resultantsof the tensile and compressive force due to flexure. It need not betaken less than the greater of 0.9de or 0.72h.
= de – a/2 a = 6.86
44.81 – (6.86/2) = 41.4 in.
0.9de = 0.9 x (44.81)

= 40.3 in.

0.72h = 0.72x (48) h = 48 in

= 34.6 in.

Therefore, use dv = 41.4 in. for Vc calculation.

β = factor indicating ability of diagonally cracked concrete to transmit tension as specified in S5.8.3.4

= for non-pre-stressed sections, β may be taken as 2.0

Vc = 0.0316 x (2)$\sqrt{3}$x(48) x (41.4)

= 217.5 k

Vs = Shear resistance due to steel (k)

= [Av x fy x dv x (cot θ + cot α)sin α]/s (S5.8.3.3-4)

where:

s = Spacing of stirrups (in.)= 7 in assume

θ = Angle of inclination of diagonal compressive stresses as determined in S5.8.3.4 (deg)

= 45 deg for non-pre-stressed members

α = Angle of inclination of transverse reinforcement to longitudinal axis (deg) = 90 deg for vertical stirrups

Av = (6 legs of #5 bars)(0.31)= 1.86 in^2

Vs = [1.86 x 60 x (41.4) x (1/tan 45)]/7= 660 k

Vp = Component in the direction of the applied shear of the effective pre-stressing force; positive if resisting the applied shear (k), not applicable in the pier cap

= 0 for non-pre-stressed members

Therefore, Vn is the lesser of:

Vn = 217.5 + 660 + 0

= 877.5 k

Or

Vn = 0.25 x f'c x bv x dv + Vp

= 0.25 x 3 x 48 x 41.4 + 0

= 1,490.40 k

Use Vn = 877.50 k

Therefore,
$V_r = \phi V_n$
$= 0.9 \times (877.5)$
$= 789.8$ k $< V_u = 798.3$ k NOT OK

5.2.8.2. Check the Minimum Transverse Reinforcement (S5.8.2.5)

A minimum amount of transverse reinforcement is required to restrain the growth of diagonal cracking and to increase the ductility of the section. A larger amount of transverse reinforcement is required to control cracking as the concrete strength is increased.

Where transverse reinforcement is required, as specified in S5.8.2.4, the area of steel must satisfy:

$$A_v = 0.0316\sqrt{f'c}\,b_v s/f_y \qquad \text{(S5.8.2.5-1)}$$

where:

b_v = width of web adjusted for the presence of ducts as specified in S5.8.2.9 (in.)
$A_v = 0.0316 \times \sqrt{3} \times (48) \times (7)/60$
$= 0.307 \text{ in}^2 < 1.86 \text{ in}^2$ provided OK

5.2.8.3. Check the Maximum Spacing of the Transverse Reinforcement (S5.8.2.7)

The spacing of the transverse reinforcement must not exceed the maximum permitted spacing, s_{max}, determined as:

If $v_u < 0.125 \times f'c$, then,

$$s_{max} = 0.8d_v \leq 24.0 \text{ in.} \qquad \text{(S5.8.2.7-1)}$$

If $v_u \geq 0.125f'c$, then,

$$s_{max} = 0.4d_v \leq 12.0 \text{ in.} \qquad \text{(S5.8.2.7-2)}$$

The shear stress on the concrete, v_u, is taken to be:

$$v_u = V_u/(\phi b_v d_v) \qquad \text{(S5.8.2.9-1)}$$

$= 798.3/[0.9 \times 48 \times (41.4)]$
$= 0.446 \text{ ksi} > 0.125 \times 3 = 0.375 \text{ ksi}$

Therefore, use Eq. S5.8.2.7-2

$s_{max} = 0.4x41.4 = 16.56$ in.

s_{max} cannot exceed 12 in., therefore, use 12 in. as maximum

Use, $S_{max} = 12$ in

$s_{actual} = 7$ in. < 12 in. OK

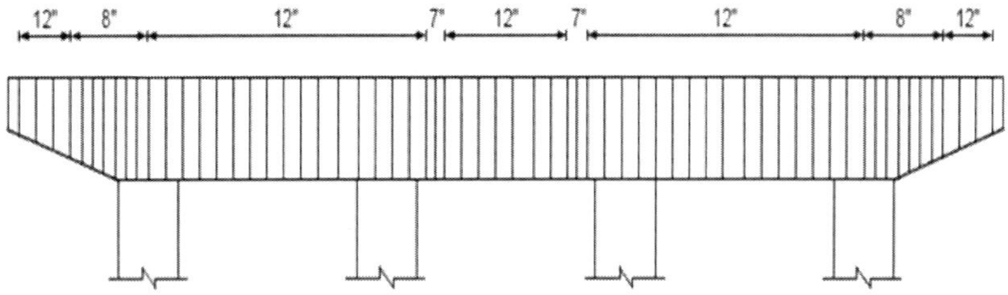

Figure 5.7. Stirrup Distribution in the Bent Cap. Source: Figure 7.2-7 from AASHTO LRFD Bridge Design Specifications, FHWA Office of Bridges and Structures, 1200 New Jersey Ave., SE, Washington, DC 20590.

5.2.9. AASHTO-LRFD Design Step 7.2.3 - Column Design

5.2.9.1. Required Information

General:

$f'c = 3.0$ ksi

$Ec = 3{,}321$ ksi (S5.4.2.4)

$n = 9$

$fy = 60$ ksi

Circular Columns:

Column diameter = 3.5 ft.

Column area, $Ag = 9.62$ ft^2

Side cover = 2 in. (AASHTO LRFD Table S5.12.3-1)

Vertical reinforcing bar diameter (#8) = 1.0 in.

Steel area = 0.79 in^2

Number of bars = 16

Total area of longitudinal reinforcement = 12.64 in^2

Type of transverse reinforcement = Ties tie spacing = 12 in.

Transverse reinforcement bar diameter (#3) = 0.375 in. (S5.10.6.3)

Transverse reinforcement area = 0.11 in^2/bar

The example bridge is in Seismic Zone 1, therefore, a seismic investigation is not necessary for the column design. Article S5.10.11 provides provisions for seismic design where applicable.

5.2.9.2. Applied Moments and Shears

The maximum biaxial responses occur on column 1 at 0.0 ft. from the bottom (top face of footing).

From the load analysis of the bent, the maximum load effects at the critical location were obtained and are listed in Table 5.2 (AASHTO-LRFD Table 7.2-2).

Table 5.2. (AASHTO-LRFD Table 7.2-2) – Maximum Factored Load Effects and the Concurrent Load Effects for Strength Limit States

Load effect maximized	**Limit State**	**M_t (k-ft)**	**M_l (k-ft)**	**P_u (k)**	**M_u (k-ft)**
Positive M_t	Strength V	342	352	1,062	491
Negative M_t	Strength V	-129	-216	682	252
Positive M_l	Strength V	174	822	1,070	840
Negative M_l	Strength V	116	-824	1,076	832
Axial Load P	Strength I	90	-316	1,293	329

where:

Mt: Factored moment about the transverse axis

Ml: Factored moment about the longitudinal axis

Pu: Factored axial load

Sample hand calculations are presented for the case of maximum positive Ml from Table 5.2 (AASHTO-LRFD Table 7.2-2).

5.2.9.3. Maximum Shear Occurs on Column 1 at 0.0 ft. from the Bottom (Top Face of Footing)

Factored shears – strength limit state:

V_t = 44.8 k (Str-V)
V_l = 26 k (Str-V)

5.2.9.4. Check Limits for Reinforcement in Compression Members (S5.7.4.2)

The maximum area of non-pre-stressed longitudinal reinforcement for non-composite compression components shall be such that:

$As/Ag \leq 0.08$ (S5.7.4.2-1)

where:
As = Area of non-pre-stressed tension steel (in^2)
Ag = Gross area of section (in^2)
= 12.64/[9.62 x (144)]
= 0.009 < 0.08 OK

The minimum area of non-pre-stressed longitudinal reinforcement for non-composite compression components shall be such that:

$As \times fy/Ag \times f'c \geq 0.135$ (S5.7.4.2-3)

= 12.64 x 60/[9.62 x (144 x 3)]
= 0.182 > 0.135 OK

Therefore, the column satisfies the minimum steel area criteria, does not use a reduced effective section. For oversized columns, the required minimum longitudinal reinforcement may be reduced by assuming the column area is in accordance with S5.7.4.2.

5.2.9.5. Strength Reduction Factor, ϕ, to Be Applied to the Nominal Axial Resistance (S5.5.4.2)

For compression members with flexure, the value of ϕ may be increased linearly from axial (0.75) to the value for flexure (0.9) as the factored axial load resistance, ϕPn, decreases from 0.10f'cAg to zero. The resistance factor is incorporated in the interaction diagram of the column shown graphically in Figure 5.8 (AASHTO-LRFD Figure 7.2-8) and in tabulated form in Table 5.3 (AASHTO-LRFD Table 7.2-3).

Table 5.3. (AASHTO-LRFD Table 7.2-3) – Column Interaction Diagram in Tabulated Form

P(k)	M (k-ft)	P (k)	M (k-ft)
		(cont.)	(cont.)
P_{max} = 2,555	764	799	1,354
2,396	907	639	1,289
2,236	1031	479	1,192
2,076	1135	319	1,124
1,917	1222	160	1,037
1,757	1291	0	928
1,597	1348	-137	766
1,437	1389	-273	594
1,278	1419	-410	410
1,118	1424	-546	212
958	1404	-683	0

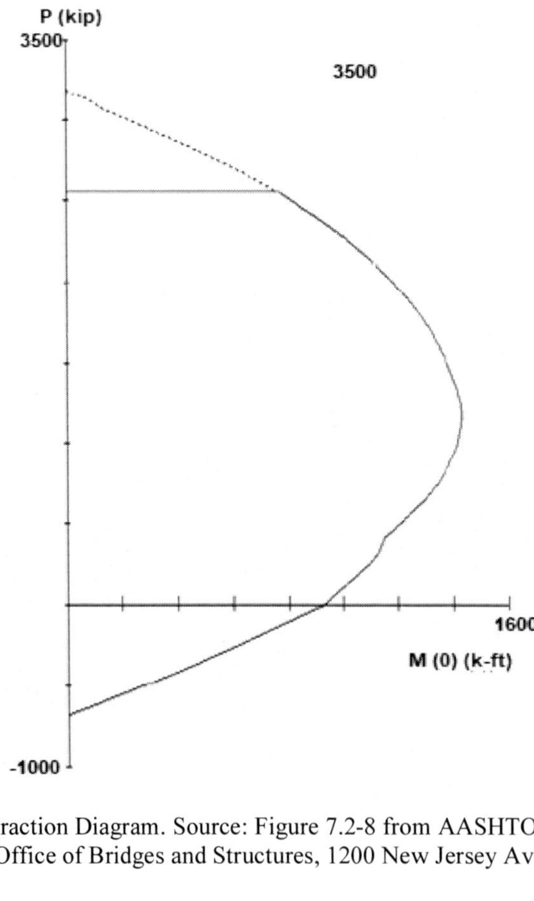

Figure 5.8. Column Interaction Diagram. Source: Figure 7.2-8 from AASHTO LRFD Bridge Design Specifications, FHWA Office of Bridges and Structures, 1200 New Jersey Ave., SE, Washington, DC 20590.

5.2.10. AASHTO-LRFD Design Step 7.2.3.1

5.2.10.1. Slenderness Effects

The effective length factor, K, is taken from S4.6.2.5. The slenderness moment magnification factors are typically determined in accordance with S4.5.3.2.2. Provisions specific to the slenderness of concrete columns are listed in S5.7.4.3.

Typically, the columns are assumed un-braced in the plane of the bent with the effective length factor, K, taken as 1.2 to account for the high rigidity of the footing and the pier cap. In the direction perpendicular to the bent K may be determined as follows:

- If the movement of the cap is not restrained in the direction perpendicular to the bent, the column is considered not braced and the column is assumed to behave as a free cantilever. K is taken equal to 2.1 (see AASHTO LRFD Table SC4.6.2.5-1)
- If the movement of the cap is restrained in the direction perpendicular to the bent, the column is considered braced in this direction and K is taken equal to 0.8 (see AASHTO LRFD Table SC4.6.2.5-1)

For the example, the integral abutments provide restraint to the movements of the bent in the longitudinal direction of the bridge (approximately perpendicular to the bent). However, this restraint is usually ignored and the columns are considered un-braced in this direction, i.e.;

K = 2.1

The slenderness ratio is calculated as K x l_u/r

where:

K = effective length factor taken as 1.2 in the plane of the bent and 2.1 in the direction perpendicular to the bent

Effective length in the plane of the bent = 1.2

Effective length in the direction perpendicular to the bent = 2.1

l_u = Un-braced length calculated in accordance with S5.7.4.3 (ft.)

= Distance from the top of the footing to the bottom of the cap

= 18 ft.

r = radius of gyration (ft.)

= ¼ the diameter of circular columns = 0.875 ft.

For a column to be considered slender, Klu/r should exceed 22 for un-braced columns and, for braced columns, should exceed 34–12(M1/M2) where M1 and M2 are the smaller and larger end moments, respectively. The term (M1/M2) is positive for single curvature flexure (S5.7.4.3)

5.2.10.2. Slenderness Ratio in the Plane of the Bent

$K \times l_u/r = 1.2 \times (18)/(0.875)$
$= 24.69 > 22$ therefore, the column is slightly slender

5.2.10.3. Slenderness Ratio out of the Plane of the Bent

$K \times l_u/r = 2.1 \times (18)/(0.875)$
$= 43.2 > 22$ therefore, the column is slender

With the column slender in both directions, effect of slenderness needs to be considered.

5.2.10.4. Moment Magnification in the Bent

Longitudinal direction:

$$Mcl = \delta bM2b + \delta sM2s \quad \text{(S4.5.3.2.2b-1)}$$

where:

$$\delta b = Cm/[1 - (Pu/\phi Pe)] \geq 1.0 \quad \text{(S4.5.3.2.2b-3)}$$

$$\delta s = 1/[1 - \Sigma Pu/\phi\Sigma Pe] \quad \text{(S4.5.3.2.2b-4)}$$

where:

Cm = parameter of the effect of moment-curvature = 1.0 for members not braced for side sway (S4.5.3.2.2b)

Pu = factored axial load for critical case, (k), see Table **5**.2 (AASHTO-LRFDTable 7.2-2)= 1,070 k

Pe = Euler buckling load (k)

$\phi = 0.75$, resistance factor for axial compression (S5.5.4.2)

M_{2b} = Moment on compression member due to factored gravity loads that result in no appreciable side sway calculated by conventional first-order elastic frame analysis, always positive (k-ft)

M_{2s}= moment on compression member due to factored lateral or gravity loads that result in side sway, Δ, greater than lu/1500, calculated by conventional first-order elastic frame analysis, always positive (k-ft)

Calculate Pe,

$Pe = \pi 2EI/(Klu)2$ (S4.5.3.2.2b-5)

where:

EI = Column flexural stiffness calculated using the provisions of S5.7.4.3 and is taken as the greater of:

$EI = [EcIg/5 + EsIs]/(1 + \beta d)$ (S5.7.4.3-1)

And

$EI = [EcIg/2.5]/(1 + \beta d)$ (S5.7.4.3-2)

where:

Ec = Modulus of elasticity of concrete per S5.4.2.4 (ksi)

$= 33{,}000 w_c^{1.5} \sqrt{f'_c}$

Wc = 0.15

f'c = 3

Ec = 3321 ksi

Ig = Moment of inertia of gross concrete section about the centroidal axis (in^4)

$= \pi r^4/4$

r = 1.75 ft

$Ig = \pi \times [1.75 \times (12)]^4/4$

= 1,52,745 in4

βd = Ratio of the maximum factored permanent load moment to the maximum factored total load moment, always positive. This can be determined for each separate load case, or for simplicity as shown here, it can be taken as the ratio of the maximum factored permanent load from all cases to the maximum factored total load moment from all cases at the point of interest.

= Ml permanent/Ml total

Ml permanent = 118.3

Ml total = 822

βd = 118.3/822

= 0.144

As a simplification, steel reinforcement in the column is ignored in calculating EI, therefore, neglect Eq. S5.7.4.3-1.

EI = [3321 x (152745)/2.5]/(1 + 0.144)
= 1.77E+08 k-in^2
K = Effective length factor per AASHTO LRFD Table SC4.6.2.5-1 = 2.1
lu = unsupported length of the compression member (in.) = 18 x 12
= 216 in.
Pe = π^2(1.77E+08)/[2.1(216)]^2
= 8508 k

Therefore, the moment magnification factors δb and δs can be calculated.

δb = 1.0/[1 – (1070/[0.75(8508)])]
= 1.2
δs = 1/[1 – ΣPu/φΣPe]

ΣPu and ΣPe are the sum of the applied factored loads and the sum of the buckling loads of all columns in the bent, respectively. For hand calculations, it is not feasible to do calculations involving several columns simultaneously. Therefore, in this example, Pu and Pe of the column being designed are used instead of ΣPu and ΣPe.

δs = 1.2

Therefore, the magnified moment in the longitudinal direction is taken as:

$Mc1 = \delta b \times M_{2b} + \delta s \times M_{2s}$ (S4.5.3.2.2b-1)

$= 1.2x(M_{2b} + M_{2s})$
= 1.2 x (total factored moment, Ml)
= 1.2 x 822= 986.4 k-ft

Transverse direction:

$Mct = \delta_b \times M_{2b} + \delta_s x M_{2s}$ (S4.5.3.2.2b-1)

Calculate Pe,

$Pe = \pi^2 EI/(K \times l_u)^2$ (S4.5.3.2.2b-5)

where:

EI = Column flexural stiffness calculated using the provisions of S5.7.4.3 and is taken as the greater of:

$EI = [EcIg/5 + EsIs]/(1 + \beta d)$ (S5.7.4.3-1)

And

$EI = [EcIg/2.5]/(1 + \beta d)$ (S5.7.4.3-2)

where:

Ec = 3,321 ksi
Ig = 1,52,745 in4
βd = Mt permanent/Mt total
Mt permanent= 101.7
Mt total = 342
βd = 101.7/342 = 0.3

For simplification, steel reinforcement in the column is ignored in calculating EI, therefore, neglect Eq. S5.7.4.3-1.

EI = [3321 x (152745)/2.5]/(1 + 0.3)
= 1.56E+08 k-in^2
K = 1.2
l_u = 216 in.
Pe = π^2(1.56E+08)/[1.2x(216)]^2
= 22,929 k

Therefore, the moment magnification factors δb and δs can be calculated.

δb = 1/[1 – (1070/[0.75 x (23064)])]
= 1.07
δs = 1/[1 – ΣPu/φΣPe]

Similar to longitudinal, use Pu and Pe instead of ΣPu and ΣPe. = 1.07
Therefore, the magnified moment in the transverse direction is taken as:

Mct = 1.07 (total factored moment, Mt)
Mt = 174
= 1.07 x 174
= 186 k-ft

The combined moment Mu is taken as:

Mu = √(Mc1 + Mct)
= √($986.4^2 + 186^2$)
= 1004 k-ft

Factored axial load on the column for the load case being checked =1070 k

By inspection, from the column interaction diagram Figure 5.8 (AASHTO-LRFD Figure 7.2-8)) or Table 5.3 (AASHTO-LRFD Table 7.2-3), the applied factored loads (M = 1004 k-ft and P = 1070 k) are within the column resistance.

5.2.11. AASHTO-LRFD Design Step 7.2.3.2 - Transverse Reinforcement for Compression Members

5.2.11.1. Transverse Reinforcement for Compression Members (S5.10.6)

Transverse reinforcement for compression members may consist of either spirals or ties. Ties are used in this example. In tied compression members, all longitudinal bars are enclosed by lateral ties. Since the longitudinal bars are #8, use #3 bars for the ties (S5.10.6.3).

The spacing of ties is limited to the least dimension of the compression member or 12.0 in., therefore, the ties are spaced at 12.0 in. center-to-center.

Ties are located vertically no more than half a tie spacing above the footing and not more than half a tie spacing below the lowest horizontal reinforcement in the cap.

Figure5.9 (AASHTO-LRFD Figure 7.2-9) shows the column cross-section.

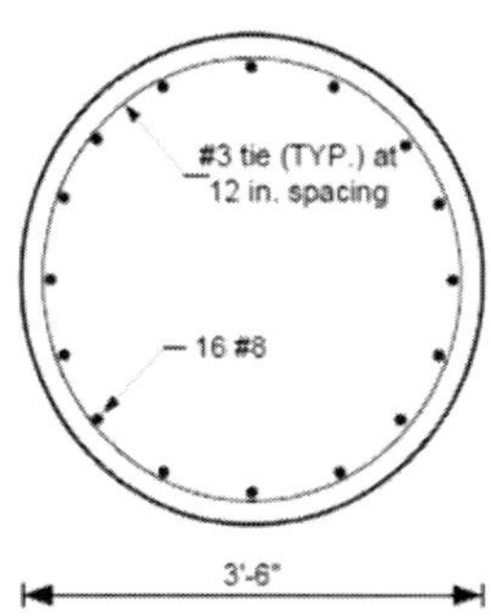

Figure 5.9. Column Cross-Section. Source: Figure 7.2-9 from AASHTO LRFD Bridge Design Specifications, FHWA Office of Bridges and Structures, 1200 New Jersey Ave., SE, Washington, DC 20590.

5.2.12. AASHTO-LRFD Design Step 7.2.4 Footing Design

Based on the intermediate bent load analysis, the critical footing is Footing 1 supporting Column 1

5.2.12.1. Required Information:

General:

f'c = 3.0 ksi

fy = 60 ksi

Side concrete cover = 3 in. (AASHTO-LRFD Table S5.12.3-1)

Top concrete cover = 3 in.

Bottom concrete cover = 3 in.

Top bars Transverse (T) or longitudinal(L) in bottom mat = L

Direction of bottom bars in bottom mat = T

A preliminary analysis of the footing yielded the following information:

Footing depth = 3 ft.

Footing width, W = 12 ft.

Footing length, L = 12 ft.

Top mat reinforcing bar diameter, #5 bars = 0.625 in.

Top mat reinforcing bar area, #5 bars = 0.31 in^2

Bottom mat reinforcing bar diameter, #9 bars = 1.128 in.

Bottom mat reinforcing bar area, #9 bars = 1.0 in^2

Number of bars = 13 bars in each direction in both the top and bottom mats

5.2.12.2. Location of Critical Sections

According to S5.13.3.6.1, the critical section for one-way shear is at a distance dv, the shear depth calculated in accordance with S5.8.2.9, from the face of the column. For two-way shear, the critical section is at a distance of dv/2 from the face of the column.

For moment, the critical section is taken at the face of the column in accordance with S5.13.3.4. For the circular column in this example, the face of the column is assumed to be located at the face of an equivalent square area concentric with the circular column in accordance with S5.13.3.4.

5.2.12.3. Determine the Critical Faces along the y-Axis for Moment

Since the column has a circular cross-section, the column may be transformed into an effective square cross-section for the footing analysis.

Critical face in y-direction = ½ footing width, W – ½ equivalent column width

Shaft area = 9.62 ft2

Equivalent column width = √shaft area

= √9.62

= 3.10 ft

Critical face in y-direction = ½ footing width, W – ½ equivalent column width
= ½ (12) – ½ (3.1)
= 4.45 ft.
Critical faces in the y-direction = 4.45 ft. and 7.55 ft.

5.2.12.4. Determine the Critical Faces along the x-Axis for Moment

For a square footing with an equivalent square column:

Critical face in the x-direction = Critical face in the y-direction
= 4.45 ft.
Critical faces in the x-direction = 4.45 ft. and 7.55 ft.

See Figure 5.10 (AASHTO-LRFD Figure 7.2-10) for a schematic showing the critical sections for moments.

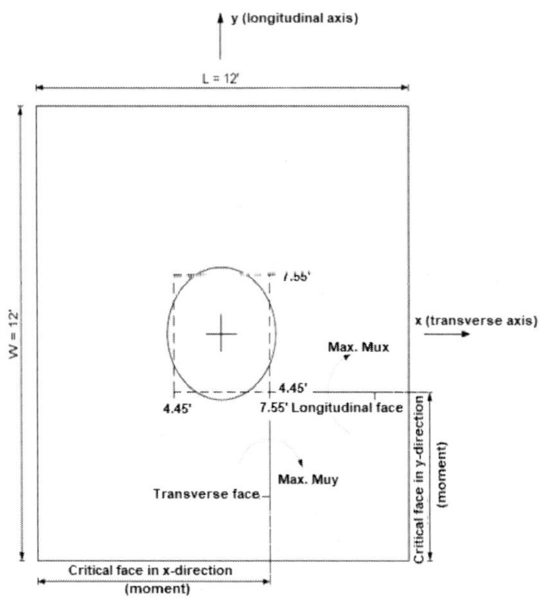

Figure 5.10. Critical Sections for Moment. Source: Figure 7.2-10 from AASHTO LRFD Bridge Design Specifications, FHWA Office of Bridges and Structures, 1200 New Jersey Ave., SE, Washington, DC 20590.

5.2.12.5. Design Factored Loads at the Critical Section

From the analysis of the intermediate bent computer program, the cases of loading that produced maximum load effects and the other concurrent load effects on the footing are shown in Table 5.4 (AASHTO-LRFD Table 7.2-4).

Each row in Table 5.4 (AASHTO-LRFD Table 7.2-4) represents the maximum value of one load effect (max. +Mt,-Ml, etc.). The corresponding concurrent load effects are also given. Many engineers design the footing for the above listed cases. However, computer design programs are able to check many more cases of loading to determine the most critical case. For example, a load case that is not an axial load or maximum moment may still produce the maximum combined effects on the footing. From the output of a footing design program, the critical case for the footing design was found to produce the following factored footing loads under Strength I limit state:

Table 5.4. (AASHTO-LRFD Table 7.2-4) – Loads on Critical Footing (Footing Supporting Column 1)

Load effect maximized	Limit State	M_t (k-ft)	M_l (k-ft)	P_u (k)
Positive M_t	Strength V	423	377	1,143
Negative M_t	Strength III	-154	-197	628
Positive M_l	Strength V	232	895	1,151
Negative M_l	Strength V	158	-897	1,157
Axial Load P	Strength V	121	-363	1,374

Pu = 1,374 k
Mt = -121 k-ft
Ml = 626 k-ft

The critical Service I loads:

Pu = 891 k
Mt,s = 176 k-ft
Ml,s = 620 k-ft

For the sample calculations below, the factored loads listed above for the critical case of loading were used.

5.2.12.6. Sample Calculations for the Critical Footing under the Critical Case of Loading

If M/P < L/6 then the soil under the entire area of the footing is completely in compression and the soil stress may be determined using the conventional stress formula (i.e., $\sigma = P/A \pm Mc/I$).

Mt/Pu = 121/1374
= 0.088 < 12/6 = 2 OK
Ml/Pu = 626/1374
= 0.456 < 12/6 = 2 OK

Therefore, the soil area under the footing is under compression.

5.2.12.7. Moment

For Mux (k-ft/ft), where Mux is the maximum factored moment per unit width of the footing due to the combined forces at a longitudinal face, see Figure 5.10 (AASHTO-LRFD Figure 7.2-10):

$$\sigma 1, \sigma 2 = P/LW \pm M_l(L/2)/(L^3W/12)$$

where:

σ1 = stress at beginning of footing in direction considered (ksf), see Figure 5.10 (AASHTO-LRFD Figure 7.2-10)
σ2 = stress at end of footing in direction considered (ksf)
P = axial load from above (k)
M1 = moment on longitudinal face from above (k-ft)
L = total length of footing (ft.)
W = total width of footing (ft.)
$\sigma 1 = 1374/[12 \text{ x } (12)] + 626 \text{ x } (12/2)/[12^3(12)/12]$
= 9.54 + 2.17
= 11.71 ksf
$\sigma 2 = 1374/[12 \text{ x } (12)] - 626 \text{ x } (12/2)/[12^3(12)/12]$
= 9.54 - 2.17
= 7.37 ksf

Interpolate to calculate σ3, the stress at critical location for moment at face of column, 4.45 ft. from the end of the footing along the width.

σ3 = 10.1 ksf

Therefore,

$$Mux = \sigma 3L1(L1/2) + 0.5(\sigma 1 - \sigma 3)(L1)(2L1/3)$$

where:

L1 = distance from the edge of footing to the critical location (ft.)

$Mux = 10.1(4.45)(4.45/2) + 0.5(11.71 – 10.1)(4.45)[2(4.45)/3]$
$= 100 + 10.63$
$= 110.63$ k-ft/ft

For Muy (k-ft/ft), where Muy is the maximum factored moment per unit length from the combined forces at a transverse face acting at 4.45 ft. from the face of the column, see Figure 5.10 (AASHTO-LRFD Figure 7.2-10):

$$\sigma5, \sigma6 = P/LW \pm Mt(W/2)/(W3L/12)$$

where:

Mt = moment on transverse face from above (k-ft)
$\sigma5 = 1374/[12 \times (12)] – (-121) \times (12/2)/[(12^3)(12)/12]$
$= 9.54- (-0.42)$
$= 9.96$ ksf
$\sigma6 = 9.54+ (-0.42)$
$= 9.12$ ksf

Interpolate to calculate σ7, the stress at critical location for moment (at face of column, 4.45 ft. from the end of the footing along the length).

$\sigma7 = 9.65$ ksf

Therefore,

$Muy = \sigma7L3(L3/2) + 0.5(\sigma5 – \sigma7)(L3)(2L3/3)$
$= 9.65(4.45)(4.45/2) + 0.5(9.96 – 9.65)(4.45)[2(4.45)/3]$
$= 95.55 + 2.05$
$= 97.6$ k-ft/ft

Factored applied design moment, Service I limit state, calculated using the same method as above:

Mux,s = 75.9 k-ft/ft
Muy,s = 72 k-ft/ft

Where Mux,s is the maximum service moment from combined forces at a longitudinal face at 4.45 ft. along the width and Muy,s is the maximum service moment from combined forces at a transverse face at 7.55 ft. along the length.

5.2.12.8. Shear

Factored applied design shear.

For Vux (k/ft), where Vux is the shear per unit length at a longitudinal face:

$$Vux = \sigma 4L2 + 0.5(\sigma 1 - \sigma 4)L2$$

where:

L2 = distance from the edge of footing to a distance dv from the effective column (ft.)

Based on the preliminary analysis of the footing, dv is estimated as 30.3 in. Generally, for load calculations, dv may be assumed equal to the effective depth of the reinforcement minus 1 inch.

Small differences between dv assumed here for load calculations and the final dv will not result in significant difference in the final results.

dv = 30.3

The critical face along the y-axis = 4.45 – 30.3/12

= 1.925 ft. from the edge of the footing

By interpolation between $\sigma 1$ and $\sigma 2$, $\sigma 4 = 11.01$ ksf

Vux = 11.01(1.925) + 0.5(11.71 – 11.01)(1.925)

= 21.19 + 0.67

= 21.9 k/ft

For Vuy (k/ft), where Vuy is the shear per unit length at a transverse face:

$$Vux = \sigma 8L4 + 0.5(\sigma 5 - \sigma 8)L4$$

where:

dv = 31.4 in. for this direction (from preliminary design). Alternatively, for load calculations, dv may be assumed equal to the effective depth of the reinforcement minus 1 inch).

The critical face along the x-axis = 4.45 – 31.4/12

= 1.83 ft. from the edge of the footing

By interpolation between $\sigma 5$ and $\sigma 6$, $\sigma 8 = 9.83$ ksf

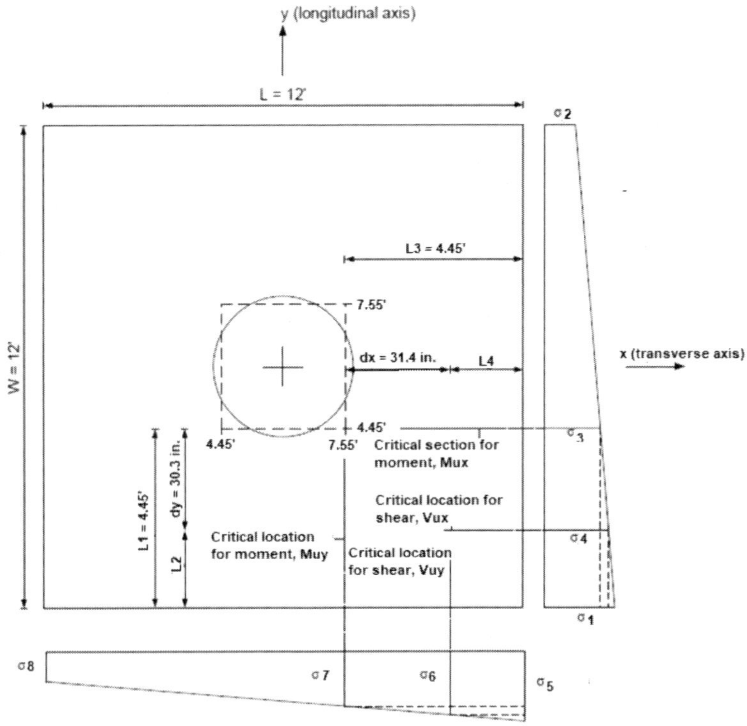

Figure 5.11. Stress at Critical Locations for Moment and Shear. Source: Figure 7.2-11 from AASHTO LRFD Bridge Design Specifications, FHWA Office of Bridges and Structures, 1200 New Jersey Ave., SE, Washington, DC 20590.

Vux = 9.83(1.83) + 0.5 x (9.96 – 9.83) x (1.83)
= 17.99 + 0.12 = 18.1 k/ft

5.2.13. AASHTO-LRFD Design Step 7.2.4.1 Flexural Resistance (S5.7.3.2)

Check the design moment strength (S5.7.3.2)

Article S5.13.3.5 allows the reinforcement in square footings to be uniformly distributed across the entire width of the footing.

Check the moment resistance for moment at the critical longitudinal face (S5.13.3.4)

The critical section is at the face of the effective square column (4.45 ft. from the edge of the footing along the width). In the case of columns that are not rectangular, the critical section is taken at the side of the concentric rectangle of equivalent area as in this example.

Mrx = ϕMnx (S5.7.3.2.1-1)

where:

$\phi = 0.9$ (S5.5.4.2.1)

Mnx = Asfy(dsx – a/2) (S5.7.3.2.2-1)

Determine dsx, the distance from the top bars of the bottom reinforcing mat to the compression surface.

dsx = footing depth – bottom cvr – bottom bar dia. – ½ top bar dia. in bottom mat
= 3x(12) – 3 – 1.128 – ½ x(1.128)
= 31.3 in.

Determine As per foot of length. The maximum bar spacing across the width of the footing is assumed to be 12.0 in. in each direction on all faces (S5.10.8.2). Use 13 #9 bars and determine the actual spacing.

Actual bar spacing = [L – 2 x (side cover) – bar diameter]/(n_{bars} – 1)
= [12 – 2(3)/12 – 1.128/12]/(13 – 1)
= 11.49 in.

As = 1(12/11.49)
= 1.04 in^2

Determine "a", the depth of the equivalent stress block.

a = Asfy/0.85f'cb (S5.7.3.1.1-4)

For a strip 12 in. wide,

b = 12 in
As = 1.04 in^2
a = 1.04x(60)/[0.85x(3)x(12)] = 2.04 in.

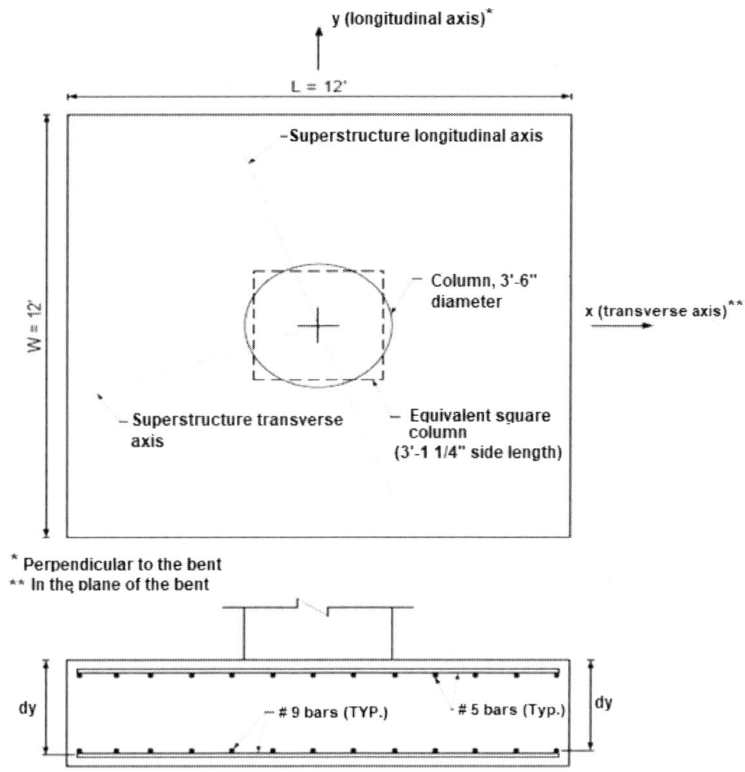

Figure 5.12. Footing Reinforcement Locations. Source: Figure 7.2-12 from AASHTO LRFD Bridge Design Specifications, FHWA Office of Bridges and Structures, 1200 New Jersey Ave., SE, Washington, DC 20590.

Calculate ϕMnx, the factored flexural resistance.

Mrx = ϕMnx

= 0.9x[1.04(60)(31.3 – 2.04/2)]/12 (S5.7.3.2.2-1)

= 141.71 k-ft/ft >Mux = 110.63 k-ft/ft OK

Applied factored moment,

Check minimum temperature and shrinkage steel (S5.10.8)

According to S5.10.8.1, reinforcement for shrinkage and temperature stresses shall be provided near surfaces of concrete exposed to daily temperature changes and in structural mass concrete. Footings are not exposed to daily temperature changes and, therefore, are not checked for temperature and shrinkage reinforcement. Nominal reinforcement is provided at the top of the footing to arrest possible cracking during the concrete early age before the footing is covered with fill.

5.2.14. AASHTO-LRFD Design Step 7.2.4.2 Limits for Reinforcement (S5.7.3.3)

5.2.14.1. Check Maximum Reinforcement (S5.7.3.3.1)

c/de = ≤ 0.42 (S5.7.3.3.1-1)

where:

c = a/β1
= 2.04/0.85
= 2.4 in/ft
c/de = 2.4/31.3
= 0.077 < 0.42 OK

Minimum reinforcement check (S5.7.3.3.2)

Unless otherwise specified, at any section of a flexural component, the amount of non-pre-stressed tensile reinforcement shall be adequate to develop a factored flexural resistance, Mr, at least equal to the lesser of:

1.2Mcr = 1.2frS

where:

fr = 0.24√f 'c (S5.4.2.6)

= 0.24x√3
= 0.42 ksi

For a 1 ft. wide strip,

3 ft. thick,

S = bh2/6
= (1x12)x(3x12)^2/6
= 2592 in3/ft

1.2Mcr = (1.2x0.42x2592)/12
= 108.9 k-ft/ft

Or

1.33Mux = 1.33x110.63
= 147.1 k-ft/ft

Therefore, the minimum required section moment resistance = 108.9 k-ft/ft. Provided moment resistance = 143 k-ft/ft > 108.9 k-ft/ft OK

5.2.14.2. Check the Moment Resistance for Moment at the Critical Transverse Face

The critical face is at the equivalent length of the shaft (7.55 ft. from the edge of the footing along the length). In the case of columns that are not rectangular, the critical section is taken at the sideof the concentric rectangle of equivalent area.

Mry = ϕMny
= ϕ[Asfy(dsy – a/2)] (S5.7.3.2.2-1)

Determine dsy, the distance from the bottom bars of the bottom reinforcing mat to the compression surface.

dsy = footing depth – cover – ½ (bottom bar diameter)
= 3x(12) – 3 – ½ x(1.128)
= 32.4 in.

Determine As per foot of length

Actual bar spacing = [W – 2(side cover) – bar diameter]/(nbars – 1)
= (12 – (2X3)/12– 1.128/12)/(13 – 1)
= 11.49 in.

As = 1X(12/11.49)
= 1.04 in^2

Determine "a", depth of the equivalent stress block.

a = As x fy/(0.85 x f'c x b)

For a strip
Width = 12 in,
b = 12 in
As = 1.04 in^2

a = (1.04X60)/(0.85X3X12) = 2.04 in.

Calculate ϕMny, the factored flexural resistance

Mry = ϕMny
= ((0.9X1.04X60)X(32.4 – 2.04/2))/12
= 146.9 k-ft/ft > Muy = 97.6 k-ft/ftOK

5.2.15. AASHTO-LRFD Design Step 7.2.4.3 Control of Cracking by Distribution of Reinforcement (S5.7.3.4)

5.2.15.1. Check Distribution about Footing Length, L

fs, allow = Z/(dcA)1/3 ≤ 0.60fy (S5.7.3.4-1)

where:

Z = 170 k/in. (moderate exposure conditions assumed, no dry/wet cycles and no harmful chemicals in the soil)

Notice that the value of the of the crack control factor, Z, used by different jurisdictions varies based on local conditions and past experience.

dc = bottom cover + ½ bar diameter
= 2+ ½(1.128)
= 2.56 in.

A = 2dc(bar spacing)
= 2X2.56X11.49
= 58.8 in^2
fs, allow = Z/[(dc x A)$^{1/3}$]
= 170/((2.56X58.8)^1/3)
= 32 ksi < 0.6X60 = 36 ksi

therefore, use fs, allow = 32 ksi

5.2.15.2. Check Actual Steel Stress, fs, Actual

For 3 ksi concrete, the modular ratio, n = 9

Maximum service load moment as shown earlier = 77.3 k-ft

The transformed moment of inertia is calculated assuming elastic behavior, i.e., linear stress and strain distribution. In this case, the first moment of area of the transformed steel on the tension side about the neutral axis is assumed equal to that of the concrete in compression.

Assume the neutral axis at a distance "y" from the compression face of the section.

Section width = bar spacing = 11.49 in.

Transformed steel area = (bar area)(modular ratio) = 1 x 9 = 9 in^2

By equating the first moment of area of the transformed steel about that of the concrete, both about the neutral axis:

9.0 x (31.3 – y) = 11.41y(y/2)

Solving the equation results in y = 6.28 in.

$I_{transformed} = A_{ts} \times (dsx - y)^2 + by^3/3$

$= 9 \times (31.3 - 6.28)^2 + 11.49 \times (6.28)^3/3$

= 6583 in4

Stress in the steel, fs, actual = (Msc/I)n, where Ms is the moment acting on the 11.41 in. wide section.

fs,actual = (77.3 x 11.49) x (31.3 – 6.28)/6583) x9

= 30.4 ksi < fs, allow = 32 ksi OK

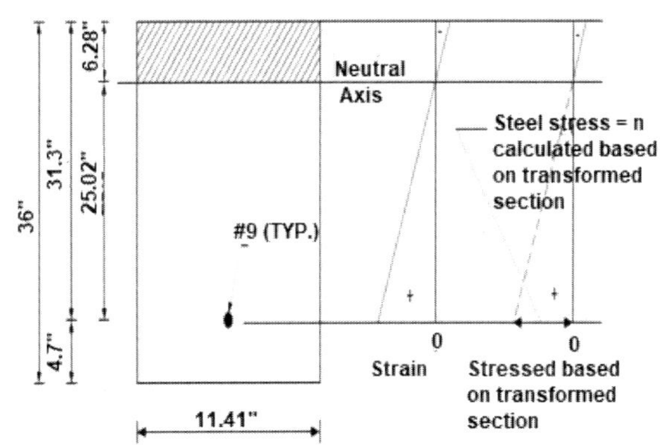

Figure 5.13. Crack Control for Top Bar Reinforcement under Service Load. Source: Figure 7.2-13 from AASHTO LRFD Bridge Design Specifications, FHWA Office of Bridges and Structures, 1200 New Jersey Ave., SE, Washington, DC 20590.

5.2.15.3. Check Distribution about Footing Width, W

This check is conducted similarly to the check shown above for the distribution about the footing length and the reinforcement is found to be adequate.

5.2.16. AASHTO-LRFD Design Step 7.2.4.4 Shear Analysis

5.2.16.1. Check Design Shear Strength (S5.8.3.3)

According to S5.13.3.6.1, the most critical of the following conditions shall govern the design for shear:

- One-way action, with a critical section extending in a plane across the entire width and located at a distance taken as specified in S5.8.3.2.
- Two-way action, with a critical section perpendicular to the plane of the slab and located so that its perimeter, bo, is a minimum but not closer than 0.5dv to the perimeter of the concentrated load or reaction area.

The subscripts "x" and "y" in the next section refer to the shear at a longitudinal face and shear at a transverse face, respectively.

5.2.16.2. Determine the Location of the Critical Face along the y-Axis

Since the column has a circular cross-section, the column may be transformed into an effective square cross-section for the footing analysis.

As stated previously, the critical section for one-way shear is at a distance dv, the shear depth calculated in accordance with S5.8.2.9, from the face of the column and for two-way shear at a distance of dv/2 from the face of the column.

Determine the effective shear depth, dvx, for a longitudinal face.

dvx = Effective shear depth for a longitudinal face per S5.8.2.9 (in.)

= dsx – a/2 (S5.8.2.9)

= 31.3– 2.04/2 = 30.3 in.

But not less than:

0.9dsx = 0.9 x 31.3 = 28.2 in.
0.72h = 0.72 x 36 = 25.9 in.

Therefore, use dvx = 30.3 in.
The critical face along the y-axis = 4.45 –30.3/12
= 1.925 ft. from the edge of the footing

5.2.16.3. Determine the Location of the Critical Face along the x-Axis

Determine the effective shear depth, dvy, for a transverse face.

dvy = effective shear depth for a transverse face per S5.8.2.9 (in.)
= dsy – a/2 = 32.4 –2.04/2 = 31.4 in.

But not less than:

0.9dsy = 0.9 x32.4 = 29.2 in.
0.72h = 0.72 x 36 = 25.9 in.
Therefore, use dvy = 31.4 in.
The critical face along the x-axis = 4.45 –31.4/12
= 1.833 ft. from the edge of the footing

See Figure 5.14 (AASHTO-LRFD Figure 7.2-14) for locations of the critical sections.

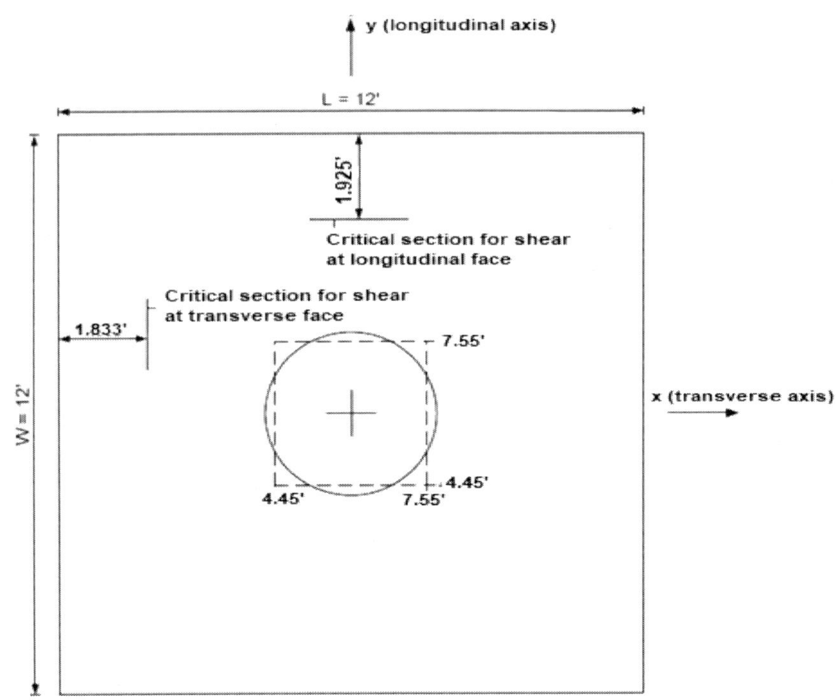

Figure 5.14. Critical Sections for Shear. Source: Figure 7.2-14 from AASHTO LRFD Bridge Design Specifications, FHWA Office of Bridges and Structures, 1200 New Jersey Ave., SE, Washington, DC 20590.

5.2.16.4. Determine One-Way Shear Capacity for Longitudinal Face (S5.8.3.3)

For one-way action, the shear resistance of the footing of slab will satisfy the requirements specified in S5.8.3.

$$Vrx = \phi Vnx \quad \text{(S5.8.2.1-2)}$$

The nominal shear resistance, Vnx, is taken as the lesser of:

$$Vnx = Vc + Vs + Vp \quad \text{(S5.8.3.3-1)}$$

Or

$$Vnx = 0.25f'cbvdvx + Vp \quad \text{(S5.8.3.3-2)}$$

$$Vc = 0.0316\beta\sqrt{f'c}bvdv \quad \text{(S5.8.3.3-3)}$$

where,

$\beta = 2$
bv = 12 in. (to obtain shear per foot of footing)
dvx = effective shear depth for a longitudinal face per S5.8.2.9 (in.)
= 30.3 in. from above
Vp = 0 k

The nominal shear resistance is then taken as the lesser of:

$Vnx = 0.0316x\ (2) \times \sqrt{3}(12) \times (30.3)$
= 39.8 k/ft and
$Vnx = 0.25 \times f'c \times bv \times$
$dv = 0.25 \times 3 \times 12 \times 30.3$
= 272.7 k/ft

Therefore, use Vnx = 39.8 k/ft
$Vrx = \phi \times$
$Vnx = 0.9 \times 39.8$
= 35.82 k/ft > applied shear Vux,
Vux = 21.9 k/ft (calculated earlier) OK

Determine one-way shear capacity for transverse face;

$$Vry = \phi \times Vny \quad \text{(S5.8.2.1-2)}$$

The nominal shear resistance, Vnx, is taken as the lesser of:

$$Vny = Vc + Vs + Vp \qquad \text{(S5.8.3.3-1)}$$

Or

$$Vny = 0.25f'cbvdvy + Vp \qquad \text{(S5.8.3.3-2)}$$

$$Vc = 0.0316 \times \beta \times \sqrt{f'c} \times bv \times dvy \qquad \text{(S5.8.3.3-3)}$$

where:

$\beta = 2$
bv = 12 in. (to obtain shear per foot of footing)
dvy = effective shear depth for a transverse face per S5.8.2.9 (in.)
= 31.4 in. from above
Vp = 0 k

The nominal shear resistance is then taken as the lesser of:

$Vcy = 0.0316(2.0)\sqrt{3}(12)(31.4)$
= 41.2 k/ft

And

$Vny = 0.25 \times f'c \times bv\ xdv$

= 0.25 x 3 x 12 x 31.4
= 282.6 k/ft

Therefore, use Vny = 41.2 k/ft
$Vry = \phi \times$
Vny = 0.9 x 41.2
= 37.1 k/ft > applied shear Vuy,
Vuy = 18.1 k/ft (calculated earlier)OK

5.2.16.5. Determine Two-Way (Punching) Shear Capacity at the Column (S5.13.3.6.3)

For two-way action for sections without transverse reinforcement, the nominal shear resistance, Vn in kips, of the concrete shall be taken as:

$Vn = (0.0636+0.126/\beta c \times \sqrt{f'c} \times bo \times dv \leq 0.126\sqrt{f'c} \times bo \times dv$

(S5.13.3.6.3-1)

where:

βc = Ratio of long side to short side of the rectangular through which the concentrated load or reaction force is transmitted = (column equivalent length)/(column equivalent width) = 3.10/3.10 = 1.0 (notice, for circular columns this ratio is always 1.0)

dv = Average effective shear depth (in.) = (dvx + dvy)/2 = (30.3 + 31.4)/2 = 30.9 in.

bo = Perimeter of the critical section (in.), the critical section is 0.5dv from the reaction area (S5.13.3.6.1). Use the circular column cross-section and cylindrical surface for punching shear.

$= 2\pi \times (42/2 + 30.9/2)$

$= 229$ in.

$Vn = (0.063 + 0.126/1.0)\sqrt{3}(229) \times (30.9) = 2{,}316$ k

The nominal shear resistance, Vn, cannot exceed $0.126\sqrt{f'c}bodv$

$Vn = 0.126\sqrt{3}(229)(30.9)$

$= 1{,}544$ k

Therefore,

$Vr = 0.9 \times 1{,}544 = 1{,}390$ k

The maximum factored vertical force for punching shear calculations equals the maximum factored axial load on the footing minus the factored weight of the footing.

P2 way = 1,374 – 1.25 x [12 x (12) x (3)] x (0.150)

= 1,293 k

The maximum shear force for punching shear calculations for a footing with the entire footing area under compression and the column at the center of the footing:

$V_{2\ way} = P_{2\ way} x$ (1 – area within punching shear perimeter/footing area)

$= 1{,}293 \times [1 - \pi((42/2 + 30.9/2)/12)2/12 \times (12)]$

$= 1{,}293 \times (1 - 0.201)$

$= 1{,}033$ k < Vr = 1,390 kOK

For footings with eccentric columns or with tension under some of the footing area, the design force for punching shear is calculated as the applied load, P2way, minus the soil load in the area within the perimeter of the punching shear failure.

5.2.17. AASHTO-LRFD Design Step 7.2.4.5 Foundation Soil Bearing Resistance at the Strength Limit State (S10.6.3)

5.2.17.1. Foundation Assumptions

Footings rest on dry cohesionless soil
Angle of internal friction of the soil (ϕf) = 32 degrees
Depth of the bottom of the footing from the ground surface = 6 ft
Soil density = 120 lb/ft3
Footing plan dimensions are 12 ft. by 12 ft

5.2.17.2. Footing Effective Dimensions

According to S10.6.3.1.1, where loads are eccentric, the effective footing dimensions L' and B', as specified in S10.6.3.1.5, shall be used instead of the overall dimensions L and B in all equations, tables, and figures pertaining to bearing capacity.

Therefore, for each load case shown in Table 5.4 (AASHTO-LRFD Table 7.2-4), a unique combination of the footing effective dimensions is used. In the following section, the case of maximum axial load on the footing will be used to illustrate the bearing capacity calculations.

The footing effective dimensions are calculated using S10.6.3.1.5 and Figure 5.15 (AASHTO-LRFD Figure SC10.6.3.1.5-1 (shown below).

$$B' = B - 2eB$$

where:

B = Eccentricity parallel to dimension B (ft.)

$$B' = 12 - 2(121/1{,}374)$$

$$= 11.82 \text{ ft.}$$

$$L' = L - 2eL \qquad \text{(S10.6.3.1.5-2)}$$

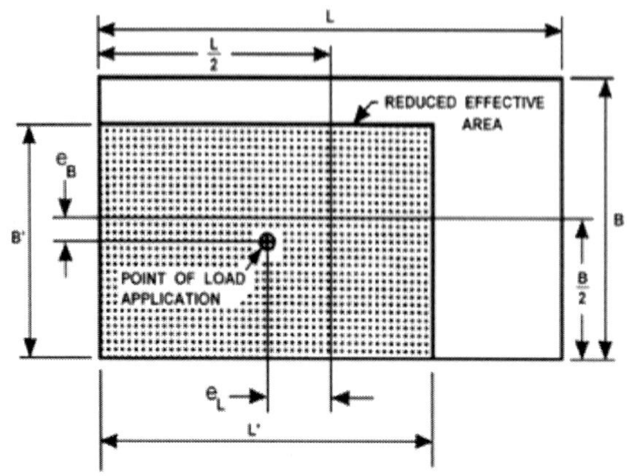

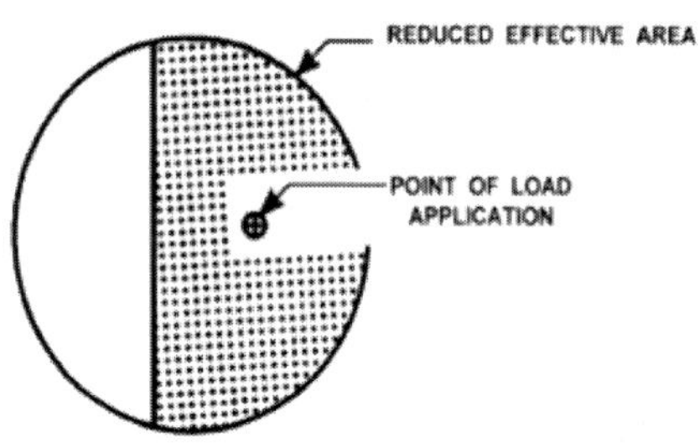

Figure 5.15. Reduced Footing Dimensions (Reproduced from the Specifications). Source: Figure SC10.6.3.1.5-1 from AASHTO LRFD Bridge Design Specifications, FHWA Office of Bridges and Structures, 1200 New Jersey Ave., SE, Washington, DC 20590.

where:

eL = Eccentricity parallel to dimension L (ft.)

L′ = 12 – 2(626/1,374) (S10.6.3.1.5-2)

= 11.09 ft.

According to S10.6.3.1.2c, for cohesionless soil, the nominal bearing resistance of a layer of the soil in TSF may be determined as:

qult = 0.5γBCw1Nγm + γCw2DfNqm (S10.6.3.1.2c-1)

where:

Df = Depth of footing from ground level (ft.)
= 6 ft.
γ = Total, i.e., moist density of sand or gravel (TCF)
= 120/2,000
= 0.06 TCF

B = footing width (ft.)= smaller of 11.82 and 11.09 ft.
= 11.09 ft.
Cw1, Cw2 = Coefficients as specified in AASHTO LRFD Table S10.6.3.1.2c-1 as a function of Dw (dimensionless)
= for dry soil with a large depth, CW1 = CW2 = 1
Dw = Depth to water surface taken from the ground surface (ft.)
= Assume a large distance relative to the footing dimensions
$N\gamma m$, Nqm = Modified bearing capacity factor (dimensionless)

Substituting in Eq. S10.6.3.1.2c-1:
qult = 0.5 x (0.06) x (11.09) x (1.0) x $N\gamma m$ + 0.06 x (1.0) x (6) x Nqm = 0.334$N\gamma m$ + 0.36Nqm

From Eqs. S10.6.3.1.2c-2 and -3

$N\gamma m = N\gamma S\gamma c\gamma i\gamma$ (S10.6.3.1.2c-2) Nqm

= NqSqcqiqdq (S10.6.3.1.2c-3)

where:
$N\gamma$ = Bearing capacity factor as specified in AASHTO LRFD Table S10.6.3.1.2c-2 for footings on relatively level ground
Nq = Bearing capacity factor as specified in AASHTO LRFD Table S10.6.3.1.2c-2 for relatively level ground
Sq, $S\gamma$ = Shape factors specified in AASHTO LRFD Tables S10.6.3.1.2c-3 and -4,respectively
cq, $c\gamma$ = Soil compressibility factors specified in AASHTO LRFD Tables S10.6.3.1.2c-5
iq, $i\gamma$ = Load inclination factors specified in AASHTO LRFD Tables S10.6.3.1.2c-7 and -8
dq= Depth factor specified in AASHTO LRFD Table S10.6.3.1.2c-9

From AASHTO LRFD Table S10.6.3.1.2c-2: $N\gamma = 30$ for $\phi f = 32$ degrees.
From AASHTO LRFD Table S10.6.3.1.2c-2: Nq = 23 for $\phi f = 32$ degrees.

$L'/B' = 11.82/11.09 = 1.07$

Interpolate between $L'/B' = 1$ and 2. However, using values corresponding to L'/B' = 1.0 will not lead to significant change because $L'/B' \approx 1.0$.

From AASHTO LRFD Table S10.6.3.1.2c-3: Sq = 1.62 for L′/B′ = 1.0 and $\phi f = 30$ degrees.

From AASHTO LRFD Table S10.6.3.1.2c-4: Sγ = 0.6 for L′/B′ = 1.0 and $\phi f = 30$ degrees.

Soil stress at the footing depth before excavation, q = 0.06(6) = 0.36 TSF

For AASHTO LRFD Tables S10.6.3.1.2c-5 and -6, either interpolate between q = 0.25 and q = 0.5 or, as a conservative approach, use the value corresponding to q = 0.5. For this example, the value corresponding to q = 0.5 TSF is used.

From AASHTO LRFD Table S10.6.3.1.2c-5: cq, cγ = 1.0 for q = 0.5 and $\phi f = 32$ degrees.

The maximum factored horizontal load on the bottom of the column from the bent analysis equals 46.0 and 26.0 kips in the transverse and longitudinal directions, respectively. In AASHTO LRFD Table S10.6.3.1.2c-7, it is intended to use the un-factored horizontal and vertical loads. However, due to the small ratio of horizontal to vertical loads, using the factored loads does not affect the results.

Horizontal-to-vertical load ratio:

H/V = 44.8/1,374
= 0.033 in the transverse direction
H/V = 26.0/1,374
= 0.019 in the longitudinal direction

AASHTO LRFD Table S10.6.3.1.2-7 lists values for iq, iγ that correspond to horizontal-to-vertical load ratios of 0.0 and 0.1. Interpolation between the two values is acceptable. A more conservative approach is to use the value corresponding to H/V = 0.1.

From AASHTO LRFD Table S10.6.3.1.2c-7: iq = 0.85 for square footing with H/V = 0.1

From AASHTO LRFD Table S10.6.3.1.2c-7: iγ = 0.77 for square footing with H/V = 0.1

AASHTO LRFD Table S10.6.3.1.2c-9 lists values for dq that correspond to a friction angle, $\phi f \geq 32$ degrees and for $Df/B \geq 1.0$. For this example, $\phi f = 30$ degrees and Df/B = 6/11.13 = 0.54

By extrapolation from AASHTO LRFD Table S10.6.3.1.2c-9, use dq = 1.05

Substituting in Eqs. S10.6.3.1.2c-2 and -3:

$$N\gamma m = N\gamma \times S\gamma \times c\gamma \times i\gamma \qquad \text{(S10.6.3.1.2c-2)}$$

$$= 30 \times 0.6 \times 1.0 \times 0.77 = 13.86$$

Nqm = Nq x Sq x cq x iq x dq (S10.6.3.1.2c-3)

= 23 x 1.62 x 1.0 x 0.85 x 1.05 = 33.3

Therefore,
qult = 0.333 x Nγm + 0.36 x Nqm = 0.333 x (13.86) + 0.36 x (33.3) = 16.6 TSF

5.2.17.3. Resistance Factor

From AASHTO LRFD Table S10.5.5-1, several resistance factors are listed for cohesionless soil (sand). The selection of a particular resistance factor depends on the method of soil exploration used to determine the soil properties. Assuming that ϕ was estimated from SPT data, the resistance factor = 0.35.

According to S10.6.3.1.1,
qR = ϕ x qn = ϕ x qult
= 0.35 x 16.6 = 5.81 TSF

Footing load resistance = (qR) x (footing effective area)
= 5.81 x 11.82 x 11.09
= 762.0 Tons
= 1,524 k > 1,374 k applied OK

The soil load resistance check may be repeated using the same procedures for other load cases.

5.3. Computer Applications for the Design of Bridge Piers in AASHTO LRFD

The computer applications for the design of bridge piers in AASHTO LRFD, BS Eurocode II and IRC 112/IRC 6are described in detail in the softcopy volume 'User's Guide for Hands on Practice with Computer Applications' for the book 'Computer Aided Bridge Engineering' is available for download from web site 'www.roadbridgedesign.com'.

REFERENCES

ASTRA Pro software, User's Guide, Tutorial Videos, *User's Manual and Design Manual*, Techsoft Engineering Services, Web site: www.techsoftglobal.com and www.roadbridgedesign.com

Bridge Engineering Handbook, Second Edition, Superstructure Design, Edited by Wai Fah Chen and Lian Duan, Published by CRC Press, Taylor & Francis Group, 6000 Broken Sound Parkway, NW, Suite 300, Boca Raton, FL 33487, Web site: www.crcpress.com

Design Specifications, Customary U.S. Units, Sixth Edition 2012, American Association of State Highway and Transportation Officials. 444 North Capitol Street, NW, Suite 249, Washington, DC 20001, Phone 202-624-5800/Fax 202-624-5806, Web site: www.transportation.org

Chapter 6

DESIGN OF PRE-STRESSED POST TENSION BOX-GIRDER BRIDGE SUPERSTRUCTURES IN AASHTO-LRFD

ABSTRACT

The efficient dispersal of congested traffic, economic considerations, and aesthetic desirability have increased the popularity of box girder bridges these days in modern highway, including urban interchanges and metro rail systems.

Box girders are prominently used in freeway and bridge systems due to their structural efficiency, serviceability, better stability, pleasing aesthetics and economy of construction. The box girders are an efficient form of construction for bridges because they minimize weight, while maximizing flexural stiffness and capacity. Box girders have high torsional stiffness and strength, compared with an equivalent member of open cross section. Although significant research has been underway on advanced analysis for many years to better understand the behavior of all types of box-girder bridges, the results of these various research works are scattered and unevaluated. Hence, a transparent understanding of more recent work on straight and curved box-girder bridges is highly desired. The main objective is to give a clear vision about non composite straight and curved box girder bridges. This study would enable bridge engineers to better understand the behavior of box girders with different variations in parameters such as curvature and shape.

In AASHTO-LRFD Standard, the design of pre-stressed concrete (PSC) box girder is done for bending moments and shear forces by dead load, super imposed dead load (SIDL) and live loads (LL). The shrinkage strain, creep strain and effect of temperature rise and fall are also determined. The design is complete for pre-stressing cables, end anchorages, and un-tensioned reinforcement.

Note: The values in 'Red Color' are Design Input Data by the User.

6.1. General

Post-tensioned concrete box girders are widely used in the highway bridges in California. Figure 6.1 shows the post-tensioned (CIP/PT) concrete box girder bridge. Post-tensioned (PT) tendons have been widely utilized in concrete bridges in the United States. The advantages of PT bridges compared to bridges constructed using conventional reinforcement include greater span length, structural efficiency, reduced materials, and a more streamlined appearance. However, PT tendons can be susceptible to corrosion and ultimately failure if physical deficiencies (PDs) or chemical deficiencies (CDs) are present. Examples of PDs include separation, segregation, presence of soft material, and free water, while an example of a CD includes concentrations of chloride that exceed the allowable limit as specified by the American Association of State Highway and Transportation Officials and other specifications. The failure of a few tendons can compromise overall structural integrity. Inspections of bridge PT tendons have revealed both PDs and CDs as well as strand tendon failures caused by corrosion have been reported.

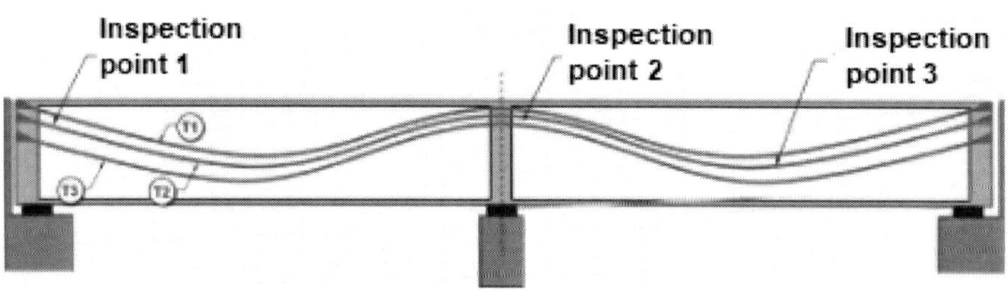

Figure 6.1. Post-tensioned concrete box girders.

6.2. Basic Concepts of Post Tension

Post tensioning is one method of pre-stressing concrete. The concrete members are cast first. Then, after the concrete has gained sufficient strength, tendons (strands of high strength steel wire) are inserted into preformed ducts and tensioned to induce compressive stresses in the expected tensile stress regions of the member. Concrete must be free to shorten under the pre-compression. The strands are then anchored and a corrosion protection such as grout or grease is installed (Gerwick, 1997).

Before further discussing pre-stressing, we should compare it with conventionally reinforced concrete. Prior to gravity loading, the stress level in conventional reinforced concrete is zero.

The reinforced steel is only activated by the placement of the gravity load. The concrete and reinforced steel act as a composite section. However, once the tensile capacity of the concrete surrounding the longitudinal reinforcement has been surpassed, the concrete cracks. Pre-stressed concrete activates the steel prior to gravity loading through pre-stressing the reinforcement. This prevents cracking at service loads in pre-stressed concrete.

Pre-stressed concrete utilizes high strength materials effectively. Concrete is strong in compression, but weak in tension. High tensile strength of pre-stressing steel and high compressive strength of concrete can be utilized more efficiently by pre-tensioning high strength steel so that the concrete remains in compression under service loads activated while the surrounding concrete is compressed. The pre-stressing operation results in a self-equilibrating internal stress system which accomplishes tensile stress in the steel and compressive stress in the concrete that significantly improves the system response to induced service loads (Collins and Mitchell, 1997).

The primary objectives of using pre-stressing is to produce zero tension in the concrete under dead loads and to have service load stress less than the cracking strength of the concrete along the cross section. Thus the steel is in constant tension. Because of this the concrete remains in compression under service loads throughout the life of the structure. Both materials are being activated and used to their maximum efficiency. Figure 6.2 shows elastic stress distribution for a pre-stressed beam after initial pre-stressing.

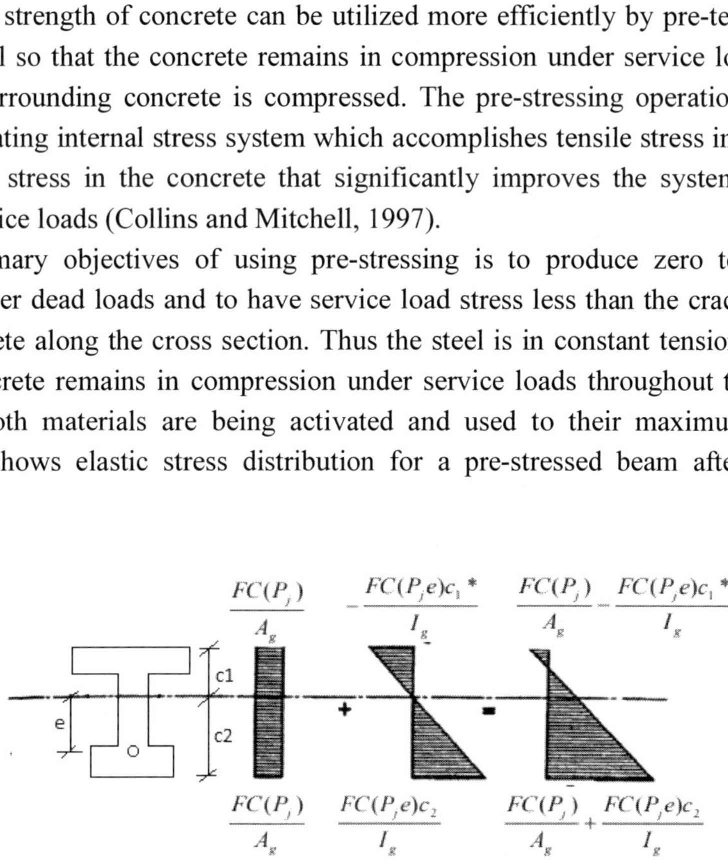

*Note: Component of equation is negative because c is on opposite side of centre of gravity from the tendon. Tension is denoted as negative (-), compression is denoted as positive (+).

Figure 6.2. Elastic stresses in an un-cracked pre-stress beam. Effects of initial pre-stress by component (Nilson, 1987).

The stress at any point of the cross-section can be expressed as:

$$f_{pe} = \frac{FC(P_j)}{A_g} \pm \left(\frac{FC(P_j e)y}{I_g} + \frac{MC_s(P_j)y}{I_g} \right) \quad (6.1)$$

where:

Ag=gGross area of section (in^2)

e=Eccentricity of resultant of pre-stressing with respect to the centroid of the cross section. Always taken as a positive (ft)

FC=Force coefficient for loss

fpe=Effective stress in the pre-stressing steel after losses (ksi)

Ig=Moment of inertia of the gross concrete section about the centroidal axis, neglecting reinforcement (in^4)

Pj=Force in pre-stress strands before losses (ksi)

MCs =Secondary moment force coefficient for loss (ft)

y=Distance from the neutral axis to a point on member cross-section (in.)

The pre-stressing force effect is accomplished by two components of the general equation shown above as Eqn-6.1 The first component is uniform compression stress due to the axial pre-stressing force. The second component is the bending stress caused by eccentricity of the pre-stressing steel with respect to the center of gravity of the cross section. This creates a linear change in stress throughout the beam cross section (Figure 6.2). It is noted that the distance from the neutral axis to the fiber in question y (y is the general term, c1 and c2 which are more specific terms shown in Figure 6.2), may result in a negative value for the bending part of the equation. It is possible the pre-stressing force will create tension across the center of gravity from the tendon, and therefore part of the beam section may be in tension prior to applying load. The use of pre-stressed concrete has its advantages and limitations. Some limitations are its low superstructure ductility, the need for higher concrete compressive strengths, and larger member sizes to accommodate ducts inside the girders.

Post-tension box girder superstructures are commonly used due to their low costs, their performance throughout the life of the structure, and contractor's experience with the structure type. Post-tensioning also allows for thinner superstructures. A continuous superstructure increases the stiffness of the bridge frame in the longitudinal direction and gives the designer the option to fix the columns to the superstructure, reducing foundation costs.

6.3. Material Properties

At first glance, pre-stressed concrete and reinforced concrete make use of the same two core materials: concrete and steel. However, the behavior of the materials vary due to usage. Conventional concrete structures use deformed bars for reinforcement. Most pre-stressed applications use tightly wrapped, low-relaxation (lo-lax) seven-wire strand. As shown in Figure 6.3, the stress strain curves for those steel are quite different.

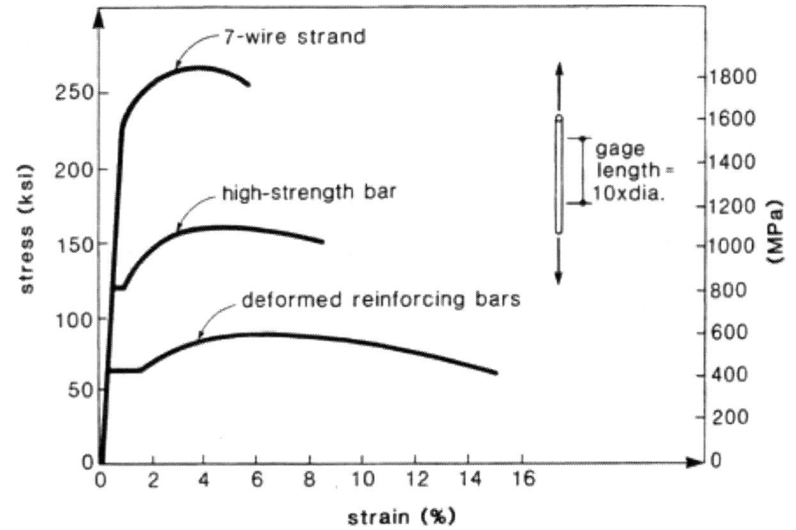

Figure 6.3. Stress-strain curves of mild steel (deformed reinforcing bars) and pre-stress steel (7-Wire Strand) (Collins and Mitchell, 1997).

6.4. Design of Pre-Stressed Post Tension Box-Girder Bridge

The span arrangement and configuration should be closely studied, first in respect to the site location of the bridge. A bridge crossing over a navigable waterway is very much dictated by the horizontal and vertical clearance required. It is also important to know the soil condition and landscapes (e.g., over water, land, valley, or mountainous area). For segmental concrete bridges, uniformity of the span lengths is critical in order to maximize the benefit of pre-casting the segments. The more uniform the span distribution, the more economical the bridge. It is also preferable to have an uneven number of spans from the architectural point of view. A range of 20 to 50 metres for PSC I-Girder and 30 to 60 metres for incremental launched box girders bridges are recommended.

It is not recommended to place many piers with short span lengths. Therefore, the number of piers should be reduced and the span length increased. The shape and size of the piers are also important. For shorter span length, the lateral pier dimensions should be slender in order to reduce the wall view effect from an oblique view. For shallow valley, it is important to consider the L/H ratio of the opening between two piers, where L is the span length and H is the pier height. It is preferable to have an L/H ratio equal to or greater than 1.5. The end spans should be less than the typical span length (60% to 80% of typical span length) in order to achieve an efficient design.

The next step is to study the structural system of the bridge. The final bridge structural system is studied carefully, particularly the stability during construction and how the static system changes from stage to stage of the construction until the bridge is completed. This is the nature of segmental bridges, unlike other types of structures such as reinforced concrete or steel conventional types of bridges. The erection and stitching of the segment pieces require many stages and the structural system is constantly changing from time to time until completion.

Segmental bridge construction is not only popular for long span bridges over rivers and canyons but also in urban and city interchanges and for transit structures such as light rail and high-speed rail guide-way structures. Aesthetic value has become one of the most important parameters in selecting the bridge structure. The owners and stakeholders are placing more and more importance on aesthetically pleasing structures. Fortunately for segmental box girder bridges, the shape of the box girder is already inherently attractive. The architect should properly design the pier shape to ensure that the overall look of the bridge is compatible and suitable for a particular location.

The design forces obtained from the planer frame analysis and 3D live load influence lines may be combined in a spreadsheet using the LRFD Service Limit State and Strength Limit State combinations. The maximum tensile and compressive stresses at each predetermined section in the top slab are summarized and compared to the stress limits at Service Limit States specified in AASHTO-LRFD Bridge Design Specifications (AASHTO 2012).

The pre-stressing force is usually estimated in preliminary calculations, and then analyzed in a planer, time-dependent run using the design software. All other loads are incorporated into the planer model, except live loads. The results are then compiled in a spreadsheet to check stresses. By varying the pre-stressing force, the combined stresses of service limit states are calculated. Using the selected tendon forces per unit length, the size and spacing of transverse tendons in the segment are determined.

The LRFD Strength Limit States may be also tabulated in a spreadsheet program of software ASTRA Pro and an envelope of maximum and minimum values is determined for each chosen section. The values of the moment envelope can then be compared to the calculated bending capacities for each of the corresponding transverse components. Concrete service stresses were computed using several AASHTO standard specifications and the current LRFD specifications for comparison purposes.

6.4.1. Service Limit State Design (3.5.5)

When checking concrete stresses, only Service Limit State I is checked with a live load factor of 1.0 for tension as well as compression. Also, a linear temperature gradient of 10° F between interior and exterior surfaces of the box is used in Service Limit State I. The current LRFD specification does not specify this loading, leaving it up to the owner or designer to establish if it should be included on a project-by-project basis. This example is based on a load factor of 0.5 for transverse temperature gradient when accompanying live load. Also, in addition to Service Limit State I, LRFD requires a check for service load stresses due to dead load and full temperature gradient. This limit state can often govern at locations where live load influences are small.

In addition to service limit states under maximum loading, temporary stresses such as those prior to barrier placement and vehicular traffic should be checked to ensure that allowable stresses are not exceeded during the construction process. Service load combinations used in this example include the following:

Service I (Tension & Compression)

1.0(DC+DW+ EL)+1.0(PT)+1.0(LL + IM)+1.0(CR + SH)+/− 0.5(TG)

Segmental Load Combination (LRFD Equation 3.4.1-2)

1.0(DC+DW+ EL)+1.0(PT)+1.0(CR + SH)+/−1.0(TG)

6.4.2. Strength Limit State Design (3.5.6.1 Flexural Strength Design Check)

For purposes of the transverse design, Strength Limit State IV is the same as Strength Limit State I without live load, with 25% more self-weight. This loading does not govern in this example.

For temperature gradient load factors, LRFD specifications suggest determining a load factor on a project-specific basis, with a recommendation of 0.0 for most instances. Since these loads are a result of restrained deformations, the loads should disappear if the reinforcement begins to yield at ultimate. In addition, the Segmental Guide Specifications do not include this component in ultimate load combinations. For these reasons, the temperature gradient was not used in the strength limit state combinations below.

The LRFD specifications require minimum reinforcement equal to that required to resist 1.2 times the cracking moment. This requirement governs only the bottom slab (soffit) design. To satisfy the minimum steel requirement, the transverse bar spacing in the bottom soffit was decreased from 12 (205)to 8 in (203 mm), which represents an increase in reinforcement of 50%.

Also, under ultimate flexure, the amount of web steel reinforcing required for transverse bending should be calculated. This should be combined in an appropriate manner with reinforcing required for longitudinal shear. Listed below is the ultimate load combination per LRFD: (Strength I)

$$\gamma_p DC + \gamma_p DW + 1.0EL + 1.75(LL + IM) + 0.5(CR + SH)$$

6.4.3. Shear Strength Design Check (3.5.6.2)

Traditionally, shear behavior has been ignored in the design of concrete decks for AASHTO bridges. Box girder decks are similar in this sense, but can often have large construction loads placed on them. In these special cases, both one-way and two-way action shear should be investigated.

6.4.4. Longitudinal Design (3.6.1 Design Methodology)

The following design example illustrates the longitudinal design process for a typical precast segmental bridge. The structure is assumed to be erected using the precast balanced cantilever construction method with a ground-based crane. Owing to changes in the static system during erection, as cantilevers are made continuous through cast-in-place closure joints, it is necessary to analyze the structure for time-dependent effects. Time-dependent analysis is a function of the segment casting date, times that the segments are incorporated into the structure, as well as dates associated with changes in the static system throughout the construction process.

It is customary to establish an assumed sequence of construction and to estimate a reasonable construction schedule. Casting and erection dates of the segments are established on the basis of the construction schedule and the segment production rate. Casting dates are a function of an assumed number of casting cells and the time required to cast each segment. For purposes of estimating these dates, the segment production rate is assumed as one typical segment per day per casting cell and one pier/expansion joint segment per week per casting cell. Segments are not to be erected earlier than one month after casting. During construction, when actual casting and erection dates become available, the stage-by-stage analysis should be rerun in order to obtain correct camber values.

Time-dependent properties of concrete are established on the basis of the environmental humidity and dimensions of the cross section, and can be adjusted for concrete composition (e.g., limestone aggregate), rate of hardening, and ambient temperature. Section properties are determined for each segment considering the effects of shear lag in the top and bottom slab.

The above information may be entered into a time-dependent analysis software. A stage-by-stage analysis may be performed using an assumed post-tensioning layout while carefully modeling appropriate boundary conditions for each step of the construction process. After the construction has been modeled, the structure may be stepped through time day 10,000 to allow all time-dependent effects to occur. It is also essential in statically indeterminate structures to sum up all locked-in forces that result from various stages of structural systems until day 10,000. Additional loads may be placed on the structure such as live load, temperature gradient, and support settlement, as appropriate, and analyzed for the initial (at end of construction) and final conditions at day 10,000.

6.4.5. Tendon Layout and Envelope (3.6.2)

An approximate tendon layout can be based on preliminary calculations for construction loading of an atypical cantilever. Span continuity tendons can be estimated by preliminary design based on the final structure's approximate creep and shrinkage effects using load factor dead and live load combinations. The assumed layout can then easily be modified during the final design to satisfy all applicable LRFD Limit State Load Combinations.

The preliminary design for this example indicates the need for twelve cantilever tendons and five bottom continuity tendons per web. On the basis of previous experience, two eight-strand continuity tendons were added in the top slab across the closure pour to control stresses resulting from temperature gradients. The final design resulted in an increase of one cantilever tendon and one bottom span continuity tendon at interior spans only.

The tendons used are based on a 12-strand system using 0.6 in. (15.24 mm) diameter strands. Only 11 strands were used for bottom continuity tendons to provide space for 5% contingency post-tensioning as required for internal tendons. One out of 12 strands will provide approximately 8% of the contingency post-tensioning if needed. An empty duct was provided for the cantilever tendons combined with an anchorage on the last segment of the cantilever to allow for contingency post-tensioning. This empty duct should be grouted if no contingency tendons are required.

6.4.6. Design of Top RCC Deck Slab

6.4.6.1. 2-Span Cast-in-Place Post-Tensioned Concrete Box Girder Bridge [CIPPTCBGB] Design

2-Span Cast-in-Place Post-Tensioned Concrete Box Girder Bridge [CIPPTCBGB] Design. This worksheet illustrates the design of a two span cast-in-place post-tensioned concrete box girder bridge. The bridge has spans of 118 feet and 130 feet. The bridge has zero skew. Standard ADOT 42-inch F-shape barriers will be used resulting in a bridge configuration of 1'-7" barrier, 12'-0" outside shoulder, two 12'-0" lanes, a 6'-0" inside shoulder and a 1'-7" barrier. The overall out-to-out width of the bridge is 45'-2". A plan view and typical section of the bridge are shown in Figures 6.4 and 6.5.

The following legend is used for the references shown in the left-hand column:

[2.2.2] LRFD Specification Article Number
[2.2.2-1] LRFD Specification Table or Equation Number
[C2.2.2] LRFD Specification Commentary
[A2.2.2] LRFD Specification Appendix
[BDG] ADOT LRFD Bridge Design Guideline

Note: The values below are Design Input Data by the User.

Bridge Geometry
Span 1 length 130 ft
Span 2 length 118 ft
Bridge width42 ft
Roadway width42.00 ft
Superstructure depth5.50 ft
Web spacing7.75 ft
Web thickness12.00 in
Top slab thickness8.00 in
Bottom slab thickness6.00 in
Deck overhang2.63 ft
Bar dia = 0.625 in

6.4.6.2. AASHTO LRFD [Table 2.5.2.6.3-1] Minimum Requirements

The minimum span to depth ratio for a multi-span bridge should be taken as 0.04 resulting in a minimum depth of 0.04 x 118= 4.72 ft. Use 5'-6".

AASHTO LRFD Cl. [9.7.1.1] [BDG]

The minimum top slab thickness shall be as shown in the LRFD Bridge Design Guidelines. For a centerline spacing of 7.75 feet, the effective length is 6.75 feet resulting in a minimum thickness of 8.00 inches. The minimum overhang thickness is 9.00 inches, one inch thicker than the interior slab.

AASHTO LRFD Cl. [5.14.1.5.1b] [BDG]

Centerline spacing = 7.75 ft.
Effective length = 6.75 ft.
Minimum thickness = 8 in.
Overhang thickness = 9 in.
Minimum thickness of the web = 12 in.

The minimum bottom slab thickness shall be the larger of:

1/30 the clear web spacing =6.75 x 12 /30 = 2.7 inches 6.0 inches.

AASHTO LRFD Cl. [C5.14.1.5.1c] [BDG]

The minimum thickness of the web shall be =12 in.

Concrete Deck Slab Minimum Requirements

Slab thickness = 8.00 in.
Top concrete cover = 2.50 in.
Bottom concrete cover = 1.00 in.
Wearing surface = 0.50 in.

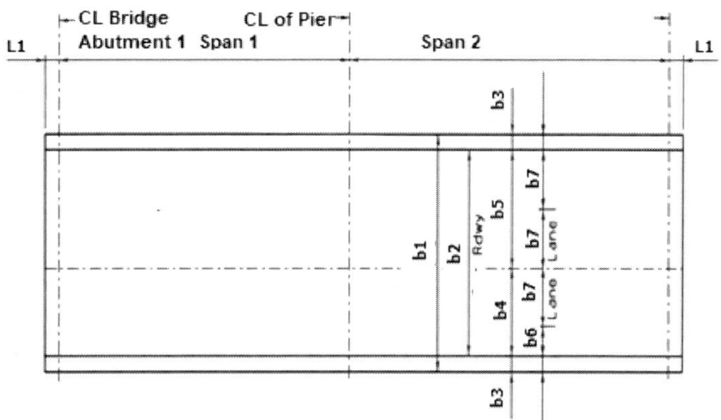

LOCATION PLAN

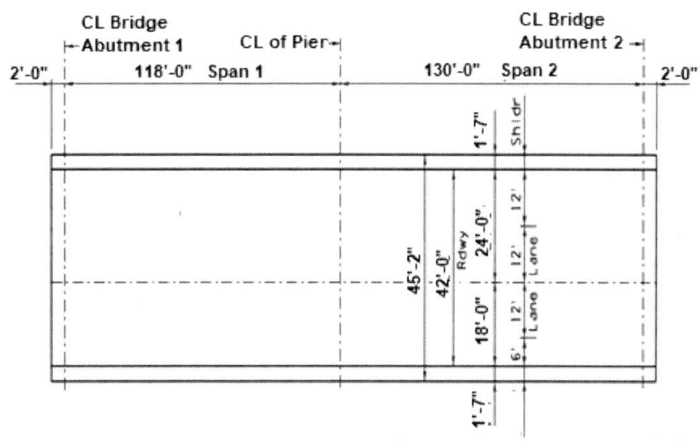

LOCATION PLAN

Figure 6.4. Concrete Deck Slab Minimum Requirements.

L1 = 2 ft.
b1 = 45.17 ft.
b2 = 42 ft.
b3 = 1.583 ft.
b4 = 18 ft.
b5 = 24 ft.
b6 = 12 ft.
b7 = 12 ft.

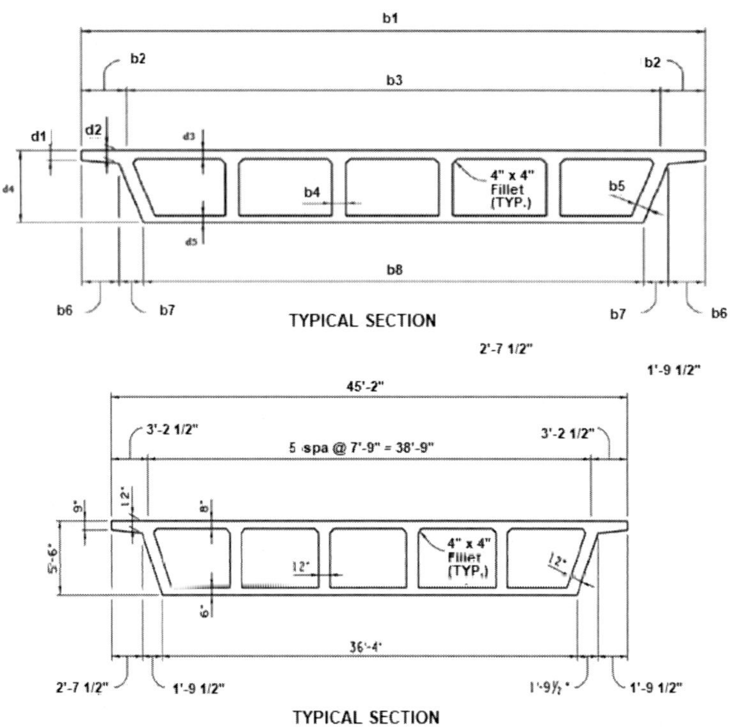

Figure 6.5. Multi-Cell Box Girder Cross Section.

b1= 542.00 in d1= 9.00 in.
b2= 38.50 in d2= 12.00 in.
b3= 465.00 in d3= 8.00 in.
b4= 12.00 in d4= 66.00 in.
b5= 12.00 in d5= 6.00 in.
b6= 31.50 in.
b7= 21.50 in.
b8= 436.00 in.

Material Properties

6.4.6.3. AASHTO Cl. [5.4.3.1] and [5.4.3.2] Reinforcing Steel

Yield Strengthfy =60 ksi
Modulus of ElasticityEs =29,000 ksi

6.4.6.4. AASHTO LRFD [Table 5.4.4.1-1] and [5.4.4.2] Pre-Stressing STRAND

Low relaxation pre-stressing strands
0.6" diameter strandAps = 0.217 in2
Tensile Strength fpu =270 ksi
Yield Strengthfpy =243 ksi
Modulus ElasticityEp =28500 ksi

6.4.6.5. AASHTO LRFD Cl. [5.4.2.1] [BDG] Concrete

The final and release concrete strengths are specified below:

Superstructure Column & Drilled Shaft

f'c = 4.5 ksi f'c =3.5 ksi
f'ci = 3.5 ksi

AASHTO LRFD [Table 3.5.1-1]

Unit weight for normal weight concrete is listed below:

AASHTO LRFD Cl. [C3.5.1]

Unit weight for computingEc =0.145 kcf
Unit weight for DL calculation=0.150 kcf

AASHTO LRFD Cl. [C5.4.2.4]

The modulus of elasticity for normal weight concrete where the unit weight is 0.145 kcf may be taken as shown below:

$E_c = 1820\sqrt{f'c} = 1820 \times \sqrt{4.5} = 3861$ ksi
$E_{ci} = 1820\sqrt{f'ci} = 1820 \times \sqrt{3.5} = 3405$ ksi

AASHTO LRFD Cl. [5.7.1]

The modular ratio of reinforcing to concrete should be rounded to the nearest whole number.

n =29000/3861=7.51 Usen =8

AASHTO LRFD Cl. [5.7.2.2]

$\beta 1$ = The ratio of the depth of the equivalent uniformly stressed compression zone assumed in the strength limit state to the depth of the actual compression zone stress block.

$\beta 1 = 0.85 - 0.05 \times [f'c-4.0/1.0] = 0.85 - 0.05 \times [4.5 - 4/1] = 0.825$

6.4.6.6. AASHTO LRFD Cl. [5.4.2.6] Modulus of Rupture

The modulus of rupture for normal weight concrete has two values. When usedto calculate service level cracking, as specified in Article 5.7.3.4 for side reinforcing or in Article 5.7.3.6.2 for determination of deflections, the following equation should be used:

$fr = 0.24\sqrt{f'c}$

Service Level Cracking

For superstructure calculations:

$fr = 0.24 \times \sqrt{4.5} = 0.509$ ksi

For substructure calculations:

$fr = 0.24 \times \sqrt{3.5} = 0.449$ ksi

When the modulus of rupture is used to calculate the cracking moment of a member for determination of the minimum reinforcing requirement as specified in Article 5.7.3.3.2, the following equation should be used:

Minimum Reinforcing

$fr = 0.37 \times \sqrt{4.5} = 0.785$ ksi

For substructure calculations:

$fr = 0.37 \times \sqrt{3.5} = 0.692$ ksi

6.4.6.7. AASHTO LRFD Cl. [1.3.2] Limit States

In the LRFD specification, the general equation for design is shown below:

$\Sigma\, \eta_i \times \gamma_i \times Q_i \leq = \phi \times Rn = Rr$

For loads for which a maximum value of γi is appropriate:

$\eta_i = \eta_D \times \eta_R \times \eta_I \geq 0.95$

For loads for which a minimum value of γi is appropriate:

$\eta_i = 1/\ \eta_D \times \eta_R \times \eta_I \leq 1.0$

6.4.6.8. AASHTO LRFD Cl. [1.3.3] Ductility

For strength limit state for conventional design and details complying with the LRFD specifications and for all other limit states:

$\eta_D = 1$

6.4.6.9. AASHTO LRFD Cl. [3.4.1] [BDG] Redundancy

For the strength limit state for conventional levels of redundancy and for all other limit states:

$\eta_R = 1$

6.4.6.10. AASHTO LRFD Cl. [1.3.4] and [1.3.5] Operational Importance

For the strength limit state for typical bridges and for all other limit states:

$\eta_I = 1$

AASHTO LRFD Cl. [3.4.1] [BDG]

For an ordinary structure with conventional design and details and conventional levels of ductility, redundancy, and operational importance, it can be seen thatηi = 1.0 for all cases. Since multiplying by 1.0 will not change any answers, the load modifier ηi has not been included in this example.

For actual designs, the importance factor may be a value other than one. The importance factor should be selected in accordance with the ADOT LRFD Bridge Design Guidelines.

6.4.6.11. Deck Design [BDG]

As bridges age, decks are one of the first elements to show signs of wear and tear. As such, ADOT has modified some LRFD deck design criteria to reflect past performance of decks in Arizona. Section 9 of the Bridge Design Guidelines provides a thorough background and guidance on deck design.

6.4.6.12. AASHTO LRFD Cl. [9.7.2.3] Effective Length [BDG]

ADOT Bridge Design Guidelines specify that deck design be based on the effective length rather than the centerline-to-centerline distance specified in the LRFD specifications. The effective length for monolithic cast-in-place concrete is the clear distance between supports. For this example with a center line-to-centerline web spacing of 7.75 feet and web width of 12 inches, the effective length is 6.75 feet. The resulting minimum deck slab thickness as per ADOT guidelines is 8.00 inches.

6.4.6.13. AASHTO LRFD Cl. [9.6.1] Method of Analysis [BDG]

In-depth rigorous analysis for deck design is not warranted for ordinary bridges. The empirical design method specified in AASHTO LRFD [9.7.2] is not allowed by ADOT Bridge Group. Therefore the approximate elastic methods specified in [4.6.2.1] will be used. Dead load analysis will be based on a strip method using the simplified moment equation of [w S2 / 10], for both positive and negative moments, where "S" is the effective length.

6.4.6.14. AASHTO LRFD Cl. [A4.1] Live Loads

The un-factored live loads found in Appendix A4.1 will be used. Multiple presence and dynamic load allowance are included in the chart. Since ADOT bases deck design on the effective length, the chart should be entered under "S" equal to the effective length of 6.75 feet rather than the centerline-to-centerline distance of 7.75 feet. Since the effective length is used the correction for negative moment from centerline of the web to the design section should be zero. Entering the chart yields the following live load moments:

Pos M =5.1 ft-k/ft
Neg M =-5.5 ft-k/ft(0 inches from centerline)

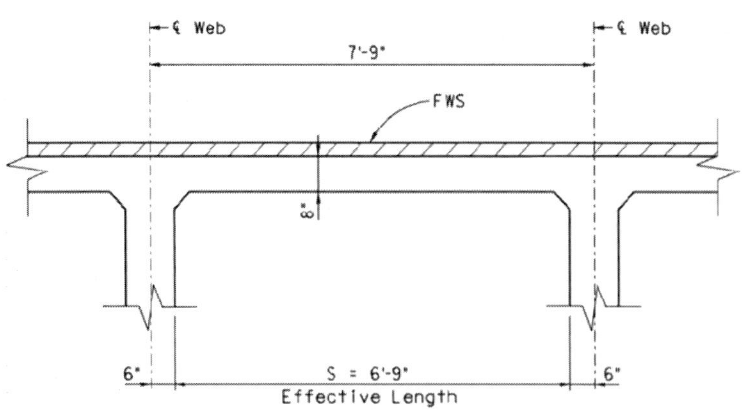

Figure 6.6. Effective Length for Live Loads.

Positive Moment Design

A summary of positive moments follows:

DC Loads

Deck 0.15x (8 / 12) x (6.75)^2÷ 10 = 0.46ft-k

DW Loads

FWS 0.025x (6.75)^2÷10=0.11ft-k

Vehicle

LL + IM = 5.10 ft-k

6.4.6.15. AASHTO LRFD Cl. [9.5.2] Service I Limit State [BDG]

Deck design is normally controlled by the service limit state. The working stress in the deck is calculated by the standard methods used in the past. For this check, Service I moments should be used.

Ms =1.0x (MDC + MDW) + 1.0x (MLL+IM)

Ms = 1 x(0.46 +0.11) +1 x (5.1) =5.67 ft-k

Try #5 reinforcing bars

ds =8–0.5ws - 1 clr – 0.625 / 2=6.19 in

Determine approximate area reinforcing as follows:

fs = 24

j = 0.9

As ≈ Ms/(fsj x ds) = (5.67) x (12) / (24) x (0.9) x (6.19) = 0.509in^2

Try #5 @7inches

Bar area= 0.31

As =(0.31) x (12 / 7) = 0.531in^2

6.4.6.16. AASHTO Cl. [9.5.2] Allowable Stress [BDG]

The allowable stress for a deck under service loads is not limited by the LRFD specifications. The 2006 Interim Revisions replaced the direct stress check with a maximum spacing requirement to control cracking. However, the maximum allowable stress in a deck is limited to 24 ksi per the LRFD Bridge Design Guidelines.

Determine stress due to service moment:

p = As/bds =0.531 / (12) x (6.19) = 0.007149

np =8x (0.007149) = 0.05719

$k=\sqrt{2np + np2} - np = \sqrt{(2x\ (0.05719) + (0.05719)^2)} - 0.05719 = 0.286$
$j = 1 - k/3 = 1 - 0.286/3 = 0.905$
$fs = Ms / Asjds = (5.67) \times (12) / (0.531) \times (0.905) \times (6.19)$
$= 22.87$ ksi

Since the applied stress is less than 24 ksi, the LRFD Bridge Practice Guideline service limit state requirement is satisfied.

6.4.6.17. Control of Cracking

AASHTO Cl. [5.7.3.4]

For all concrete components in which the tension in the cross-section exceeds 80 percent of the modulus of rupture at the service limit state load combination the maximum spacing requirement in equation 5.7.3.4-1 shall be satisfied.

Modulus of rupture = 80% = 0.8
$fsa = 0.80fr = 0.80 \times (c\ 0.24\ f'c) = 0.8 \times 0.509 = 0.407$ ksi
Wearing surface = 0.5 in slab thickness = 8.00 in.
$Scr = (12) \times (8-0.5)^2 \div 6 = 112.5$in4
$fs = Ms / Scr = (5.67) \times (12) / 112.5 = 0.605$ ksi>fsa = 0.407 ksi

Since the service limit state stress exceeds the allowable, the spacing, s, of mild steel reinforcing in the layer closest to the tension force shall satisfy the following:

AASHTO Cl. [5.7.3.4-1]

$s \leq 700\gamma e / (\beta s \times fs) - 2dc$ where
$\gamma e = 0.75$ for Class 2 exposure condition for decks
$dc = 1$ clear $+ 0.625 \div 2 = 1.31$ inches
$fs = 22.87$ ksi
hnet =7.50 inches
$\beta s = 1 + dc / 0.7 \times (h-dc) = 1+(1.31/(0.7 \times (7.5-1.31))) = 1.3$
$s \leq ((700 \times 0.75) / (1.3 \times 22.87)) - (2 \times 1.31) = 15.04$ in.

Since the spacing of 7 inches is less than 15.04, the cracking criteria is satisfied.

6.4.6.18. AASHTO LRFD [Table 3.4.1-1] Strength I Limit State

Factored moment for Strength I is as follows:

$\gamma_{Dc} = 1.25$
$\gamma_{DW} = 1.5$

$Mu = \gamma_{Dc}(M_{DC}) + \gamma_{DW}(M_{DW}) + 1.75(M_{LL+IM})$

$Mu = 1.25x\ 0.46 + 1.5x\ 0.11 + 1.75\ x\ 5.1 = 9.67$ ft-k

6.4.6.19. AASHTO LRFD [5.7.3] and [5.7.3.2.2-1] Flexural Resistance

The flexural resistance of a reinforced concrete rectangular section is:

$Mr = \varphi\ Mn$

$= \varphi Asfy\ (ds - a/2)$

AASHTO LRFD [5.7.3.1.1-4]

$fy = 60$

$c = Asfy / 0.8f'c\beta 1b = 0.531x\ 60 / 0.85x\ 4.5x0.825\ x\ 12$

$= 0.841$ in.

AASHTO LRFD [5.7.3.2.3]

$a = \beta 1c = 0.825\ x\ 0.841 = 0.69$ in.

The tensile strain must be calculated as follows:

$\varepsilon T = 0.003\ x\ (dt/c - 1) = 0.003\ x\ (6.19/0.841 - 1) = 0.019$

AASHTO Cl. [5.5.4.2.1]

Since $\varepsilon T > 0.005$, the member is tension controlled and $\phi = 0.90$. $Mr = 0.9\ x\ 0.531\ x\ 60\ x\ (6.19 - (0.69/2)) \div 12 = 13.97$ ft-k

Since the flexural resistance, Mr, is greater than the factored moment, Mu, the strength limit state is satisfied.

6.4.6.20. AASHTO Cl. [5.7.3.3.1] Maximum Reinforcing

The 2006 Interim Revisions eliminated this limit. Below a net tensile strain in the extreme tension steel of 0.005, the factored resistance is reduced as the tension reinforcement quantity increases. This reduction compensates for the decreasing ductility with increasing over strength.

6.4.6.21. AASHTO Cl. [5.7.3.3.2] Minimum Reinforcing

The LRFD specification specifies that all sections requiring reinforcing must have sufficient strength to resist a moment equal to at least 1.2 times the moment that causes a concrete section to crack or 1.33 Mu. A conservative simplification for positive moments is to ignore the 0.5 inch wearing surface for this calculation. If this check is satisfied there are no further calculations required. If the criteria is not satisfied one check should be made with the wearing surface subtracted and one with the full section to determine which of the two is more critical.

Sc = (12) x (8)^2 /6 = 128in3
fr = 0.37 $\sqrt{f'c}$ = 0.37x$\sqrt{4.5}$= 0.785 ksi
1.2 Mcr = 1.2 frSc = 1.2x 0.785x 128 ÷ 12
= 10.05 ft-k
1.2 Mcr = 10.05 ≤Mr =13.97ft-k

Hence, the minimum reinforcement limit is satisfied.

6.4.6.22. AASHTO Cl. [9.5.3] & [5.5.3.1] Fatigue Limit State

Fatigue need not be investigated for concrete deck slabs in multi-girder applications. The interior deck is adequately reinforced for positive moment using #5 @ 7 inches.

6.4.6.23. AASHTO Cl. [9.7.3.2] Distribution Reinforcement

Reinforcement shall be placed in the secondary direction in the bottom of slabs as a percentage of the primary reinforcement for positive moments as follows:

220 /$\sqrt{S}$ < 67 percent maximum
220 /$\sqrt{6.75}$< 85 percent maximum
Usc 67% Maximum = 0.67
As = 0.67 x0.531 = 0.356 in^2
Use #5 @ 10" ⇒ As = 0.31x12/10 = 0.372 in^2

The LRFD specification does not allow for a reduction of this reinforcing in the outer quarter of the span as was allowed in the standard specifications.

6.4.6.24. AASHTO Cl. [9.7.1.3] Skewed Decks [BPG]

For bridges with skews less than 20 degrees, the ADOT LRFD Bridge Design Guidelines specifies that the primary reinforcing shall be placed parallel to the skew. For the zero degree skew in this example, the transverse deck reinforcing is placed normal to the webs.

6.4.6.25. Negative Moment Design

A summary of negative moments follows:
DC Loads

Deck0.15x 8 /12 x (6.75)^2 ÷ 10 =-0.46 ft-k

DW Loads

FWS0.025 x (6.75)^2÷10 = -0.11ft-k

Vehicle

LL + IM = -5.50ft-k

6.4.6.26. Service Limit State

AASHTO LRFD [9.5.2] [BDG] & [Table 3.4.1-1]

Deck design is normally controlled by the service limit state. The working stress in the deck is calculated by the standard methods used in the past. For this check, Service I moments should be used.

Ms = 1.0 (MDC) + (MDW) + 1(MLL+IM) Ms =1x(0.46 +0.11) + 1x(5.5) = 6.07 ft-k

Try #5 reinforcing bars

ds = 8 –2.5clear –0.625/2 = 5.19 inches

Determine approximate reinforcing area as follows:

As ≈Ms/fsjds = 6.07 x 12/ 24x 0.9 x 5.19 = 0.65in2
Try #5 @ 5 ½ inches
As =0.31 x12/5.5 = 0.676 in2

6.4.6.27. Allowable Stress

Determine stress due to service moment:

p =As/bds = 0.676/12x 5.19 = 0.01085
np = 8x0.01085= 0.0868
k=√2np + np2 - np = √(2 x0.0868 + (0.0868) ^2) -0.0868
= 0.339
j = 1- k/3 =1-0.339 /3= 0.887
fs = Ms / Asjds = 6.07 x12 / 0.676x 0.887x 5.19
= 23.41 ksi ≤ 24.0ksi

AASHTO Cl. [9.5.2] [BDG]

Since the applied stress is less than the allowable specified in the LRFD Bridge Design Guidelines, the service limit state stress requirement is satisfied.

6.4.6.28. Control of Cracking

AASHTO Cl. [5.7.3.4]

The deck must be checked for control of cracking. For all concrete components in which the tension in the cross section exceeds 80 percent of the modulus of rupture at the service limit state load combination, the maximum spacing requirement in Equation 5.7.3.4-1 shall be satisfied.

$fsa = 0.80fr = 0.80(0.24 \sqrt{f'c}) = 0.8 \times 0.509 = 0.407$ ksi
$Scr = (12) \times (7.5)^2 \div 6 = 112.5$ in^3
$fs = Ms / Scr = 6.07 \times 12 / 112.5 = 0.647$ ksi $> fsa = 0.407$ ksi

Since the service limit state stress exceeds the allowable, the spacing, s, of mild steel reinforcing in the layer closest to the tension force shall satisfy the following:

AASHTO Cl. [5.7.3.4-1]

$s \leq 700\gamma e / \beta sfs - 2dc$
$\gamma e = 0.75$ for Class 2 exposure condition for decks
$dc = 2.5$ clear $+ 0.625 \div 2 = 2.81$ inches
$fs = 23.41$ ksi
$h = 8.00$ inches
$\beta s = 1 + dc/0.7(h-dc) = 1 + (2.81 / (0.7 \times (8-2.81))) = 1.77$
$s \leq 700 \times 0.75 / 1.77 \times 23.41 - 2 \times 2.81$
$= 7.05$ in

Since the spacing of 5.5 inches is less than 7.05 the cracking criteria is satisfied.

6.4.6.29. AASHTO Cl. [3.4.1] Strength I Limit State

Factored moment for Strength I is as follows:

$Mu = \gamma_{Dc} (M_{DC}) + \gamma_{DW}(M_{DW}) + 1.75(M_{LL+IM})$
$Mu = 1.25 \times 0.46 + 1.5 \times 0.11 + 1.75 \times 5.5 = 10.37$ ft-k

6.4.6.30. AASHTO Cl. [5.7.3] Flexural Resistance

The flexural resistance of a reinforced concrete rectangular section is:

AASHTO Cl. [5.7.3]

Mr =φ x Mn = φ x As x fy x (ds -a/2)

AASHTO Cl. [5.7.3.1.1-4]

c =Asfy / 0.8/5f'cβ1b = (0.676) x (60) / (0.85) x (4.5)x(0.825) x (12) = 1.071 in

AASHTO Cl. [5.7.3.2.3]

a = β1c =0.825 x 1.071 = 0.88 inches
εT =0.003 x (dt/c -1) = 0.003 x (5.19/1.071-1) = 0.012

AASHTO Cl.[5.5.4.2.1]

Since εT > 0.005, the member is tension controlled and ϕ = 0.9
Mr =0.9x0.676 x 60) x(5.19-(0.88/2))÷ 12= 14.45 ft-k

Since the flexural resistance, Mr, is greater than the factored moment, Mu, the strength limit state is satisfied.

6.4.6.31. AASHTO Cl. [5.7.3.3.2] Minimum Reinforcing

The LRFD specification specifies that all sections requiring reinforcing must have sufficient strength to resist a moment equal to at least 1.2 times the moment that causes a concrete section to crack or 1.33 Mu. The most critical cracking load for negative moment will be caused by ignoring the 0.5 inch wearing surface and considering the full depth of the section.

Sc =12x (8)^2 ÷ 6
= 128 in3
12 Mcr =1.2frSc = 1.2x 0.785 x 128 ÷ 12
= 10.05 ft-k
1.2Mcr =10.05≤ Mr =14.45 ft-k

Therefore, the minimum reinforcement limit is satisfied.

6.4.6.32. AASHTO Cl. [9.5.3] & [5.5.3.1] Fatigue Limit State

Fatigue need not be investigated for concrete deck slabs in multi-girder applications. The interior deck is adequately reinforced for negative moment using #5 @ 5 ½ inches.

6.4.6.33. AASHTO Cl. [C4.6.2.1.6] Shear

Past practice has been not to check shear in typical decks. For a standard concrete deck shear need not be investigated.

6.4.6.34. Overhang Design

AASHTO Cl. [Appendix A13] and AASHTO Cl. [Article A13.4.1]

Design Case 1

The overhang shall be designed for the three design cases described below:

Design Case 1: Transverse forces specified in [AASHTO LRFD Table A13.2-1] Extreme Event Limit Combination II Limit State

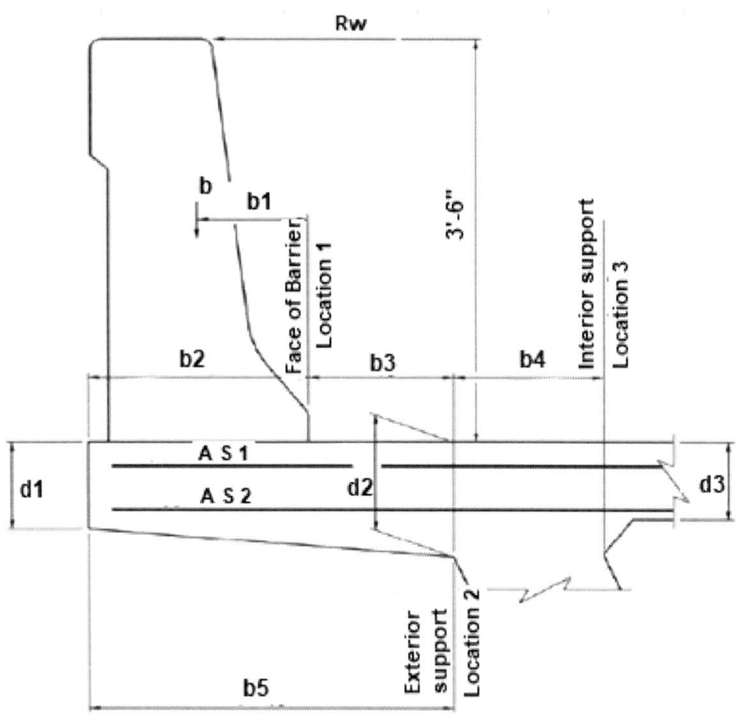

Figure 6.7. Overhang Design – Case 1

b= 0.538 ft 6.456 in

b1= 0.946 ft 11.352in

b2= 1.58 ft 18.96 in

b3= 1.04 ft 12.48 in

b4= 1.083 ft 12.996in

b5= 2.625 ft 31.5 in

d1= 0.75 ft 9 in

d2= 1 ft 12 in
d3= 0.67 ft 8 in

The deck overhang must be designed to resist the forces from a railing collision using the forces given in Section 13, Appendix A, AASHTO LRFD. TL-5 rail will be used with a 42 inch height. A summary of the design forces is shown below:

Table 6.1. Summary of the Design Forces

Design Forces	Forces	Units
F_t, Transverse	124.0	Kips
F_l, Longitudinal	41.0	Kips
F_v, Vertical Down	80.0	Kips
L_t and L_l	8.0	Feet
L_v	40.0	Feet
H_e Minimum	42.0	Feet

The philosophy behind the overhang analysis is that the deck should be stronger than the barrier. This ensures that any damage done to the barrier is easier to repair and that the assumptions made in the barrier analysis are valid. The forces in the barrier must be known to analyze the deck.

Rw = total transverse resistance of the railing.
Lc = critical length of yield line failure. See Figures 6.8 and 6.9.

For impacts within a wall segment:

AASHTO Cl. [A13.3.1-1]

Rw= (2/2Lc - Lt)(8Mb + 8Mw + McLc2/H)

AASHTO Cl. [A13.3.1-2]

Lc =Lt/2 + √(Lt/2)2 + 8H(Mb + Mw)/Mc

AASHTO Cl. [Section 9] [BDG]

Required design values for the ADOT 42-inch F-shape barrier shown in SD1.02 are published in the Bridge Design Guidelines and are repeated below:

Mb =0.00 ft-k
Mc =15.16ft-k

Mw = 56.42 ft-k

Rw =129.6k

Lc =8/2 √(8/2)^2+8x (3.5) x (0+56.42)/15.16= 14.96 ft

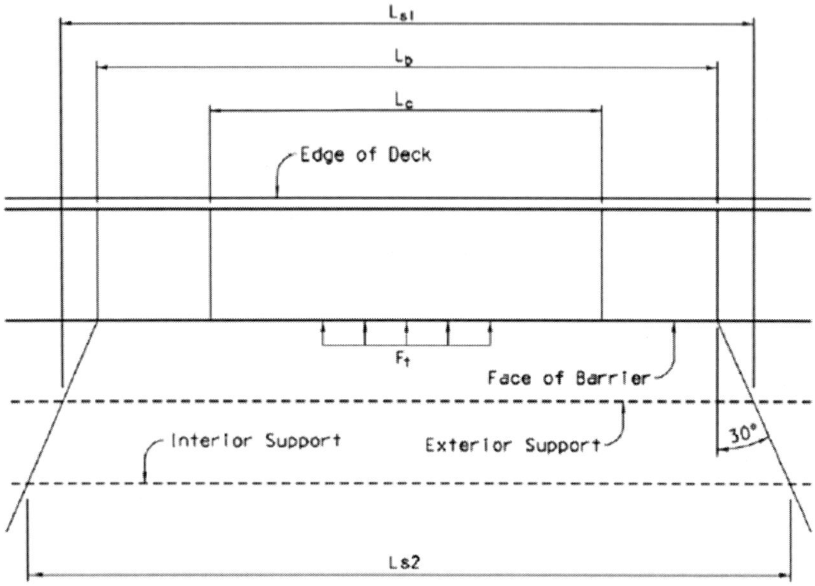

Figure 6.8. Railing Barrier in PLAN.

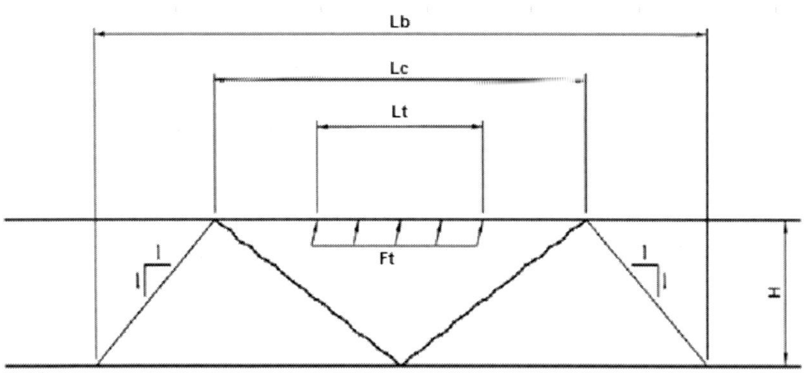

Figure 6.9. Railing Barrier in ELEVATION.

Since the railing resistance to transverse load, Rw = 129.60 kips, is greater than the applied load, Ft = 124.00 kips, the railing is adequately designed for the test level specified.

Barrier Connection to Deck

The strength of the attachment of the barrier to the deck must also be checked. The deck will only see the lesser of the strength of the barrier or the strength of the connection. For the 42 inch barrier, #5 reinforcing at 12 inches connects the:

Barrier to the deck

Bar dia = 0.625 in
$As=0.310in^2$
ds=17.25 –1 ½ clear – 0.625 /2 = 15.44 inches

Flexure

For a reinforcing bar not parallel to the compression face, only the parallel component is considered. The #5 reinforcing is oriented at an angle of 26 degrees.

$\theta = 26$ degree
$f'c = 4$ ksi
$\beta 1 = 0.85$
$c=Asfy\cos\theta / 0.85f'c\beta 1b = (0.31) \times (60)\times Cos(26) /0.85 \times (4) \times (0.85) \times (12) = 0.482$ in
$a = \beta 1c =0.85\times 0.482 = 0.41$ inches
Mn=As x fy x cos (θ) x (ds - a/2)ft-k
Mn=(0.31) x (60)x cos (26) x (15.44-0.41/2)/12 = 21.22 ft-k
$\phi = 1$
ϕMn =1 x 21.22 = 21.22 ft-k
ϕPu =21.22x 12 ÷ 42 = 6.06 k/ft

6.4.6.35. Shear

The barrier to deck interface must also resist the factored collision load. The normal method of determining the strength is to use a shear friction analysis. However, in this case with sloping reinforcing, the horizontal component of reinforcing force will also directly resist the horizontal force.

$Rn =As\ fy\ \sin\theta = (0.31 \times 60)\times \sin (26) = 8.15$ k/ft

The strength of the connection is limited by the lesser of the shear or flexural strength. In this case, the resistance of the connection is controlled by flexure with a value equal to 6.063 k/ft.

The design of the deck overhang is complicated because both a bending moment and a tension force are applied. The problem can be solved using equilibrium and strain compatibility. In lieu of that more complex method, a simpler method will be demonstrated here.

Face of Barrier
Location 1
Figure 6.7

The design horizontal force in the barrier is distributed over the length Lb equal to Lc plus twice the height of the barrier. See Figures 6.8 and 6.9.

Lb =14.96+2x 3.5 = 21.96 ft
Pu =129.6 /21.96=5.9 k/ft< 6.06k/ftper connection strength Dimensions
h =9+ (12-9) x 1.58 / 2.625=10.81 in =0.901ft
d1 =10.81 – 2.5 clr – 0.625/ 2 = 8 in

6.4.6.36. Moment at Face of Barrier

Deck =0.15 x9 / 12 x (1.58)^2 ÷ 2=0.14 ft-k
and 0.15x (10.81-9)/ 12 x (1.58)^2 ÷ 6 =0.01 ft-k
and 0.15 ft-k

AASHTO Cl. [3.5.1] [BDG]

Barrier = 0.538 x0.946=0.51 ft-k
Collision =5.9x[3.5+0.901 /2]=23.31ft-k

[Appendix A 13] AASHTO Cl. [A13.4.1]

The load factor for dead load shall be taken as 1.

Extreme Event II

Mu =1x (0.15 +0.51) + 1x 23.31=23.97 ft-k

AASHTO LRFD [Table 3.4.1-1]

e =Mu / Pu = 23.97 x12 / 5.9 =48.75 in
Assume the top layer of reinforcing yields and fs = 60 ksi
Determine resulting force in the reinforcing (#5 @ 5 ½")
T1 = 0.676 x 60 =40.56k

6.4.6.37. Simplified Method

The simplified method of analysis will be used. If only the top layer of reinforcing is considered in determining strength, the assumption can be made that the reinforcing will yield. By assuming the safety factor for axial tension is 1.0 the strength equation can be solved directly. This method will determine whether the section has adequate strength. However, the method does not consider the bottom layer of reinforcing does not maintain the required constant eccentricity and does not determine the maximum strain. For an in-depth review of the development of this equation refer to Appendix A of these guidelines.

ϕMn =ϕ[T1(d1 -a/2) - Pu(h/2 - a/2)]

where,

a=T1 - Pu / 0.85 f'c b

= 40.56 -5.9 / (0.85) x (4.5) x (12) = 0.76 in

ϕMn =(1) x [(40.56) x (8-0.76/2) - (5.9) x (10.81/2 - 0.76/2)] ÷ 12

ϕMn =23.28 ft-k

Since ϕ Mn =23.28 < Mu =23.97 ft-k, the overhang does not have adequate strength. Note that the resulting eccentricity equals (23.28) x (12) ÷5.9 = 47.35 inches compared to the actual eccentricity of 48.81 inches that is fixed by the constant deck thickness, barrier height and dead load moment. Reinforcing in the deck should be increased since ϕ Mn < Mu.

6.4.6.38. Development Length AASHTO Cl. [5.11.2]

The reinforcing must be properly developed from the barrier face towards the edge of the deck. The available embedment length equals 19 inches minus 2inches clear or 17 inches. For the #5 transverse reinforcing in the deck, the required development length is as follows:

AASHTO Cl. [5.11.2.1.1]

For No. 11 bar and smaller: 1.25Ab fy / √f'c = (1.25) x (0.31) x (60) /√4.5 = 10.96 in

But not less than 0.4 db fy = 0.4 x 0.625 x60 = 15 in

Since the available length is greater than the required length, the reinforcing is adequately developed using straight bars.

Exterior Support, Location 2, Figure 6.8

The deck slab must also be evaluated at the exterior overhang support. At this location the design horizontal force is distributed over a length Ls1 equal to the length Lc plus twice the height of the barrier plus a distribution length from the face of the barrier to the exterior support. See Figures 6.7, 6.8 and 6.9. Using a distribution of 30 degrees from the face of barrier to the exterior support results in the following:

LS1 =14.96+2x(3.5) + (2)xtan(30)x(1.04) =23.16 ft

Pu =129.6 / 23.16 = 5.6k/ft

Dimensions

h = 12.00 in
d1 = 12– 2.5 clr – 0.625 /2= 9.19in

Moment at Exterior Support

DC Loads

Deck=0.15x (9 / 12) x (2.63)^2/ 2 =0.39 ft-k
= 0.15 x (3 /12) x (2.63)^2/6 =0.04 ft-k
Barrier = 0.538x (0.946+1.04)= 1.07 ft-k
DC=1.50ft-k

DW Loads

FWS=0.025 x (1.04)^2/ 2= 0.01 ft-k
Collision =5.6 x [3.5 + (12 /12) / 2]= 22.4 ft-k

Extreme Event II AASHTO Cl. [A13.4.1][3.4.1]

The load factor for dead load shall be taken as 1.

Mu=1 x 1.5 + 1x0.01+1x22.4 = 23.91 ft-k

Solve for the strength of the deck section:

T1= 0.676 x 60 =40.56 k
a=T1 -Pu / 0.85 f'c b = 40.56 -5.6/ (0.85 x 4.5 x 12) = 0.76 in
ϕMn =ϕ[T1(d1 - a/2) - Pu(h/2 - a/2)]
ϕMn =(1) x [(40.56) x (9.19 -0.76/2) - (5.6) x (12/2 -0.76/2)] ÷ 12
ϕMn = 27.16 ft-k >Mu = 23.91 ft-k

Therefore, the section has adequate strength.

Interior Support Location 3 Figure 6.7

Design Case 1

The deck slab must also be evaluated at the interior point of support. At this location, the design horizontal force is distributed over a length Ls2 equal to the length Lc plus twice the height of the barrier plus a distribution length from the face of the barrier to the interior support. See Figures 6.7, 6.8 and 6.9. Using a distribution of 30 degree from the face of the barrier to the interior support results in the following:

$LS2 = 14.96 + 2x3.5 + (2)x\tan(30) \times (2.13) = 24.42$ ft

$Pu = 129.6 / 24.42 = 5.31$k/ft

Dimensions

h =8.00 in

d1 =8– 2.5 clr – 0.625 / 2= 5.19 in

Moment at Interior Support

For dead loads, use the maximum negative moments for the interior cells used in the interior deck analysis:

DC =0.46ft-k

DW =0.11ft-k

Collision = 5.31 x [3.5 + (8/ 12) / 2] = 20.36 ft-k

Extreme Event II AASHTO Cl. [A13.4.1] and [3.4.1]

The load factor for dead load shall be taken as1.

Mu = 1x0.46 + 1 x0.11 + 1x 20.36 =20.93 ft-k

Solve for the strength of the deck section:

T1=0.676 x60= 40.56 k

a=T1 -Pu / 0.85f'c b = 40.56 -5.31 / 0.85 x 4.5 x12= 0.77 in

$\phi Mn = \phi[T1(d1 - a/2) - Pu(h/2 - a/2)]$

$\phi Mn = (1) \times [(40.56) \times (5.19 - 0.77/2) - (5.31) \times (8/2 - 0.77/2)] \div 12$

$\phi Mn = 14.64$ ft-k< Mu =20.93 ft-k So, Section has inadequate strength.

[3.6.1]

Deck reinforcement should be revised.

The deck is not adequately reinforced for Design Case I.

Extreme Event II AASHTO Cl. [A13.4.1] and [3.4.1] Design Case 2: Vertical forces specified in AASHTO Cl. [A13.2] and Extreme Event in Limit State AASHTO Cl. [A13.4.1]

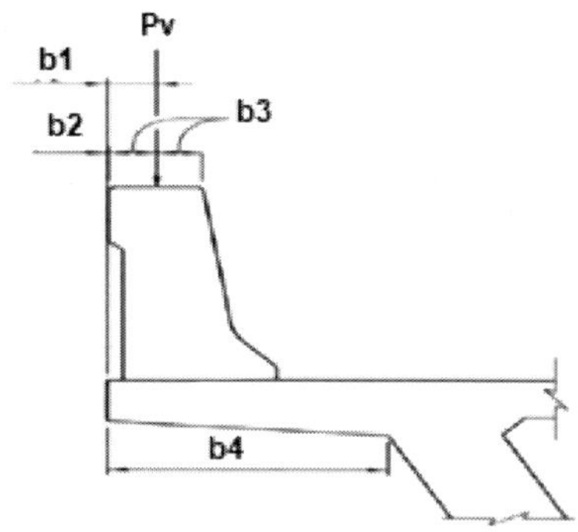

Figure 6.10. Overhang Design – Case 2.

b1 = 5.5 in 0.458 ft
b2 = 0.25 in 0.021 ft
b3 = 5.25 in 0.438 ft
b4 = 31.5 in 2.625 ft

AASHTO Cl. [A13.2-1]

DC Dead Loads 1.5ft-k

DW Dead Load

Vehicle Collision = [80/40]x [2.625 - (5.5/12)] = 4.33 ft-k

The load factor for dead load shall be taken as 1.

Mu =1 x 1.5+1 x 0.01 +1 x 4.33 = 5.84 ft-k

Flexural Resistance AASHTO Cl. [5.7.3.2]

The flexural resistance of a reinforced concrete rectangular section is:

Mr =φ Mn = φAsfy (d -a/2)

Try #5 reinforcing bars

ds =12 – 2.5 clr –0.625/2= 9.19 inches

Use #5 @ 5 ½", the same reinforcing required for the interior span and overhang Design Case 1.

AASHTO Cl. [5.7.3.1.1-4]

$c = A_s f_y / 0.85 f'c \beta_1 b = 0.676 \times 60 / 0.85 \times 4.5 \times 0.825 \times 12 = 1.071\text{in}$
$a = \beta_1 c = (0.825) \times (1.071) = 0.88\text{inches}$
$\varepsilon T = 0.003 \times (dt/c - 1) = 0.003(9.19/1.071 - 1) = 0.023$

AASHTO Cl. [C 5.5.4.2.1]

Since $\varepsilon T > 0.005$ the member is tension controlled.
$Mn = (0.676) \times (60) \times (9.19 - 0.88/2) \div 12 = 29.58$ ft-k

AASHTO Cl. [1.3.2.1]

$\phi = 1.00$
$Mr = \varphi Mn = (1) \times (29.58) = 29.58\text{ft-k}$

Since the flexural resistance, Mr, is greater than the factored moment, Mu, the extreme limit state is satisfied. The LRFD specification requires that all sections requiring reinforcing must have sufficient strength to resist a moment equal to at least 1.2 times the moment that causes a concrete section to crack or 1.33 Mu.

LL Distribution AASHTO Cl. [BDG]

$Sc = bh2 / 6 = (12) \times (12)^2 / 6 = 288\text{in}3$

AASHTO LRFD [Table 4.6.2.1.3-1]

$1.2Mcr = 1.2frSc = 1.2 \times (0.785 \times 288)/12 = 22.61\text{ft-k} < Mr = 29.58\text{ft-k}$

AASHTO LRFDIM [3.6.2]

Since the strength of the section exceeds 1.2 Mcr, the minimum reinforcing criteria is satisfied.

Multiple Presence Factor AASHTO Cl. [3.6.1.1.2-1]

Strength Limit State. Design Case 3: The loads specified in [3.6.1] that occupy the overhang Strength and Service Limit State Design Case 3

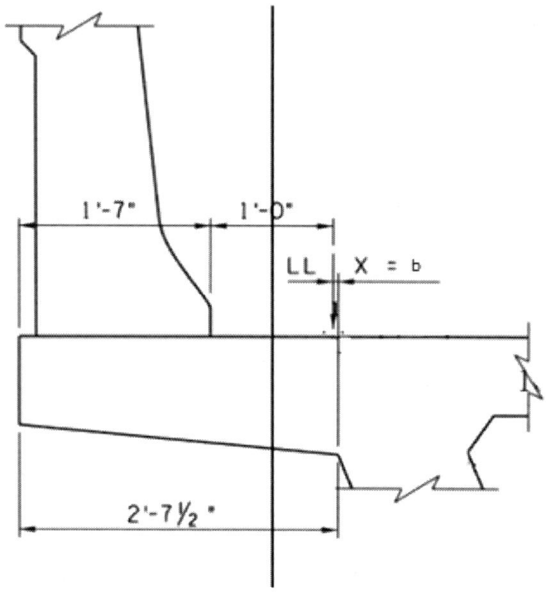

Figure 6.11. Overhang Design – Case 3.

b = 0.5 in 0.042 ft

Service I Limit State AASHTO Cl.[3.4.1]

At the face of exterior support:

DC Dead Loads = 1.5 ft-k
DW Dead Loads = 0.01 ft-k

6.4.1.39. Allowable Stress [BDG]

While the LRFD specification allows use of a uniform load of 1.00 kip/ft for service limit state where the barrier is continuous, ADOT does not. Therefore, also use the live load distribution for strength limit state for the service limit state. For a cast-in-place concrete deck overhang, the width of the primary strip is 45.0 + 10.0 X where X equals the distance from the point of load to the support.

Width Primary Strip (inches) = 45 +10x(0.042) =45.42 in = 3.79 ft
Dynamic Load Allowance, IM

For all states other than fatigue and fracture limit state, IM = 33%= 0.33.
A multiple presence factor must also be applied. Since one vehicle produces the:

Critical load, m =1.2
LL + IM = [16 x (1.33 x1.2 x 0.042)] / 3.79 = 0.28 ft-k
Mu = 1.25x1.5 + 1.5 x 0.01 + 1.75 x0.28 = 2.38ft-k
p = As / bds = 0.676 / (12x9.19) = 0.00613
np = 8x(0.00613) = 0.049
k = √2np + np2 - np = √2x(0.049)+(0.049)^2 - 0.049 = 0.266
j = 1 - k/3 = 1 - 0.266/3 = 0.911
Ms = 1 x (1.5 + 0.01) +1 x (0.28) =1.79 ft-k
fs = Ms / As jds = (1.79) x (12) / (0.676) x (0.911) x (9.19) = 3.80ksi≤ 24ksi

6.4.1.40. Control of Cracking

The flexural resistance was previously calculated for Design Case 2. Since the member is tension controlled, ϕ = 0.90. Since Mr = (0.9) x (29.58) = 26.62ft-kis greater than Mu the deck is adequately reinforced for strength.

Since the applied stress is less than the allowable specified by ADOT, the service limit state requirement is satisfied.

For all concrete components in which tension in the cross section exceeds 80 percent of the modulus of rupture at the service limit state load combination, the maximum spacing requirement in Equation 5.7.3.4-1 shall be satisfied.

fsa =0.80fr = 0.80(c 0.24 f 'c) = 0.8x(0.509) = 0.407 ksi
Scr =(12) x (11.5)^2 ÷6 = 264.5 in3
fs =Ms / Scr = (1.79)x(12) / 264.5 = 0.081ksi
0.081ksi<fsa = 0.407 ksi
ds =12– 2.5 clr – 0.625/2= 9.19 in
As (#5 @ 5 ½”) = 0.676in^2

Bottom Slab Reinforcing [5.14.1.5.2b]

The maximum allowable stress in a deck is limited to 24 ksi per the LRFD Bridge Design Guidelines.

Determine stress due to service moment:

Placed in the longitudinal direction in the bottom slab at a maximum 18 inch spacing.

As =(0.004) x (6) x (12) = 0.288 in^2 Use #5 @ 12”

6.4.1.41. Temperature and Shrinkage Reinforcing AASHTO Cl. [5.10.8]

A minimum of 0.5% reinforcement shall be placed in the transverse direction in the bottom slab at a maximum 18 inch spacing. The reinforcement shall extend to the exterior face of the outside web and be anchored by a 90° hook.

AASHTO Cl. [5.10.8-1]

As =0.005 x 6 x 12 =0.360 in^2 Use #5 @ 9" (As = 0.413 in2)

AASHTO Cl. [5.10.8-2]

Temperature and shrinkage reinforcement requirements were changed in the 2006 Interim Revisions. The required area reinforcement for the section follows:

Exterior Web

b =(5.5 – 1x12) ÷ cos(21.8) = 58.2 inches

h = 12.00

fy = 60

$As \geq 1.30bh / 2.(b + h)fy = 0.11 \leq As \leq 0.60$

$\theta = 21.8$

$As \geq 1.3 \times (58.2) \times (12) / 2x(58.2+12)x(60) = 0.108\ in^2/ft$

Top Slab

B =6.75x12 = 81.00 inches

$As \geq 1.3\ x81x8\ /2x81+ 8x\ 60 = 0.079 in^2/ft$

Since the service limit state stress is less than the allowable, the control of cracking requirement is satisfied.

Bottom Slab

$As \geq 1.3x(81)x(6) / 2x(81 +6)\ x(60) = 0.061 in^2/ft$

A minimum of 0.4% reinforcement shall be Use #5 @ 12 in^2/ft minimum reinforcement.

6.4.1.42. Superstructure Design - Section Properties

The section properties have been calculated by subtracting the ½-inch wearing surface from the top slab thickness. However, this wearing surface has been included in weight calculations. The bridge has a uniform cross section except where the web flares from 12 inches to 18 inches starting 16 feet from the face of the abutment diaphragms. A summary of section properties follows:

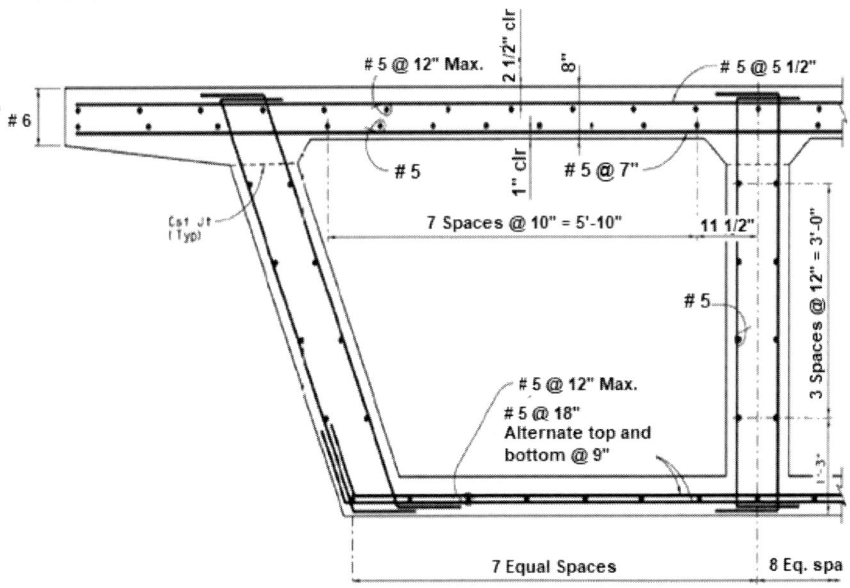

Figure 6.12. Box Girder Cross Section and Reinforcement Details.

Table 6.2. Section Properties

	12" Web	18" Web	
Yb	36.33	35.99	In
Yt	28.87	29.57	
Inertia	65,96,207	70,63,707	In^4
Area	10,741	12,660	In^2

The typical pier section is shown in Figure 6.13. To model the longitudinal frame (Figure 6.14), the equivalent length of the drilled shaft is required. This can be a complex problem. However, for a uniform layer of dense sand, the problem can be simplified by using the relative stiffness factor T. The relative stiffness factor reflects the relative ratio of flexural rigidity of the shaft to stiffness of the soil. For a dense sand with n = 0.200 kci.

AASHTO LRFD [10.7.4.2]

T =(EI/n)1/5

EI = (3405 x (π x (84)^4/ 64) = 8321548701

T = (8321548701/ 0.2)^(1/5)÷12= 11.08 ft

For drilled shafts with an embedment depth at least three times T, the equivalent length of the shaft may be taken as follows:

Le =1.8T=1.8 x 11.08 = 19.94 ft

The top 5 feet of the column/shaft embedment is ignored as described in more detail in the substructure design. Since two feet of the column is embedded, the top three feet of the shaft is ignored for lateral support. Therefore, the length of the shaft used in the structural analysis is as shown below:

L =19.94+3=22.94 feet

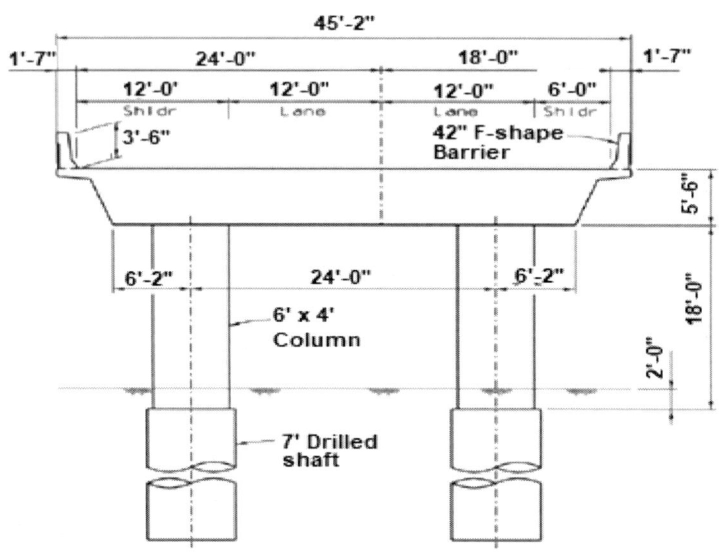

Figure 6.13. Superstructure Cross Section with Dimension Details.

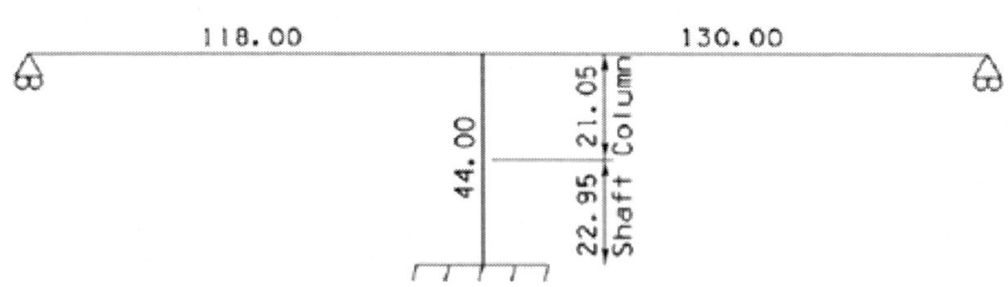

Figure 6.14. Longitudinal Frame with Dimension Details.

The equivalent length of the shaft is the length, which, when fixed at the base with the soil removed, will produce the same deflection and rotation at the top. This can be verified by using a computer program such as L-Pile. For more complex soil types that are usually encountered, L-Pile or a similar soil structure interaction program may be used to determine the deflection at the top of the column/shaft. For a prismatic column/shaft, the equivalent length can be solved for directly by using the following deflection formula:

Δ =Pl3 /3EI or solving for l yields: l =3$\sqrt{}$3EIΔ / P

In the above formula, P is the applied load at the column top and Δ is theresulting deflection.

For a non-prismatic member, the length of the shaft can be determined by trial and error equating the two deflections. This length can then be used in the frame analysis to determine moments and shears at the top of the column. Values at the base will overestimate the true magnitude of the loads. These values can be used in the design, especially if the forces are low and the 1 percent reinforcing requires controls. Otherwise, the top moments and shears should be used as loads applied to a soil structure interaction program.

The longitudinal analysis was performed using a model as shown in Figure 6.14. For this example, moments, shears and stresses were calculated at tenth points using computer software programs. A summary of the moments at critical locations is shown in Table 6.1.

Dead Load AASHTO Cl. [3.5.1]

In LFRD design, the dead load must be separated between DC loads and DW loads since their load factors differ. The DC loads include uniform loads from the self-weight of the superstructure plus 0.010 ksf for lost deck formwork, and the barriers plus the concentrated load from the intermediate diaphragm. The DW load includes the 0.025 ksf Future Wearing Surface and any utilities.

Live Load AASHTO Cl. [3.6]

The HL-93 live load in the LRFD specification differs from the HS-20-44 load in the Standard Specifications. For an in-depth discussion of live loads, refer to the computer software used to generate the live loads for the analysis. The example values shown in Table 6.3 are for one vehicle but include dynamic load allowance.

6.4.1.43. Pre-Stress Secondary Moment

The exact calculation of the secondary pre-stress moment is very difficult and time consuming since the moment is a function of the pre-stress force along the entire cable path, not just at one point. The elastic shortening would have to be calculated at each point and the resulting secondary moment recalculated. Fortunately, design does not require this level of refinement. The computer software automatically calculated the secondary moment based on uniform long-term losses along the span.

6.4.1.44. Live Load Distribution

AASHTO Cl. [4.6.2.2.1]

The LRFD specification has made major changes to the live load distribution factors. However, for a cast-in-place concrete box girder bridge, a unit design is allowed by multiplying the interior distribution factor by the number of webs. The live load distribution factor for moment for an interior web with one lane loaded is:

Nc= number of cells=5
S= web spacing (ft)=7.75 ft
L= span length of beam (ft). For negative moment use the average of the adjacent spans.

AASHTO Cl. [4.6.2.2.2b-1]

LL Distribution= (1.75 + S/3.6)(1/L)0.35 x (1/Nc)0.45
Span 1 LL Distribution= (1.75 +7.75/3.6)x(1/130)^0.35 x (1/5)^0.45= 0.344
Span 2 LL Distribution= (1.75 + 7.75/3.6)x(1/118)^0.35 x (1/5)^0.45 = 0.356
Negative LL Distribution= (1.75 + 7.75/3.6)x(1/124)^0.35 x (1/5)^0.45 = 0.35
(Consider average span)

The distribution factor for moment with two or more lanes loaded is:

LL Distribution= (13/Nc)0.3 x (S/5.8) x (1/L)0.25
Span 1 LL Distribution= (13/5)^0.3 x (7.75/5.8) x (1/130)^0.25 = 0.527
Span 2LL Distribution= (13/5)^0.3 x (7.75/5.8) x (1/118)^0.25 = 0.54
Negative LL Distribution= (13/5)^0.3 x (7.75/5.8) x (1/124)^0.25 = 0.533
Span 1 LL Distribution=(0.527x 6webs) = 3.162
Span 2 LL Distribution=(0.54x6webs) = 3.24
Negative LL Distribution=(0.533x6 webs) = 3.198

6.4.1.45. Skew Reduction

Since the bridge is right angle, the live load skew reduction factor is not applied.

6.4.1.46. Dynamic Load Allowance

AASHTO Cl. [3.6.2.1-1]

The dynamic load allowance IM equals 33% for strength and service limit states.

Dynamic load allowance applies to the truck or tandem but not to the design lane load. The dynamic load allowance has been included in the summation of live loads for one vehicle.

A summary of un-factored moments at critical locations from the computer program follows. The negative moments at the pier shown below have been reduced from the centerline moments.

Table 6.3. Un-factored moments (kip-ft)

	0.4 Span 1	**1.0 Span 1**	**0.0 Span 2**	**0.6 Span 2**
Superstructure	10990	-20545	-20909	15044
Barrier	975	-1834	-1867	1340
DC	11965	-22379	-22776	16384
DW (FWS)	951	-1790	-1822	1308
Vehicle LL+ IM	2677	-2614	-2736	3023
LL Distribution	3.240	3.198	3.198	3.162
Total LL + IM	8673	-8360	-8750	9559
Secondary M	2497	6243	7041	2817

The LRFD specification has made major changes to the group load combinations contained in [AASHTO LRFD Table 3.4.1-1]. There are several limit states that must be considered in design of the superstructure. The secondary moments caused by the pre-stressing must be considered in the Strength Limit States. These moments are permanent loads classified as locked-in erection stresses(EL) with a load factor = 1.0. limit states for this problem are as follows:

STRENGTH I – Basic load combination relating to the normal vehicular use of the bridge without wind.

$$Mu = 1.25 \times (DC) + 1.50 \times (DW) + 1.75 \times (LL + IM) + 1.0 \times (EL)$$

Span 1

$$0.4\ Mu = 1.25\times(11965) + 1.5\times(951) + 1.75\times(8673) + 1\times(2497) = 34058 \text{ ft-k}$$
$$1.0\ Mu = 1.25\times(-22379) + 1.5\times(-1790) + 1.75\times(-8360) + 1\times(6243) = -39046 \text{ ft-k}$$

Span 2

$$0.0\ Mu = 1.25\times(-22776) + 1.5\times(-1822) + 1.75\times(-8750) + 1\times(7041) = -39475 \text{ ft-k}$$
$$0.6\ Mu = 1.25\times(16384) + 1.5\times(1308) + 1.75\times(9559) + 1\times(2817) = 41987 \text{ ft-k}$$

AASHTO Cl. [C5.7.1]

For service limit states, the secondary moment is considered as part of the resisting pre-stress force and will be included as a resistance rather than a load. This simplifies the calculations since the secondary moment is a function of the amount of pre-stressing.

AASHTO LRFD [Table 3.4.1-1]

SERVICE I – Load combination relating to normal operational use of the bridge including wind loads to control crack width in reinforced concrete structures.

Ms =1.0 x (DC + DW) + 1.0 x (LL + IM)

Span 1

0.4 Ms = 1x (11965 + 951) + 1x 8673= 21589ft-k
1.0 Ms =1x (-22379 +(-1790)) +1 x (-8360)= 32529ft-k

Span 2

0.0 Ms =1x (-22776 +(-1822)) + 1x (-8750) = -33348ft-k
0.6 Ms =1 x (16384+ 1308) +1 x (9559) = 27251 ft-k

SERVICE II – Load combination relating only to tension in pre-stressed concrete superstructures with the objective of crack control.

Ms =1.0 x (DC + DW) + 0.80 x (LL + IM)

Span 1

0.4 Ms =1x (11965 + 951) +0.8x (8673) = 19854ft-k
1.0 Ms =1x (-22379 +(-1790)) +0.8x (-8360) =-30857 ft-k

Span 2

0.0 Ms =1 x (-22776 +-1822)) +0.8x (-8750) =-31598 ft-k
0.6 Ms =1 x (16384 +1308) +0.8 x (9559) = 25339 ft-k

SERVICE III – Load combination relating only to tension in pre-stressed concrete superstructures with the objective of crack control.

Ms =1.0 x (DC + DW) + 0.80 x (LL + IM)

Span 1

0.4 Ms =1 x (11965 + 951) + 0.8x (8673) = 19854 ft-k
1.0 Ms =1x (-22379 +(-1790)) + 0.8x (-8360) = -30857ft-k

Span 2

0.0 Ms =1 x (-22776 +(-1822)) +0.8x (-8750) = -31598 ft-k
0.6 Ms =1 x (16384+1308) + 0.8 x (9559) = 25339 ft-k

6.4.2. Pre-Stress Design

The design of a post-tensioned concrete bridge involves making assumptions, calculating results, comparing the results to the assumptions and reiterating the process until convergence. Computer software was used to achieve this. Note that only 199 strands are required, but 200 strands are specified to allow for a symmetric pattern. A summary of results follows:

fpj =0.77
fpu = 270
No. Strands =200 (0.6" ϕ strands)
Aps =0.217 x200 =43.4 in2
Pj =0.77 x 270x 43.4 =9023 k

6.4.2.1. Step 1 – Assume Cable Path

The first step in design is to assume a cable path as shown in Figure 6.15. The location of the center of gravity at the ends is very important for the anchor zone design. Placing the center of gravity at the neutral axis results in a uniform stress distribution at the ends but the top tendons will probably be too high to have sufficient top edge clearance. Placing the cable path near the geometric center of the section is usually a good compromise. For the long span, the cable path should be as low as possible. For the shorter span the cable path is often raised to prevent upward deflection growth. At the pier the cable path should be located as high as possible consistent with deck and pier cap reinforcing requirements. Care must be taken to ensure that the cable path can be physically located where assumed. A check on the center of gravity at the critical locations is required once the area of pre-stressing steel is determined.

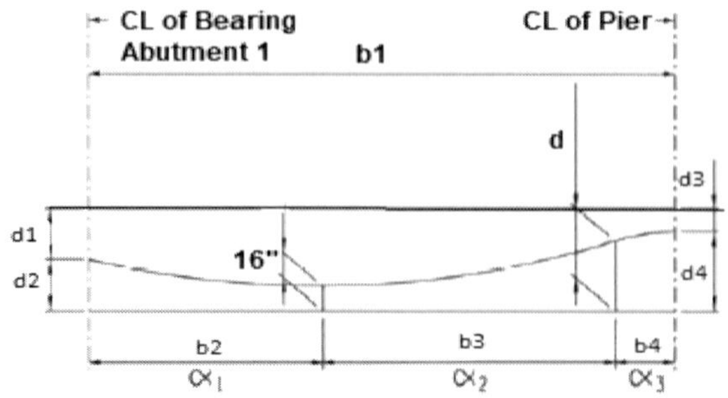

Figure 6.15 (a) – Cable Path Span 1.

Span 1

b1 = 118 ft 1416
b2 = 47.2 ft 566.4
b3 = 59 ft 708
b4 = 11.8 ft 141.6
d = 3.76 ft 45.1667
d1 = 2.75 ft 33
d2 = 2.75 ft 33
d3 = 1.25 ft 15
d4 = 4.25 ft 51

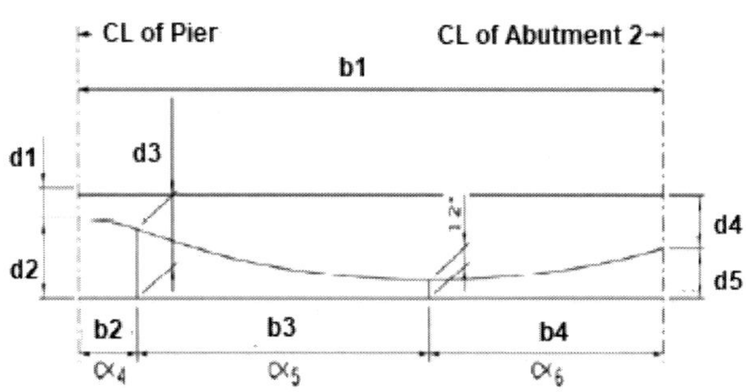

Figure 6.15 (b) - Cable Path Span 2.

Span 2

b1 = 130 ft 1560
b2 = 13 ft 156
b3 = 65 ft 780
b4 = 52 ft 624
d1 = 1.25 ft 15
d2 = 4.25 ft 51
d3 = 3.71 ft 44.52

d4 =	2.75	ft	33
d5 =	2.75	ft	33

6.4.2.2. Step 2 – Verify Cable Path AASHTO Cl. [C5.9.1.6]

For 200 strands, use two tendons per web. One duct will hold 17 strands and the other 16 strands for four webs. The other two webs will have 17 strands in each duct. The ducts must clear the three layers of #5 reinforcing in the bottom slab. From manufacturers' literature, the outside diameter of the duct will be 4 3/8" maximum. When the strands are pulled, they will rise at the low points and lower at the high points and not be located in the center of the duct. To estimate this effect, the variable Z is used. For ducts over 4 inch diameter, Z = 1 inch.

Tendons per web = 2 nos.
Diameter of the duct= 4.375 in
No of layers= 3 layers
Z=1 inch.
One duct= 17 strands
Other duct= 16 strands

At the low point:

DuctC.G.=4.375+ 0.5 = 4.88 inches
C.G. ducts=2clr + 3x(0.625) +4.88 = 8.76 inches
Strand C.G.= [((16x 4.375)÷2) + (17x(4.375+ 1 + (4.375÷2)))]÷33 = 4.96 inches
C.G. strands= 8.76 + (4.96– 4.88) + Z = 9.84 inches

At the pier: (Allow 2" for #11 pier cap reinforcing)

C.G. ducts=7+ 2+4.88 =13.88 inches
Strand C.G.= [((17x 4.375)÷2) + (16x(4.375+ 1 + (4.375÷2)))]÷33 = 4.79 inches
C.G. strands=13.88 + (4.79– 4.88) + Z = 14.79 inches

Since there are many possible combinations of size of ducts and different suppliers, one should be conservative in estimating the cg of the strands. Therefore use 12 inches for the location of the cg of the strands at low point of span 2 and 15 inches for the location of the cg of the strands at the pier.

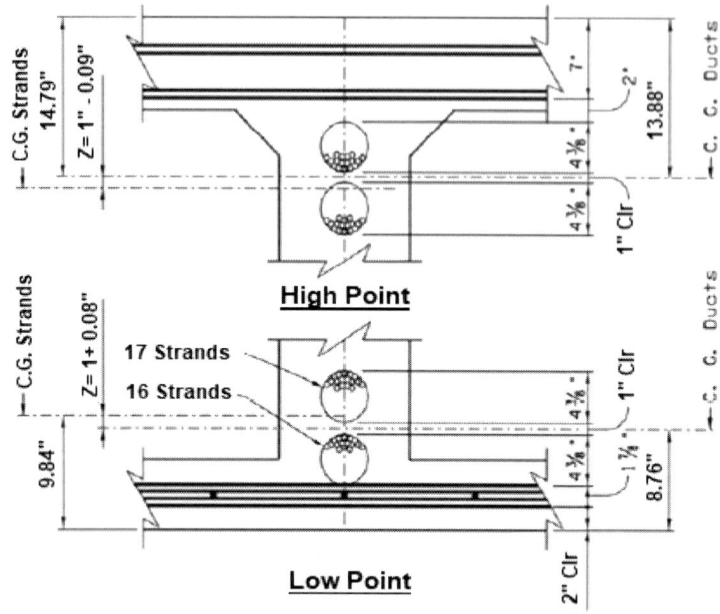

Figure 6.16. Strand C.G.

6.4.2.3. Step 3 – Calculate Friction Losses AASHTO Cl. [5.9.5] [BDG]

Total losses in pre-stress are due to friction loss, anchor set loss, elastic shortening, shrinkage, creep and relaxation. When the strands are pulled through the ducts, friction losses occur. Some loss is due to a uniform friction along the length of the path and some is due to angle changes in the cable path. Figure 2.14 is a diagram showing a summary of the friction losses, anchor set losses, elastic shortening losses and time dependent losses. Friction coefficients for rigid galvanized metal ducts are as follows:

k =0.0002
μ =0.25

For an un-symmetric two span bridge, the cable should be jacked from the long end. The increased friction losses in the shorter span should not shift the critical location to the short span but the stresses in the short span should also be checked. The bridge will be jacked from the Abutment 2 end only.

The angle change for various segments (See Figure 2.12) is shown below:

$\alpha 1= 2x\ (33\text{-}16) \div 12 / 47.2=0.060028$ radians
$\alpha 2= \alpha 3=2x(45.1667\text{-}16) \div 12/59=0.082392$ radians

$\alpha4 = \alpha5 = 2$ x (44.52-12)/12 ÷65 =0.083385 radians
$\alpha6 =$ 2x (33-12)/12 ÷ 52 = 0.067308 radians

Computer software calculated the friction losses at the tenth points along the span. The friction loss calculations will be shown for the critical locations along the span.

AASHTO Cl. [5.9.5.2.2b-1]

$\Delta f_{pF} = f_{pj} \times (1 - e^{-(Kx + \mu\alpha)})$

0.6 Span 2

$kx + \mu\alpha = (0.0002 \times 52) + (0.25) \times [0.067308] = 0.027227$
$\Delta f_{pF} = f_{pj} \times (1 - e^{(-0.027227)}) = 0.0269$ fpj

End Seat Loss 92.09 ft (See Step 4)

$kx + \mu\alpha = (0.0002 \times 92.09) + (0.25) \times [(92.09-52)/65 \times 0.083385+0.067308] = 0.048102$
Δf_{pF} = fpj x (1-e^(-0.048102)) = 0.0470fpj

Span 2

$kx + \mu\alpha = (0.0002 \times 130) + 0.25 \times [2 \times (0.083385) + 0.067308] = 0.08452$
ΔfpF =fpj x $(1 - e^{(-0.08452)}) = 0.081$ fpj

0.4 Span 1

$kx + \mu\alpha = (0.0002) \times (130+11.8+59) + (0.25) \times [2 \times (0.082392) + 2 \times (0.083385) + 0.067308] = 0.139876$
ΔfpF =fpj x $(1 - e^{(-0.139876)}) = 0.1305$ fpj

Non-Jacking End

$kx + \mu\alpha = 0.0002 \times (118+130) + (0.25) \times [0.060028 + 2 \times (0.082392) + 2 \times (0.083385) + 0.067308]$
$= 0.1643225$

ΔfpF =fpj x (1-e^(-0.1643225)) = 0.1515 fpj

A summary of friction losses at critical locations is shown in Figure 6.17.

6.4.2.4. Step 4 – Calculate Anchor Set Losses

The anchor set losses can be determined by the simplified method assuming straight line friction and anchor set losses. However, for this structure computer software was used that determined the losses using more sophisticated methods accounting for the curved loss path. The structure will be jacked from the Abutment 2 end only with an anchor set of 0.375 inches. A summary of the anchor set losses follows:

Anchor Set Length=92.09 feet
Anchor Set Loss=19.522 ksi

Jacking End:

Δf_{pA} = 19.522/ 0.77 x 270= 0.0939fpj

0.6 Span 2

Δf_{pA} = 0.0939 – 2x (0.0269) = 0.0401fpj

The initial force coefficients including both friction and anchor set losses at critical locations are as follows:

Jacking End FCi=1 –0.0939 =0.9061 fpj
0.6 Span 2 FCi=1– 0.0269 – 0.0401=0.933 fpj
End Seating FCi=1 – 0.047 =0.953 fpj
Pier FCi=1– 0.081 =0.919 fpj
0.4 Span 1 FCi= 1 –0.1305 =0.8695 fpj
Non-Jacking End FCi= 1–0.1515=0.8485 fpj

A summary of the friction and anchor set losses is shown in Figure 6.17 below:

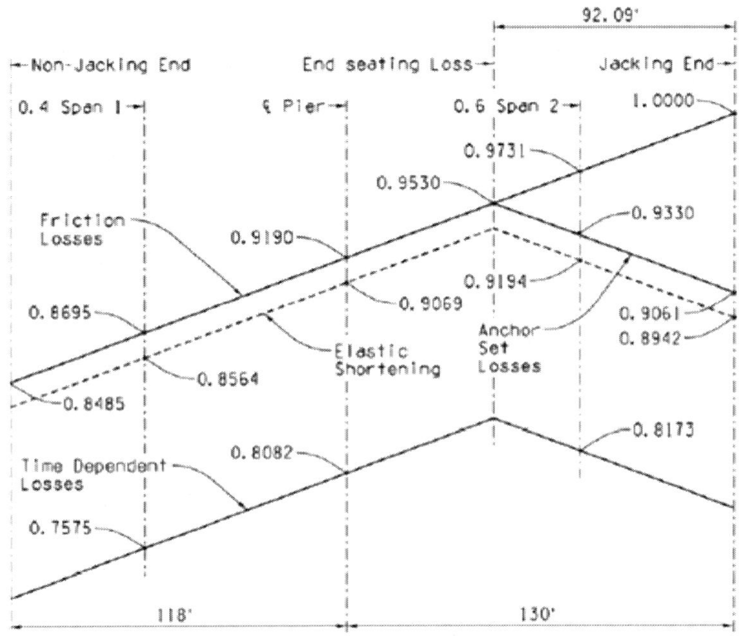

Figure 6.17. Stress Diagram.

6.4.2.5. Step 5 – Calculate Pre-Stress Losses

Elastic Shortening AASHTO Cl. [C 5.9.5.2.3b-1]

Elastic shortening losses require that the number of tendons in the bridge be known. The estimate of 12 tendons (2 tendons per web) will be used to calculate the elastic shortening losses. Elastic shortening losses can be calculated directly with a rather lengthy equation in lieu of a trial and error method. The equation for calculation of elastic shortening in the LRFD commentary [C5.9.5.2.3b-1] is incorrect. The equation also does not directly consider the effect of the secondary pre-stress moment. The correct formula is given below:

$$\Delta f_{pES} = A_{ps}\,(FCi)\,fpj\,(I+Aem^2) - e_m\,(M_g + M_{p/s})A\,/\,A_{ps}(I+Ae^2{}_m) + A.I.Eci/E_p \times 2N/(N-1)$$

This equation can be modified by dividing both the numerator and denominator by A and substituting r2 for the ratio I / A. This version of the equation produces more manageable numbers.

$$\Delta f_{pES} = (FCi)f_{pj}A_{ps}(r2+e_m2) - e_m\,(M_g + M_{p/s})\,/\,A_{ps}(r2+e2_m) + E_{ci}/E_p \times 2N/(N-1)$$

I = 65,96,207

A = 10,741

r^2 = I / A = 6596207 / 10741 = 614.11 in^2

The problem of determination of elastic shortening losses is more difficult than first viewed. The secondary moment is a function of the stress in the tendons along the entire cable path. The secondary moment could be calculated considering the elastic shortening losses along the span but this refinement is not considered necessary considering the accuracy of the loss calculations. The secondary moment considering final losses will be used to determine the elastic shortening loss. The losses will be calculated for the three critical locations.

Aps = 43.4
Eci = 3405
Ep = 28500
N = 12
Mg = 15044
Mp/s = 2817

Pre-Stress Loss

0.6 Span 2

e =36.63 – 12 =24.63 in
Aps ($r^2 + e_m^2$) = (43.4) x((614.11)+ (24.63)^2) = 52980
I.Eci / Ep . 2.N / (N-1)= [(6596207 x3405)/28500]x[(2x12)/ (12-1)]= 1719432

Elastic Shortening

Δf_{pES} =[(52980x0.933 x 0.77x270)- (24.63 x (15044+ 2817) x 12)]/ (52980 + 1719432)
= 2.82 ksi
FC_{ES} = 0.933 -2.82/ (0.77 x 270) = 0.9194

Calculate f_{cgp} and verify the elastic shortening by substituting into AASHTO [Eqn. 5.9.5.2.3b-1].

fcgp= 43.4 x [0.933 x 0.77 x270 -2.82] x (1/10741 + (24.63)^2 / 6596207) - (15044+ 2817) x 12 x 24.63/ 6596207
= 0.735 ksi

AASHTO Cl. [5.9.5.2.3b-1]

Δf_{pES} =N-1/2NXE_P/E_{ci}Xf_{cgp}

=[(12-1)/(2x 12)] x [28500/3405] x (0.735) = 2.82 ksi OK

Time-Dependent Losses AASHTO Cl. [5.9.5] [BDG]

The method of loss calculation contained in the 2006 LRFD specification shall not be used for post-tensioned box girder bridges. AASHTO LRFD Specifications Third Edition, 2004 shall be used as specified in the Bridge Design Guidelines.

Shrinkage AASHTO Cl. [5.9.5.] [BDG]

For Arizona, most locations have an average relative humidity of approximately 40%. The equation for shrinkage losses follows:

H= 40

ΔfpSR=(13.5 – 0.123 H) = 13.5 - (0.123 x 40) = 8.58 ksi

Creep AASHTO Cl. [5.9.5] [BDG]

The equation for creep follows:

ΔfpCR =12.0 fcgp – 7.0 Δfcdp

where fcgp has been previously calculated in the determination of elastic shortening losses and Δfcdp equals the change in concrete stress due to externally applied dead loads excluding self weight.

Δfcgp =(1340 +1308)x (12 x 24.63)/ 6596207= 0.119 ksi

ΔfpCR = 12 x (0.735) – 7 x (0.119) =7.99ksi

Relaxation AASHTO Cl. [5.9.5] [BDG]

For low relaxation strands, the relaxation in the pre-stressing strands equals 30% of the equation shown below:

ΔfpR2 =20.0 –0.3 ΔfpF – 0.4 ΔfpES –0.2 x (ΔfpSR + ΔfpCR)

where ΔfpF = the friction loss below 0.70 fpu at the point under consideration.

At 0.6 Span 2, the friction stress is 0.9330fpj or (0.9330) (0.77) fpu = 0.7184 fpu. Since this value is greater than 0.70,ΔfpF = 0

Δf_{pR2} =0.3x [20–0.3x (0) – 0.4x (2.82) –0.2 x (8.58 + 7.99)] = 4.67 ksi

Total Losses

The final total loss excluding friction and anchor set loss is:

Final Loss=2.82+8.58 + 7.99 + 4.67= 24.06ksi

FC_f= 0.933-24.06/ (0.77x 270)= 0.8173

Refer to Figure 6.17 for an overall view of losses.

Pre-Stress Loss

0.4 Span 1

e = 36.63 – 16 = 20.63 in

A_{ps} (r2 + e_m2) = (43.4) x(614.11+ (20.63)^2) = 45123

I . E_{ci} / E_p x 2.N/(N-1)= (6596207) x (3405)/28500x2x 12/ (12-1) = 1719432

Elastic Shortening

Δf_{pES}= (45123) x (0.8695) x (0.77 x 270) - (20.63) x (10990+ 2497) x 12 / (45123+1719432)

Δf_{pES} = 2.73 ksi

FC_{ES} = 0.8695- 2.73/ (0.77x 270) = 0.8564

Calculate f_{cgp} and verify the elastic shortening by substituting into [AASHTO Eqn. 5.9.5.2.3b-1].

fcgp=(43.4) x [(0.8695) x (0.77) x (270) - 2.73] x (1/10741+ (20.63)^2/6596207 - (10990 + 2497) x 12 x (20.63) /6596207

=0.712 ksi

[AASHTO Equation 5.9.5.2.3b-1]

ΔfpES=N-1/2N x E_p/E_{ci} x f_{cgp}

= (12-1)/(2x12)x (28500/3405)x0.712= 2.73 ksiOK

Shrinkage

The equation for shrinkage losses follows:

Δf_{pSR} =(13.5 – 0.123 H) = 13.5- 0.123 x40 = 8.58 ksi

Creep

The equation for creep follows:

Δf_{pCR} =12.0 f_{cgp} – 7.0 Δf_{cdp}

where f_{cgp} has been previously calculated in the determination of elastic shortening losses and Δf_{cdp} equals the change in concrete stress due to externally applied dead loads excluding self weight.

Δf_{cgp} =(975 + 951) x (12) x (20.63) / (6596207) = 0.072 ksi
Δf_{pCR} = 12x0.712 – 7x 0.072=8.04 ksi

Relaxation

For low relaxation strands, the relaxation in the pre-stressing strands equals 30% of the equation shown below:

$\Delta f_{pR2} = 20.0 - 0.3\ \Delta f_{pF} - 0.4\ \Delta f_{pES} - 0.2 \text{ x } (\Delta f_{pSR} + \Delta f_{pCR})$

where Δf_{pF} = the friction loss below 0.70 fpu at the point under consideration.

At 0.4 Span 1 the friction stress is0.8695 fpj
or 0.8695x0.77 =0.6695 fpu

Since this value is less than 0.70, ΔfpF = (0.7 - 0.6695x 270) = 8.23 ksi.

Δf_{pR2} = 0.3 x [20–(0.3x 8.23) –(0.4 x 2.73)– (0.2 x (8.58+ 8.04))] = 3.93ksi

The final loss excluding friction and anchor set loss is:

Final Loss = 2.73 + 8.58+8.04+3.93=23.28 ksi
FC_f=0.8695- (23.28 /(0.77x270))= 0.7575
Refer to Figure 6.17 for an overall view of losses.

Pre-Stress Loss, Pier Span 2, and Elastic Shortening

e = 51- 36.63 = 14.37 in
$A_{ps}(r^2 + e_m^2)$ = (43.4) x (614.11 + $(14.37)^2$) = 35614
I .E_{ci} / E_p . 2.N / (N-1) = (6596207) x (3405) / 28500x 2x12/ (12-1)= 1719432
Δf_{pES} =[(35614 x 0.9194 x 0.77 x 270) - (14.37 x (20909 -7041) x 12)]/ (35614+ 1719432)
Δf_{pES} =2.52 ksi
FC_{ES} = 0.9194 - (2.52/ (0.77x 270)) = 0.9073

Calculate f_{cgp} and verify the elastic shortening by substituting into [Eqn.5.9.5.2.3b-1].

fcgp= (43.4) x ((0.9194x0.77x 270)-2.52)x ((1/10741) + ((14.37)^2 / 6596207)) - (((20909 - 7041) x (12 x 14.37)) /6596207)
= 0.656 ksi
Δf_{pES} = N - 1 / 2N . Ep / Eci . Fcgp = ((12-1)/(2x12))x (28500 / 3405)x 0.656= 2.52 ksiOK

Shrinkage

The equation for shrinkage losses follows:

Δf_{pSR} = (13.5 – 0.123 H) = 13.5- (0.123) x (40)
= 8.58 ksi

Creep

The equation for creep follows:

$\Delta f_{pCR} = 12.0\ f_{cgp} - 7.0\ \Delta f_{cdp}$

where fcgp has been previously calculated in the determination of elastic shortening losses and Δfcdp equals the change in concrete stress due to externally applied dead loads excluding self weight.

Δfcgp =(1867 +1822) x12 x14.37/ (6596207) – 0.096 ksi
ΔfpCR = 12x (0.656) – 7 x (0.096) = 7.2 ksi

Relaxation

For low relaxation strands, the relaxation in the pre-stressing strands equals 30% of the equation shown below:

$\Delta f_{pR2} = 20.0 - 0.3\ \Delta f_{pF} - 0.4\ \Delta f_{pES} - 0.2(\Delta f_{pSR} + \Delta f_{pCR})$

where Δf_{pF} = the friction loss below 0.70 fpu at the point under consideration.
At the pier, the friction stress is 0.9190fpj or (0.9194) x (0.77)fpu.= 0.7079 fpu.
Since this value is greater than 0.70, ΔfpF =0ksi.

ΔfpR2 = 0.3 x [20 –0.3x(0) –0.4 x (2.52) –0.2 x (8.58 +7.2)] = 4.75 ksi

The final loss excluding friction and anchor set loss is:

Final Loss = 2.52+8.58 + 7.2 +4.75 = 23.05 ksi
FCf = 0.9194 - 23.05 / 0.77 x270 = 0.8085

Refer to Figure 6.17 for an overall view of losses.

6.4.2.6. Step 6 –Check Allowable Stress in Strands

There are four limits for stress in pre-stressing strands. The first allowable limit is prior to seating. Bridge Group has modified the LRFD allowable of 0.9 fpy.

= (0.9 x0.9) fpu = 0.81fpu to a maximum of 0.78fpu
(1) fpj = 0.77 fpu, <0.78 fpuOK

The second stress limit is 0.7fpu at anchorages immediately after anchor set.

At this time friction losses and anchor set losses have occurred. This criteria will usually limit the allowable jacking stress.

At jacking end:

(2) Strand stress = 0.9061 fpj =(0.9061 x 0.77) = 0.698fpu< 0.7fpu

At non-jacking end:

(2) Strand stress = 0.8485 fpj =0.8485x 0.77) = 0.653fpu< 0.7fpu

The third stress limit to be checked occurs at the end of the seating loss zone immediately after anchor set.

(3) Strand stress = 0.9530fpj = (0.953 x 0.77)= 0.734 fpu<0.74fpu

The fourth stress limit is a service limit state after all losses. The maximum pre-stress strand stress occurs at the end of the seating loss zone. The maximum composite dead load and live load plus dynamic load allowance occurs at the 0.6 Span 2. Since this criteria rarely controls, adding the two maximum values, even though they do not occur at the same place and using the losses at 0.6 Span 2, is a reasonable and conservative simplification.

fpe = [0.953- (0.933 - 0.8173)]X(0.77) = 0.6447 fpu after all losses

At the service limit state, dead load and live load, plus dynamic allowance stresses are added to the strand stress since the strands are bonded through the grouting process.

f service= (1340 + 1308 +9559) x 12 x (24.63) / 6596207x 28500/3861 = 4.037 ksi
Strand stress= 0.6447+ (4.037) / (270) = 0.66 fpu
(4) Strand stress = 0.66 fpu< 0.8fpy

= 0.8 x (0.9)
= 0.72 fpu

Since the four criteria for stress in the strand are met, the jacking coefficient of 0.77 is satisfactory.

6.4.2.7. Step 7 – Verify Initial Concrete Strength

Once the amount of pre-stressing steel is determined from tension criteria, the resulting concrete stress and required concrete strength can be determined. Service I limit state is used to determine the initial concrete compressive stress. The concrete stress in compression before time dependent losses is limited to:

AASHTO Cl. [5.9.4.1.1]

f'ci = 3.5

Allowable Compression = 0.60 x f'ci =0.6x 3.5 = 2.1 ksi

The basic equation for stress in concrete follows:

$f = Pj \times FC_{ES} / A + (Pj \times FC_{ES} \times e + M_{sec})y / I + \Sigma(yM)y / I$

0.6 Span 2

Bottom Fiber

Pj = 9023
Y = 36.63

fb = (9023) x (0.9194) / 10741+ [(9023) x (0.9194) x (24.63) - (2817) x (12)] x (36.63) / 6596207-(15044) x 12x (36.63) / 6596207
= 0.717 ksi$\leq$2.1 ksi

Top Fiber

y = 28.87

ft = (9023) x 0.9194/10741 - [9023x 0.9194x24.63 - 2817x 12] x 28.87/ 6596207+ 15044 x12 x 28.87/ 6596207 = 0.816 ksi $\leq$ 2.1 ksi

0.4 Span 1

Bottom Fiber

fb = (9023) x (0.8564) / 10741+ [(9023) x (0.8564) x (20.63) - (2497) x (12)] x (36.63) / 6596207- ((10990) x 12x(36.63)/6596207
= 0.706ksi $\leq$2.1ksi

Top fiber

ft = (9023) x (0.8564) / 10741 + [(9023) x (0.8564) x (20.63) - (2497) x (12)] x (28.87) / 6596207 + ((10990) x 12 x (28.87) / 6596207= 0.73 ksi $\leq$2.1ksi

Pier Span 2

Bottom Fiber

fb = (9023) x (0.9073) /10741 - [(9023) x (0.9073) x (14.37) + (7041) x (12)] x (36.63) / 6596207 + ((20909) x12x (36.63) /6596207= 1.033ksi $\leq$2.1ksi

Top fiber

ft = (9023) x (0.9073) / 10741+ [(9023) x (0.9073) x (14.37) + (7041) x (12)] x (28.87) / 6596207 - ((20909) x12x (28.87) /6596207= 0.549ksi $\leq$2.1ksi

Jacking End

Independent analysis provides an elastic shortening loss at the support of 0.0119 Pj.

Bottom fiber

fb = (9023) x (0.8942) /10741 + [(9023) x (0.8942) x (3.63) x (36.63)+ /6596207= 0.914 ksi

The initial concrete stresses are less than the allowable compressive stress. Therefore f'ci = 3.5 ksi is acceptable. The initial concrete stress must also be checked in the design of the anchor zone. This check may control the required initial strength.

6.4.2.8. Step 8 – Temporary Tension at Ends AASHTO Cl. [5.9.4.1.2]

The ends of the structure should be checked to ensure that the end eccentricity has been limited so as to keep any tension within the allowable.

ft= (9023) x (0.8942) /10741 - (9023) x (0.8942) x (3.63) x (28.87) /6596207 = 0.623 ksi $\geq$0

Since there is no tension at the ends, the criteria is met.

6.4.2.9. Step 9 – Determine Final Concrete Strength AASHTO Cl. [5.9.4.2.1]

The required final concrete strength is determined after all pre-stress losses. Service I load combination is used.

0.6 Span 2

Case I – Permanent loads plus effective pre-stress

f'c = 4.5

Allowable Compression = 0.45 f'c = (0.45x 4.5) = 2.025 ksi

0.8173

17692

ft = (9023) x (0.8173) / 10741 - [(9023) x (0.8173) x (24.63) - (2817) x (12)] x (28.87) /6596207+ (17692) x 12x (28.87) / 6596207= 0.969ksi < 2.025 ksi

Case II – One-half the Case I loads plus LL + IM

Allowable Compression = 0.40 f'c = (0.4 x4.5) = 1.8 ksi 9559

ft = 1/2 x [0.969] + (9559x 12x 28.87) / 6596207= 0.987ksi< 1.8 ksi allowable OK

Case III – Effective pre-stress, permanent loads and transient loads

AASHTO Cl. [5.7.4.7.2c-1]

ϕw =1

Allowable Compression = 0.60 ϕw f'c = 0.6 x 1x 4.5 = 2.7 ksi

The reduction factor ϕw shall be taken equal to 1.0 when the wall slenderness ratio λw is not greater than 15. The critical ratio involves the bottom slab.

AASHTO Cl. [5.7.4.7.1]

t = 0.5

λw = Xu/t = (7.75 - 1) /0.5 = 13.5 ≤15

Since the ratio is less than the allowable, the equivalent rectangular stress block can be used.

0.8173
27251

ft = (9023) x (0.8173)/10741 - [(9023) x (0.8173) x (24.63) - (2817) x (12)] x (28.87)/6596207+ (27251) x 12 x (28.87) / 6596207
= 1.471 ksi ≤27 ksi

0.4 Span 1

Case I – Permanent Loads Plus Effective Pre-Stress

Allowable Compression =0.45 f'c =(0.45x 4.5) =2.025 ksi
12916
ft = (9023) x (0.7575) / 10741 - [(9023) x (0.7575) x (20.63) - (2497) x (12)] x (28.87)/ 6596207+ (12916) x12 x (28.87) / 6596207
= 0.829 ksi < 2.025 ksi

Case II – One-Half the Case I Loads Plus LL + IM

Allowable Compression = 0.40 f'c = (0.4 x 4.5) = 1.8 ksi
8673
ft =(1/2) x [0.829] + (8673) x12 x (28.87) /6596207= 0.87 ksi < 1.8ksiallowableOK

Case III – Effective Pre-Stress, Permanent Loads and Transient Loads

Allowable Compression = 0.60 ϕw f'c = 0.6 x 1 x4.5 = 2.7 ksi
21589
ft = (9023) x (0.7575)/10741-[(9023)x(0.7575)x(20.63)-(2497)x(12)]x(28.87)/6596207+(21589)x12x(28.87)/ 6596207=
1.284 ksi ≤2.7ksi

Pier Span 2

Case I – Permanent Loads Plus Effective Pre-Stress

Allowable Compression = 0.45 f'c = (0.45x 4.5) = 2.025 ksi
0.8085
24598
fb = (9023) x (0.8085)/10741 - [(9023) x (0.8085) x (14.37) - (7041) x (12)] x (36.63)/6596207+ (24598) x 12x (36.63) / 6596207= 1.267 ksi < 2.025 ksi

Case II – One-Half the Case I Loads Plus LL + IM

Allowable Compression = 0.40 f'c = (0.4x 4.5) = 1.8 ksi

8750

fb = (1/2) x [1.267] + (8750) x12x (36.63) / 6596207= 1.217 ksi < 1.8ksiAllowableOK

Case III – Effective Pre-Stress, Permanent Loads and Transient Loads

Allowable Compression = 0.60 ϕw f'c = 0.6x 1x4.5 = 2.7 ksi

33348

fb = (9023) x (0.8085)/ 10741 - [(9023) x (0.8085) x (14.37) + (7041) x (12)] x (36.63)/ 6596207 + (33348) x 12x (36.63) /6596207

= 1.85 ksi ≤2.7 ksi

6.4.2.10. Step 10 – Determine Final Concrete Tension [BPG]

Determination of the tension in the concrete is a Service III Limit State. The allowable tension after all losses is limited to:

Allowable Tension = 0.0948√f'c = 0.0948 x√4.5 = 0.201 ksi

The basic equation for stress in concrete follows:

$f = P_j \text{ x } FC_f / A + (P_j \text{ x } FC_{f\,x}\, e + M_{p/s}) \text{ x } y / I + \Sigma(\gamma \text{ x } M) y / I$

Bottom Fiber at 0.6 Span 2

25339

fb = (9023) x (0.8173)/10741+ [(9023) x (0.8173) x (24.63) - (2817) x (12)] x (36.63)/6596207 -(25339) x 12x (36.63) / 6596207

= -0.181 ksi ≥-0.201ksi

Bottom Fiber at 0.4 Span 1

19854

fb = (9023) x (0.7575)/10741+ [(9023) x (0.7575) x (20.63) - (2497) x (12)] x (36.63)/6596207- (19854) x 12 x (36.63) / 6596207

= -0.07 ksi ≥-0.201ksi

Top Fiber at Pier Span 2

31598

ft = (9023) x (0.8085)/10741 + [(9023) x (0.8085) x (14.37) + (7041) x (12)] x (28.87)/6596207- (31598) x 12x (28.87) /6596207

= -0.152 ksi ≥-0.201ksi

[BPG]

The member must also be checked to ensure that there is no tension under dead load and effective pre-stress.

Bottom fiber at 0.6 Span 2

fb = 0.687 +0.821- 17692 x 12x 36.63 /6596207 =0.329 ksi≥0

Bottom fiber at 0.4 Span 1

fb = 0.636 + 0.617- 12916 x 12 x 36.63/ 6596207 = 0.392 ksi≥0

Top fiber at Pier Span 2

ft = 0.679+ 0.828 - 24598 x 12x 28.87 /6596207 =0.215 ksi≥0

Since both the above criteria are met, the superstructure has adequate pre-stress reinforcement for serviceability.

Fatigue Limit State AASHTO Cl. [5.5.3.1]

Fatigue of the reinforcement need not be checked for fully pre-stressed components designed to have extreme fiber tensile stress due to Service III.

Limit State within the tensile stress limit specified in AASHTO LRFD Table 5.9.4.2.2-1.

6.4.2.11. Step 11 – Flexural Resistance

The flexural resistance of the structure must exceed the factored loads. Strength I Limit State should be compared to the resistance. The Strength I Limit State includes the secondary moment from pre-stressing.

$Mr = \phi Mn < \Sigma \gamma M$

0.6 Span 2

STRENGTH I: $\Sigma\gamma M$ = 41,987 ft-k
Aps = 0.217 x200 = 43.4in2

AASHTO Cl. [5.7.3.1.1-1]

fps = fpu(1-k.c/dp)

AASHTO Cl. [5.7.3.1.1-2]

fpy = 243
k = 2 x 1.04 - fpy / fpu)= 2x (1.04- 243 /270) = 0.28
For a rectangular section without mild reinforcing steel:

dp = 65.5 – 12= 53.5 inches

AASHTO Cl. [5.7.3.1.1-4]

β1 = 0.825
b = 542.00
c = Aps x fpu / 0.85 xf'c x β1 x b + kAps x fpu/dp= (43.4x 270)/[(0.85 x4.5 x 0.825x 542) + (0.28 x 43.4 x (270/53.5))]– 6.61 < tslab = 7.5 in

AASHTO Cl. [5.7.3.2.3]

Since the stress block depth is less than the slab, the section is treated as a rectangular section:

a = c β1 = (6.61 x0.825) =5.45 in
fps = 270x (1-0.28 x 6.61 / 53.5) =260.66 ksi

AASHTO Cl. [5.7.3.2.2-1]

Mn = Aps x fps x (dp - a/2)
= (43.4) x (260.66) x (53.5 - 5.45/2) ÷ 12
= 47867 ft-k

Determine the tensile strain as follows:

εT = 0.003 x (dT/c -1) =0.003 x (53.5/6.61-1) =0.021

Since εT >0.005, the member is tension controlled.

AASHTO Cl. [5.5.4.2] [BDG]

The resistance factor $\phi = 0.95$ for flexure of cast-in-place pre-stressed concrete.

$\phi Mn = (0.95 \times 47867) = 45{,}474$ ft-k> 41,987ft-k

Therefore, the section is adequate for flexural strength at 0.6 Span 2.

Maximum Reinforcing AASHTO Cl. [5.7.3.3.1]

The 2006 Interim Revisions eliminated the maximum reinforcing requirement replacing it with the strain limitations associated with the phi factors.

Minimum Reinforcing AASHTO Cl. [5.7.3.3.2]

There is also a minimum amount of reinforcement that must be provided in a section. The amount of pre-stressed and non-pre-stressed tensile reinforcement shall be adequate to develop a factored flexural resistance at least equal to the lesser of:

1.2 Mcr

$1.33\ \Sigma\gamma M = 1.33\ Mu$

The cracking moment is determined on the basis of elastic stress distribution and the modulus of rupture of the concrete.

AASHTO Cl. [5.7.3.3.2-1]

$$M_{cr} = S_c\,(f_r + f_{cpe}) - M_{dnc}(S_c/S_{nc} - 1) \geq S_c f_r$$

Since the structure is designed for the monolithic section to resist all loads, Snc should be substituted for Sc resulting in the second term equaling zero.

Sc = Snc = I/yb = 6596207/ 36.63 = 180077 in3

Since the effect of the secondary moment is included in the applied loads, it is not repeated in the stress equation.

$f_{cpe} = P_j\, xFC_f / A + P_j xFC_f\, xe_m\, xy_b / I$

$= (9023) \times (0.8173) / 10741 + [(9023) \times (0.8173) \times (24.63) + (36.63) / 6596207 =$ 1.695 ksi

0.785

$1.2\ M_{cr} = (1.2 \times 180077) \times (0.785 + 1.695) / 12 = 44659$ ft-k

Since 1.2 Mcr = 44,659<ϕMn = 45,474the minimum reinforcing limit is satisfied.

Flexural Resistance

0.4 Span 1

STRENGTH I:ΣγM =34,058 ft-k
Aps = (0.153 x 282) = 43.4 in^2

AASHTO Cl. [5.7.3.1-1]

fps = fpu = (1 - k.c/dp)

For a rectangular section without mild reinforcing steel:

dp = 65.5– 16 = 49.50 inches

AASHTO Cl. [5.7.3.1.1-4]

c= Aps x fpu / 0.85 f'c x β1 x b + kAps x fpu/dp
= 43.4 x 270 / (0.85 x4.5 x0.825x542)+(0.28 x43.4x (270/49.5)) = 6.6 <t_{slab} = 7.5

Since the stress block depth is less than the slab, the section is treated as a rectangular section:

a =c β1= (6.6) x(0.825) = 5.45 in
fps = 270 x (1-0.28x 6.6 / 49.5) = 259.92 ksi

AASHTO Cl. [5.7.3.2.2-1]

Mn = Apsfps (dp - a/2)

Mn = 43.4X 259.92 X (49.5 -5.45 / 2 ÷ 12= 43971 ft-k

Determine the tensile strain as follows:

εT = 0.003 x (dT/c -1) = 0.003 x(49.5 /6.6 - 1) = 0.02

Since εT >0.005, the member is tension controlled.

[BDG]

The resistance factor $\phi = 0.95$ for flexure of cast-in-place pre-stressed concrete.

$\phi Mn = (0.95 \text{ x}43971) = 41772\text{ft-k} > 34{,}058\text{ft-k}$

Therefore, the section is adequate for flexural strength at 0.4 Span 1.

Maximum Reinforcing AASHTO Cl. [5.7.3.3.1]

The 2006 Interim Revisions eliminated the maximum reinforcing requirement replacing it with the strain limitations associated with the phi factors.

Minimum Reinforcing AASHTO Cl. [5.7.3.3.2]

There is also a minimum amount of reinforcement that must be provided in a section. The cracking moment is determined on the basis of elastic stress distribution and the modulus of rupture of the concrete.

AASHTO Cl. [5.7.3.3.2-1]

$Mcr = Sc\ (fr + fcpe) - Mdnc\ (Sc/Snc - 1) \geq Scfr$

$Sc = Snc = I/yb = 6596207 / 36.63 = 180077 \text{ in}^3$

$Fcpe = (9023) \text{ x } (0.7575) / 10741 + [(9023) \text{ x } (0.7575) \text{ x } (20.63) + (36.63) / 6596207 =$ 1.419 ksi

$1.2\ Mcr = 1.2\text{x}180077\text{x } 0.785 + 1.419 \div 12$

$= 39{,}689 \text{ ft-k} < 41{,}772 \text{ ft-k}$

Therefore, the minimum reinforcing limit is satisfied.

Flexural Resistance AASHTO Cl. [5.7.3]

The flexural resistance of the structure must exceed the factored loads. Strength I Limit State should be compared to the resistance.

Pier Span 2

STRENGTH I: $\Sigma\gamma M = -39{,}475$ ft-k

$Aps = 0.217 \text{ x}200 = 43.4 \text{ in}^2$

AASHTO Cl. [5.7.3.1.1-1]

$fps = fpu \text{ x } (1 - k.c/dp)$

For a rectangular section without mild reinforcing steel:

dp = 66– 15 = 51.00 inches

AASHTO Cl. [5.7.3.1.1-4]

$\beta 1 = 0.825$
b = 436.00
C =Aps fpu/(0.85 f'c β1 b + k Aps (fpu/dp))
=43.4 x270/ 0.85x4.5 x0.825 x436+0.28x 43.4 x 270/51= 8.14>t_{slab}=6 in

Since the stress block is deeper than the slab, the section must be treated as a T-section:

AASHTO Cl. [5.7.3.1.1-3]

bw = 73.83
C =Aps x fpu- 0.85 f'c x (b - bw) x hf /0.85 x f'c x β1 x bw + k x Aps x fpu/dp=43.4 x 270-0.85 x 4.5 x436-73.83 x 6 / (0.85 x 4.5x 0.825 x73.83+ 0.28x43.4x 270 / 51) = 11.46 in

AASHTO Cl. [5.7.3.2.3]

a = cβ1 = 11.46x0.825 = 9.45 in
fps = 270x (1 -0.28 x 11.46 / 51) =253.01 ksi

AASHTO Cl. [5.7.3.2.2-1]

Mn =Aps x fpu x (dp- a/2) + 0.85 x f'c x (b-bw) x hfx (a/2 -hf/2) =43.4 x253.01 x 51-9.45 /2+0.85 x 4.5 x 436- 73.83 x6x (9.45 /2-6/2) = 522467 in-k

Determine the tensile strain as follows:

εT = 0.003 (dT /c -1) = 0.003 x (51/ 11.46-1) = 0.01

Since εT >0.005, the member is tension controlled.

[BDG]
The resistance factor ϕ = 0.95for flexure of cast-in-place pre-stressed concrete.

ϕMn =(0.95 x 522467) / 12= 41362 ft-k>Mu = 39,475 ft-k

Therefore, the section is adequately reinforced for flexure.

Minimum Reinforcing AASHTO Cl. [5.7.3.3.2]

There is also a minimum amount of reinforcement that must be provided in a section.

AASHTO Cl. [5.7.3.3.2-1]

The cracking moment is determined on the basis of elastic stress distribution and the modulus of rupture of the concrete.

Mcr =Sc(fr + fcpe) - Mdnc(Sc/Snc -1) $\geq$ Scfr

Since the structure is designed for the monolithic section to resist all loads, Snc should be substituted for Sc, resulting in the second term equaling zero.

Sc = Snc = I/yb = 6596207 / 28.87 = 228480 in^3

fcpe= ((9023 x 0.8085)/ 10741) + (9023 x 0.8085 x 14.37 x28.87) / 6596207

= 1.138 ksi

1.2 Mcr =1.2 x 228480 x(0.785 + 1.138)÷ 12 = 43937 ft-k

1.33Mu = 1.33x 39475 = 52502ft-k

Since neither criteria is satisfied, add mild steel over the pier for additional resistance.

ds = 66– 2.5cl – 0.625 – 0.625 / 2= 62.56 in

Try #5 @ 12 inches

As = 0.31 x 45= 13.95 in^2

fy = 60

bw = 362.17

c = (43.4x 270)+(13.95 x 60)-(0.85 x 4.5x362.17 x6)/ ((0.85 x4.5x0.825x 73.83)+(0.28 x 43.4)x (270 / 51))

= 14.27 in

AASHTO Cl. [5.7.3.1.1-3]

a = 0.825 x14.27 = 11.77 in

fps =270x (1-0.28x(14.27 /51)) = 248.85 ksi

Mn = Apsfps (dp - a/2) + Asfy (ds - a/2) + 0.85f'c (b-bw) hf(a/2-hf/2)

$Mn = 43.4x\ 248.85x\ (51 - 11.77/2) + 13.95 \times 60x\ (62.56 - 11.77/2) + 0.85 \times 4.5x\ 362.17x6x\ (11.77/2 - 6/2)$
$= 558663$ in-k

Determine the tensile strain as follows:

$\varepsilon T = 0.003(dT/c - 1) = 0.003 \times (62.56/14.27 - 1) = 0.01$

Since $\varepsilon T > 0.005$, the member is tension controlled.

[BDG]

The resistance factor $\phi = 0.95$ for flexure of cast-in-place pre-stressed concrete.

$\phi Mn = (0.95 \times 558663)/12 = 44227$ ft-k$>1.2Mcr = 43{,}937$ft-k

Therefore, the minimum reinforcing criteria is satisfied.

6.4.2.12. Step 12 – Shear Design (LRFD Method) AASHTO Cl. [5.8]

The LRFD method of shear design is a complete change from the methods specified in the Standard Specifications and that used by ADOT. For this example, an in-depth shear design will be performed at the critical location in Span 2 near the pier.

AASHTO Cl. [5.8.3.2] dv [5.8.2.9] Critical Section

The critical location is located a distance dv from the face of the support. This creates a problem in that dv is largest of three values, two of which are a function of distance from the pier. To eliminate the iterative process in determining the critical shear location, a simplification is required. It is recommended that the equation, dv = 0.72 h be used to determine the critical shear location. Since dv is the larger of the three values determined in Step 3,

using $dv = 0.72h = (0.72 \times 65.5)/12 = 3.93$ feet will be conservative. Since the pier diaphragm is 6 feet wide, the critical shear is located 3.00 + 3.93 =6.93feet from the centerline of the pier.

Step 1 – Determine Shear

The shears are determined from computer analysis as follows:

Table 6.4. Design Shears

	2.0	**2.1**	**Units**
Super	958.2	804.7	k
Barriers	86.1	72.1	k
DC	1044.3	876.8	k
DW (FWS)	84.1	70.4	k
LL + IM vehicle	143.9	130.4	k
Secondary M	-54.2	-54.2	k

SuperVCrit =958.2 - (958.2- 804.7) x (6.93) / 13 = 876.4kips
BarrierVCrit =86.1 – (86.1- 72.1) x (6.93) / 13 =78.6kips
DCVCrit =876.4+ 78.6=955 kips
DWVCrit =84.1 – (84.1 – 70.4) x (6.93) / 13 =76.8 kips
VehicleVCrit =143.9 – (143.9– 130.4) x (6.93) / 13 =136.7 kips

Live Load Distribution AASHTO Cl. [4.6.2.2.1]

The live load distribution factor for shear will be determined based on the provisions for a whole width design. The distribution of live load per lane for shear for an interior beam in Span 2 for one design lane loaded is:

AASHTO LRFD [Table 4.6.2.2.3a-1]

S = 7.75
d = 65.5
L = 130

LL Distribution = (S/9.5)0.6 x (d/12.0L)0.1
= (7.75 /9.5)^0.6 x (65.5 / 12x130)^0.1= 0.645

The distribution for two or more design lanes loaded is:

LL Distribution = (S/7.3)0.9 x (d/12.0L)0.1
= (7.75 / 7.3)^0.9 x (65.5 / 12x 130)^0.1= 0.769

For a whole width bridge, LL Distribution = 0.769X6 = 4.614 webs

Skew Effect AASHTO LRFD [Table 4.6.2.2.3c-1]

For skewed bridges, the shear shall be adjusted to account for the effects of the skew. For a right angle bridge the correction factor equals one.

LL+IM VCrit = 136.7 x4.614 x1 = 630.7 kips
Vu = $\Sigma\gamma_i V_i$ = 1.25x 955 + 1.5 x76.8 + 1.75x 630.7 +1 x (-54.2) = 2358 kips

Step 2 – Determine Analysis Model

Sectional Model AASHTO Cl. [5.8.3]

The sectional model of analysis is appropriate for the design of typical bridge webs where the assumptions of traditional beam theory are valid. Where the distance from the point of zero shear to the face of the support is greater than 2d the sectional model may be used. Otherwise, the strut-and-tie model should be used.

From Computer output:

Point of Zero Shear to Face of Support =79.70 - 3.00 = 76.70 ft
d = 5.5
2d =2x5.5=11 ft< 76.70ft

Therefore, the sectional model may be used.

Step 3 – Shear Depth, dv

The shear depth is the maximum of the following criteria:

1) dv = 0.9 de where,
   de = Aps x fps x dp + As x fy x ds / Ap x s x fp x s + As x fy = dp when As = 0
   From Figure 6.15 (a) and (b):
   dp = 51- (51 -44.52) x $(6.93 / 13)^2$ = 49.16 in
   dv= 0.9dp = 0.9 x (49.16) = 44.24 in
2) 0.72h = 0.72x 65.5= 47.16 in
3) dv= Mn / Asfy + Apsfpu

For a T-Section at the critical location:

AASHTO Cl. [5.7.3.1.1-3]

c=(43.4 x270) -0.85 x4.5x (436 - 73.83) x 6) / 0.85 x4.5 x0.825 x 73.83 +0.28x 43.4 x270/ 49.16= 11.36 in
a=cβ_1 =11.36 x0.825 = 9.37 in

AASHTO Cl. [5.7.3.2.3]

fps= (270) x [1-0.28x11.36/ 49.16] =252.53 ksi

AASHTO Cl. [5.7.3.1.1-1]

Mn= 43.4x 252.53x (49.16 - 9.37/2) + 0.85x 4.5 x (436-73.83) x6 x (9.37/2- 6/2) = 501443in-k

Mn= 501443 /12 = 41,787 ft-k

dv= 41787 x 12 /(0+43.4x 252.53)= 45.75 in

Based on the above, the shear depth, dv, equals 47.16 inches.

Step 4 – Calculate, Vp

Due to the cable curvature, some of the pre-stress force is in the upward vertical direction and directly resists the applied shear. See Figure 6.15 for the angle of the cable path and Figure 6.17 for the stress diagram as shown below:

α = 2 x(51 -44.52)/(12 x 13)x(6.93 /13) = 0.0443 rads

FC_f=(0.953- 0.919) x (6.93) /37.91+0.8085 = 0.8147

Vp =(9023) x (0.8147) x (0.0443) = 326 kips

Step 5 – Check Shear Width, bv

The LRFD specification requires that web width be adjusted for the presence of voided or grouted ducts. For un-grouted ducts, 50% of the width should be subtracted from the gross width and for grouted ducts, 25% should be subtracted. When the structure is first pre-stressed, the ducts are un-grouted.

For this condition of dead load and pre-stressing, the shear should be checked with the 50% reduction for ducts. For the final condition, the ducts are grouted and only the 25% reduction is required.

For un-grouted ducts under DC dead load of superstructure and diaphragm:

AASHTO Cl. [5.8.3.3-2]

$V_u <= \phi \times V_n = \phi \times (0.25 \times f'c \times bv \times dv + V_p)$

$\phi = 0.9$

Required

$$bv = \frac{\frac{Vu}{\varphi} - Vp}{0.25 \times dv \times f'c} = (1.25 \times 876.4/0.9 - 326)/\ 0.25 \times 47.16 \times 4.5 = 16.8 \text{ inches}$$

Available bv =73.83-0.5 x (4.375) x (6webs) = 60.71 inches, OK

For grouted ducts under full load:

Required bv =((2358/0.9)-326)/(0.25 x47.16 x (4.5))= 43.24 inches,
Available bv =73.83-0.25 x (4.375) x (6webs) = 67.27 inches, OK
Use bv = 67.27 inches for all future calculations involving shear.

Step 6 – Evaluate Shear Stress

AASHTO Cl. [5.8.2.9-1]

vu= Vu - ϕVp / ϕbvdv
=(2358- 0.9 x326)/(0.9 x 67.27x 47.16)= 0.723 ksi
vu / f'c=0.723/4.5= 0.161

Step 7 – Estimate Crack Angle θ

The LRFD method of shear design involves several cycles of iteration. The first step is to estimate a value of θ, the angle of inclination of diagonal compressive stress. Since the formula is not very sensitive to this estimate assume that θ = 26.5 degrees. This simplifies the equation somewhat by setting the coefficient 0.5 cotθ = 1.0.

Step 8 – Calculate Strain, εx

There are two formulae for the calculation of strain for sections containing at least the minimum amount of transverse reinforcing. The first formula is used for positive values of strain indicating tensile stresses, while the second formula is used for negative values of strain indicating compressive stresses.

Formula for εx for positive values:

εx=[|Mu| / dv + 0.5Nu + 0.5|Vu - Vp| Cot θ - Apsfpo/ 2(EsAs + EpAps]

Formula for ex for negative values:

AASHTO Cl. [5.8.3.4.2-3]

εx=[|Mu| / dv + 0.5Nu + 0.5|Vu - Vp| Cot θ - Apsfpo/ 2(EcAc + EsAs + EpAps]

where:

Ac=area of concrete on the flexural tension side of the member.
Ac=10741 - (436) x (6) -2x(0.5) x (6)x(2.4) – (26.75) x (73.83)
=6136 in^2
Aps=area of pre-stressing steel on the flexural tension side of the member.
Aps=43.4in^2 since all the strands are located above mid-depth.
As=area of non-pre-stressed steel on flexural tension side of the member.

As =0

fpo = a parameter taken as the modulus of elasticity of pre-stressing tendons multiplied by the locked-in difference in strain between the pre-stressing tendons and the surrounding concrete. For the usual levels of pre-stressing, a value of 0.7 for fpu will be appropriate.

fpo =0.7x270 = 189 ksi

Nu =factored axial force taken as positive if tensile. Nu = 0 kips

Vu =factored shear force.

Vu =2358 kips

Mu = factored moment but not to be taken less than Vudv.

The moment at the critical shear location is required. The moment at the pier centerline is used for interpolation and not the reduced moment that was used in the calculation of stresses and flexural capacity. A summary of moments follows:

Table 6.5. Moment at the Critical Shear Location

	2.0	2.1	Units
Super	-23,554	-12,095	ft-k
Barriers	-4158	-2125	ft-k
DC	-27,712	-14,220	ft-k
DW (FWS)	-2736	-1656	ft-k
LL + IM vehicle	-8750	-5299	ft-k
Secondary M	7041	6337	ft-k

Super MCrit = -23,554+ ((-12,095+ 23554)x6.93) / 13= -17445ft-k

Barrier MCrit= -4158+ (-2125 + 4158 x6.93) / 13= -3074 ft-k

MCrit= -17445- 3074=-20519 ft-k

DW MCrit= -2736 + (-1656+ 2736)x(6.93) / 13 = -2160 ft-k

LL+IM MCrit= -8750 + (-5299+8750 x 6.93) /13 = -6910 ft-k

Secondary MCrit= 7041+ (6337 - 7041 x6.93) / 13 = 6666ft-k

MCrit= 1.25x (-20519) + 1.5x (-2160) + 1.75 x (-6910) + 1x (6666) = -34315ft-k

Mu =-34,315 ft-k > Vudv =(2358) x (47.16) / 12= 9267ft-k

Determine strain as follows:

E = 29000

εx =[|34315x12|/47.16 +0+1 x |2358 -326|-(43.4) x (189) /2x(29000x0+28500 x 43.4)] =0.00104

Since the value is positive the first formula must be used.

AASHTO Cl. [5.8.3.4.2-1]

Now go into [AASHTO LRFD Table 5.8.3.4.2-1] to read the values for θ and β. From the previously calculated value of vu/f'c = 0.161, enter the ≤ 0.175 row and the ≤ 1.00 column even though the value is larger than one. The estimate for values is shown below:

θ =36.8 degrees
β =1.96

With the new value of θ, the strain must be recalculated.

εx=[34315x12|/47.16 +0+0.5 x |2358 - 326|cot(36.8) -43.4 x (189)]/ 2473800 =0.00076

With this new estimate for strain, reenter the table and determine new values for θ and β. Since our new values are the same as assumed, our iterative portion of the design is complete.

Step 9 - Calculate Concrete Shear Strength, Vc

AASHTO Cl. [5.8.3.3-3]

The nominal shear resistance from concrete, Vc, is calculated as follows:

Vc=0.0316β√f'cbvdv
=0.0316 x 1.96 x√4.5 x67.27x 47.16
= 417 kips

Step 10 - Determine Required Vertical Reinforcement, Vs

AASHTO Cl. [5.8.3.3-2]

Vn=Vc + Vs + Vp
Vn=0.25f'cbvdv + Vp

AASHTO Cl. [5.8.3.3-1]

Vs=Av x fy x dv (cotθ + cot α)sinθ / s = Av x fv x dv cotθ / s where α=90°
Vu≤VR=φ Vn=φ(Vc + Vs + Vp)
Vs=Vu/φ - Vc - Vp=Avfydv cot θ / s

Av= 3.72
fy= 60
vc= 417
s=Av x fy x dv cot θ / Vu/φ - Vc - Vp
= 3.72x 60x47.16x cot(36.8)/ 2358/0.9 - 417- 326 = 7.5 in

Use #5 stirrups at 7 inch spacing

Vs=3.72x60x 47.16cot(36.8) / 7=2010 kips

The shear strength is the lesser of:

Vn=[417 +2010+ 326]= 2753 kips
Vn=[0.25 x 4.5 x 67.27 x47.16+ 326]= 3895 kips
φVn=(0.9 x 2753)= 2478k> Vu = 2358 k

Step 11 - Longitudinal Reinforcement

In addition to vertical reinforcement, shear requires a minimum amount of longitudinal reinforcement.

Table 6.6. Summary of Critical Shear Values at the Tenth Points of Span 2

	2.1	**2.2**	**2.3**	**2.4**	**2.5**
Vu	2200	1852	1497	1139	779
Vp	612	493	372	248	123
Vc	578	636	668	551	498
Reqd Spa	18.0	25.8	40.8	33.8	61.4
Max Spa	12.0	24.0	24.0	24.0	24.0
Spa Used	12.0	24.0	24.0	24.0	24.0
Vs	2050	1881	999	1059	745
φVn	2709	2916	1835	1671	1229
	2.6	2.7	2.8	2.9	d crit
Vu	431	776	1145	1530	1772
Vp	0.0	122	242	360	426
Vc	511	531	551	771	1269
Reqd Spa	100	79.3	43.8	47.2	99.9
Max Spa	24.0	24.0	24.0	24.0	24.0
Spa Used	24.0	24.0	24.0	24.0	24.0
Vs	854	1168	1168	2238	2285
φVu	1228	1764	1764	3369	3980

The requirement for longitudinal reinforcement follows:

Asfy + Aps|fps ≥ |Mu| / dvφf+ 0.5 Nu/φc + (|Vu / φv - Vp|-0.5Vs)cotθ

Considering only the pre-stressing steel yields the following:

Mu= 34,315

(43.4x252.53)≥ |34315x12/ (47.16) x (0.95) + (|2358 / 0.9 - 326| - 0.5x 2010)xcot(36.8) 10960 ≥10914 kips

Therefore, the pre-stressing strands are adequate for longitudinal reinforcement without additional mild reinforcing. Note that the specification states that the area of longitudinal reinforcement need not be greater than the area required to resist the maximum moment alone.

A summary of critical shear design values at the tenth points of Span 2 is shown in Table 6.6.

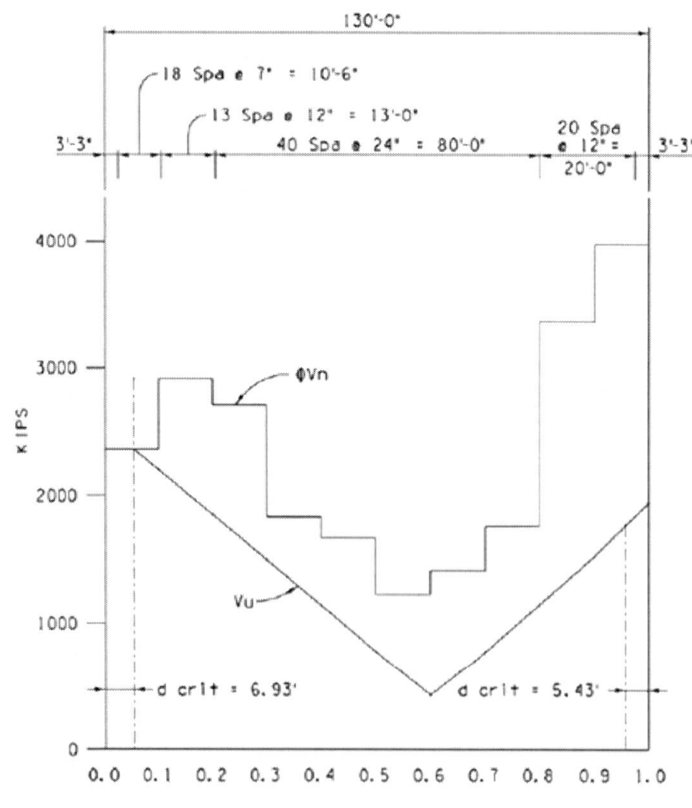

Figure 6.18. Stirrup Spacing.

[BDG]

Note that while the required spacing is low at the abutment, the stirrups are spaced at 12 inches. ADOT guidelines require the 12 inch maximum spacing within 20 feet of the abutment diaphragms to allow for flaring of the tendons without the need to change the stirrup spacing.

6.4.3. Computer Applications for the Design of PSC Box-Girder Bridge Superstructure in AASHTO LRFD

The Computer Applications for the design of PSC Box-Girder Bridge Superstructure in AASHTO LRFD and IRC 112/IRC 6 are described in detail in the softcopy volume 'User's Guide for Hands on Practice with Computer Applications' for the book 'Computer Aided Bridge Engineering' is available for download from web site 'www.roadbridgedesign.com'.

REFERENCES

ASTRA Pro software, User's Guide, Tutorial Videos, *User's Manual and Design Manual,* Techsoft Engineering Services, Web site: www.techsoftglobal.com and www.roadbridgedesign.com

Bridge Engineering Handbook, Second Edition, Superstructure Design, Edited by Wai Fah Chen and Lian Duan, Published by CRC Press, Taylor & Francis Group, 6000 Broken Sound Parkway, NW, Suite 300, Boca Raton, FL 33487, Web site: www.crcpress.com

Design Specifications, Customary U.S. Units, Sixth Edition 2012, American Association of State Highway and Transportation Officials. 444 North Capitol Street, NW, Suite 249, Washington, DC 20001, Phone 202-624-5800 / Fax 202-624-5806, Web site: www.transportation.org

LRFD Bridge Design Manual, MnDOT Bridge Office, Minnesota Department of Transportation. Minnesota Department of Transportation, 3485 Hadley Avenue, North Mail Stop 610, Oakdale, MN 55128-3307, Phone: 651/366-4500.

CONCLUSION

This book belongs to the book series of 'Computer Aided Bridge Engineering' for the design of Pre-stressed Concrete (PSC) I-Girder (I-Beam) and PSC Box-Girder Bridges. In this volume, the real project design calculations for deck-girder superstructure are presented along with the design of Abutment and Pier with Pile Foundation as the bridge substructure.

The book is proposed to be read in association with processing the design work by using the computer software ASTRA Pro as referred to in the book. The book describes two essential facets of the work, which are 'Analysis of the Grillage Model of the Deck-Girder Superstructure' and the subsequent 'Design of Deck Slab and PSC I-Girder'. Whereas the software provides three facets of the work, first is the 'Analysis of the Grillage Model of the Deck-Girder Superstructure', second is the 'Design of Deck Slab and PSC I-Girder, Abutment, Piers along with Pile Foundation' and the third is a 'Set of Sample Editable CAD Drawings for the work. The drawings may be modified as per the design work and be submitted as required for the construction. The drawings are containing information on dimensions, structural detailing, bar-bending schedules, pre-stressing details and construction guides.

ABOUT THE AUTHOR

Sandipan Goswami completed his graduate studies from the University of Calcutta and Post Graduate studies from Indian Institute of Technology-Kharagpur in 1985. Over the span of 36 years he became well known for his technical contribution to the Academic Institutes and Project Authorities to make various preparatory work along with project planning, design, estimating, funding and successful implementation for various state and central government projects. He has remained a member of various professional societies and institutions which include, Member of the IRC, Member and Fellow of the Institution of the Engineers, Chartered Engineer, India, Professional Engineer (PE) as Foreign Engineer, Board of Engineers, Malaysia. He published the widely appreciated technical paper on "Project Development – Development of Rural Road Infrastructure, IRC" and many articles in ResearchGate (Germany) and Academia (USA).

He has written the book "Computer Aided Highway Engineering" with his co-author Prof. (Dr.) P. K. Sarkar and published by CRC Press of Taylor and Francis group.

He was visiting faculty in National Institute for Training of Highway Engineers (NITHE), which is currently named as IAHE, under the 'Ministry of Road Transport and Highways'.

In his 36 years of professional career in India, Malaysia and Bangladesh has worked as Project Manager in various construction supervision projects, as Team Leader in

Detail Project Report (DPR), Feasibility Studies Report (FSR) and Construction Supervision projects, as Senior Design Engineer in various highway, bridge and tunnel projects including 256 kilometers long West Coast Expressway in Malaysia and vehicular tunnels in Kuala Lumpur City Centre, Malaysia. He was assigned to review of various DPRs, design audits, safety audits for highways, and bridges in various states highway projects assigned by 'The World Bank'. He also worked for the design and build of various national and state highway projects, bored tunnel for metro rails projects, airport planning and design projects etc. He also worked as team leader of project management consultant for urban infrastructure improvement, institutional development, capacity building with e-governance, construction management of state government funded by 'The World Bank'.

E-mail: sandipanmails@gmail.com

INDEX

D

E

F

G

H

I

J

L

M

N

O

P

R

S

T

U

V

W